OXFORD MATHEMATICAL MONOGRAPHS

Editors

I. G. MACDONALD   R. PENROSE

APPLIED SEMIGROUPS AND EVOLUTION EQUATIONS

ALDO BELLENI-MORANTE

CLARENDON PRESS - OXFORD

1979

Oxford University Press, Walton Street, Oxford OX2 6DP
OXFORD LONDON GLASGOW
NEW YORK TORONTO MELBOURNE WELLINGTON
IBADAN NAIROBI DAR ES SALAAM LUSAKA CAPE TOWN
KUALA LUMPUR SINGAPORE JAKARTA HONG KONG TOKYO
DELHI BOMBAY CALCUTTA MADRAS KARACHI

ISBN 0 19 853529 5

Printed in Great Britain by
Thomson Litho Ltd
East Kilbride, Scotland

To

Sara

PREFACE

This book is mainly intended for applied mathematicians, physicists, and engineers and, as such, it gives a self-contained introduction to the theory of semigroups and of linear and semilinear evolution-equations in Banach spaces, with particular emphasis on applications to concrete problems from mathematical physics.

Since the only prerequisite is a good knowledge of classical differential and integral calculus, the first three chapters give a 'compact picture' of Banach and Hilbert spaces and introduce the basic notions of abstract differential and integral calculus. Chapters 4 and 5 deal with semigroups and with their applications to linear and semilinear evolution equations. In Chapter 6, a detailed discussion is presented on how a problem of evolution in a given Banach space can be approximated by means of a sequence of problems in the same space or in different spaces. The relationships between the spectral properties of generators and those of semigroups are discussed in Chapter 7. Definitions and theorems of Chapters 1-7 are always supplemented with several examples completely worked out. Finally, each of the final six chapters is devoted to a complete study of a problem from applied mathematics, by using the techniques developed in the previous chapters.

This book is based on lectures given by the author to final-year undergraduates and to first-year graduates of the Mathematical Schools of Bari University and of Florence University, and on seminars given in the mathematical department of Oxford University.

Readers who want to arrive quickly at 'where the action is' may skip (in a first reading) Sections 1.4, 1.5 on Sobolev spaces (and all the Examples that deal with these spaces), Chapters 6 and 7, and some of the proofs in Chapters 2-5.

I would like to thank Dr. J.D. Murray of Oxford University and the colleagues of the mathematics departments of

Bari and Florence Universities for the many useful suggestions. I am also grateful to the Science Research Council of Great Britain because most of the final work on this book was carried out during the author's stay at the Mathematical Institute of Oxford University with a senior visiting fellowship.

Florence, 1978. A.B.-M.

CONTENTS

INTRODUCTION 1

1. BANACH AND HILBERT SPACES
   1.1. Banach and Hilbert spaces 7
   1.2. Examples of Banach and Hilbert spaces 12
        Example 1.1. R^n 12
        Example 1.2. l^p, $p \geq 1$ 13
        Example 1.3. $C([a,b])$, $-\infty < a < b < +\infty$ 15
        Example 1.4. $L^p(a,b)$, $p \geq 1$, $-\infty \leq a < b \leq +\infty$ 16
        Example 1.5. Closed linear subsets of a B-space X 20
        Example 1.6. $X = X_1 \times X_2$. 21
   1.3. Generalized derivatives 21
        Example 1.7. A function belonging to $C_0^\infty(-2,+2)$ 22
        Example 1.8. The generalized derivative of $f(x) = |x|$ 23
   1.4. Sobolev spaces of integer order 25
        Example 1.9. $W_0^{m,p}(\Omega) \to W^{m,p}(\Omega) \to L^p(\Omega)$ 29
        Example 1.10. $L^{p''}(\Omega) \to L^{p'}(\Omega)$, $1 \leq p' \leq p'' \leq \infty$ 30
        Example 1.11. $W^{1,2}(R^1) \to C_B(R^1)$ 31
   1.5. Sobolev spaces of fractional order 32
   Exercises 35

2. OPERATORS IN BANACH SPACES
   2.1. Notation and basis definitions 38
        Example 2.1. The operator $B(f) = f^2$ and its inverse 38
   2.2. Bounded linear operators 39
   2.3. Examples of bounded linear operators 44
        Example 2.2. Operators on R^n 44
        Example 2.3. Operators on l^1 45
        Example 2.4. The Fredholm integral operator on $L^2(a,b)$ 46
        Example 2.5. The operator $A = d/dx$, with $D(A) = W^{1,2}(R^1)$ and $R(A) \subset L^2(R^1)$ 47

Example 2.6. Extension of a densely defined bounded operator 48

Example 2.7. Linear functionals on $L^2(a,b)$ 50

2.4. Lipschitz operators 51

Example 2.8. The operator $A(f) = f^2$ with $D(A) = \{f:f\in C([a,b]);\ \|f\| \le r\}$ and $R(A) \subset C([a,b])$ 52

Example 2.9. The operator $A(f) = \mu f^2/(1+f^2)$ with $D(A) = L^2(a,b)$, $R(A) \subset L^2(a,b)$ 52

Example 2.10. The operator $A(f) = f^2$ with $D(A) = \{f:f\in W^{1,2}(R^1);\ \|f-f_0\|_{1,2} \le r\}$ and $R(A) \subset W^{1,2}(R^1)$ 53

Example 2.11. The operator $F(f) = (Af)(Bf)$ with A and B belonging to $\mathscr{B}(X,C([a,b]))$ 56

Example 2.12. Contraction mapping theorem 57

2.5. Closed operators 60

2.6. Self-adjoint operators 63

Example 2.13. The Fredholm integral operator 68

Example 2.14. The heat-diffusion operator $Af = kf''$ with $D(A) = \{f:f\in L^2(a,b);\ f''\in L^2(a,b); f(a) = f(b) = 0\}$ 68

Example 2.15. The operator $Af = \mathscr{F}^{-1}[-ky^2\mathscr{F}f]$, with $D(A) = L^{s+2,2}(R^1)$, $R(A) \subset L^{s,2}(R^1)$ 70

2.7. Spectral properties: basic definitions 73

2.8. Spectral properties: examples 77

Example 2.16. Resolvent of $A\in\mathscr{B}(X)$ 77

Example 2.17. Spectral properties of an operator in l^1 79

Example 2.18. The convection operator $A = -v\mathrm{d}/\mathrm{d}x$ with $D(A) = \{f:f\in L^2(a,b); f'\in L^2(a,b); f(a) = 0\}$, $R(A) \subset L^2(a,b)$ 83

Example 2.19. The heat-diffusion operator 87

Example 2.20. Resolvent of $A+B$ with $A\in\mathscr{C}(X)$ and $B\in\mathscr{B}(X)$ 94

Example 2.21. Resolvent set of a self-adjoint operator 95

Exercises 98

3. ANALYSIS IN BANACH SPACES

3.1. Strong continuity 101

Example 3.1. Strong continuity in R^n 102

Example 3.2. Strong continuity in l^1 102

Example 3.3. Strong continuity in $C([a,b])$ 103

Example 3.4. Strong continuity in $L^1(a,b)$ 104

Example 3.5. Strong continuity in $\mathscr{B}(X,Y)$ 104

3.2. Strong derivative 106

Example 3.6. Strong derivative in $C([a,b])$ 107

3.3. Strong Riemann integral 110

Example 3.7. The operator $Aw = \mathrm{d}w/\mathrm{d}t$ with $D(A) = \{w: w\in C([0,t_0];X);$ $\mathrm{d}w/\mathrm{d}t \in C([0,t_0];X);\ w(0) = \theta_X\}$, $R(A) \subset C([0,t_0];X)$ 113

Example 3.8. Differentiation under the integral sign 114

3.4. The differential equation $\mathrm{d}u/\mathrm{d}t = F(u)$ 114

3.5. Holomorphic functions 120

Example 3.9. The resolvent of a closed operator operator 122

3.6. Fréchet derivative 123

Example 3.10. F derivative of $F(f) = f_0+Bf$ with $B\in\mathscr{B}(X)$ 125

Example 3.11. F derivative of $F(f) = f^2$ with $D(F) = L^2(a,b)$, $R(F) \subset L^1(a,b)$ 125

Exercises 126

4. SEMIGROUPS

4.1. Linear initial-value problems 128

4.2. The case $A\in\mathscr{B}(X)$ 130

Example 4.1. $A\in\mathscr{B}(R^n)$ 135

Example 4.2. An integro-differential system 136

Example 4.3. Best approximation of a non-linear problem by means of a linear problem 138

4.3. The case $A\in\mathscr{C}(X)$ 139

Example 4.4. The convection operator 141
Example 4.5. The heat-diffusion operator 141
Example 4.6. An operator in l^1 141
Example 4.7. The case $A \in \mathscr{B}(X)$ 146
Example 4.8. The case $A_1 = iA$ with A self-adjoint 146
4.4. The case $A \in \mathscr{G}(1,0;X)$: two preliminary lemmas 146
4.5. The semigroup generated by $A \in \mathscr{G}(1,0;X)$ 152
4.6. The cases $A \in \mathscr{G}(M,0;X)$, $\mathscr{G}(M,\beta;X)$, $\mathscr{G}'(M,\beta;X)$ 158
Example 4.9. Conservation of the norm of $u(t) = \exp(tA_1)u_0$ with $A_1 \in \mathscr{G}'(1,0;X)$ 161
4.7. The homogeneous and the non-homogeneous initial-value problems 162
Example 4.10. Discretization of the time-like variable 164
Example 4.11. Oscillating sources 171
Example 4.12. Periodic solutions 172
Example 4.13. Oscillating heat sources 175
Exercises 176

5. PERTURBATION THEOREMS
5.1. Introduction 178
5.2. Bounded perturbations 179
Example 5.1. An integral perturbation of the heat-diffusion operator 182
Example 5.2. $A \in \mathscr{G}(1,0;l^1)$, $B \in \mathscr{B}(l^1)$ 184
Example 5.3. $A \in \mathscr{G}(M,\beta;X)$, $B = zB_0$ with $B_0 \in \mathscr{B}(X)$ and $z \in \mathbb{C}$ 184
5.3. The cases $B = B(t) \in \mathscr{B}(X)$ and B relatively bounded 187
Example 5.4. The integral operator $B(t)$ of Exercise 3.3 189
Example 5.5. Relative boundedness of the convection operator with respect to the heat-diffusion operator 190
5.4. The semilinear case 191
Example 5.6. The non-linear temperature-dependent source $F(\tau) = \mu\tau^2(x;t)\{1+\tau^2(x;t)\}^{-1}$ 192

5.5. Global solution of the semilinear problem (5.31) 205
Example 5.7. The case $\mathrm{Re}(F(f),f) \le \lambda\|f\|^2$ $\forall f \in X$, with X = a Hilbert space 211
Example 5.8. $F(f) = (1+f^2)^{-1}$, $\tilde{F}(f) = f^2[1+f^2]^{-1}$ with $D(F) = D(\tilde{F}) = C([a,b])$ 212
Example 5.9. $F(f) = (Af)(Bf)$ with A and B belonging to $\mathscr{B}(C([a,b]))$ 214
Example 5.10. A linearization procedure 215
Exercises 217

6. SEQUENCES OF SEMIGROUPS
6.1. Sequences of semigroups $\exp(tA_j) \in \mathscr{B}(X)$ 219
Example 6.1. $\exp(tA_j)f \to \exp(tA)f$ with $A_j \in \mathscr{B}(X)$ 223
6.2. Sequences of Banach spaces 225
Example 6.2. A sequence of B-spaces approximating $C([a,b])$ 225
Example 6.3. A sequence of B-spaces approximating $L^1(a,b)$ 227
6.3. Sequences of semigroups $\exp(tA_j) \in \mathscr{B}(X_j)$ 228
Example 6.4. Galerkin method 233
Example 6.5. Discretization of the operator $-v\mathrm{d}/\mathrm{d}x$ 237
Exericses 241

7. SPECTRAL REPRESENTATION OF CLOSED OPERATORS AND OF SEMIGROUPS
7.1. Introduction 244
7.2. Projections 245
Example 7.1. Projections in R^n 247
Example 7.2. Projections on a subspace of a Hilbert space 248
Example 7.3. A projection operator in L^1 248
7.3. Isolated points of the spectrum of $A \in \mathscr{C}(X)$ 249
7.4. Laurent expansions of $R(z,A)$ 253
7.5. Isolated eigenvalues 256
Example 7.4. Spectral properties of an operator in $\mathbb{C}^2$ 258

Example 7.5. Spectral properties of an operator in l^1 262
Example 7.6. Spectral properties of the heat-diffusion operator 263

7.6. Spectral representation of A and of $\exp(tA)$ 265
Example 7.7. Spectral representation of $\exp(tA)$ with $A\in\mathscr{B}(\mathbb{C}^2)$ 271
Example 7.8. Spectral representation of $\exp(tA)$ with $A\in\mathscr{G}(1,0;l^1)$ 271
Example 7.9. Spectral representation of the heat-diffusion operator 273
Example 7.10. Relationships between $\sigma(A)$ and $\sigma(\exp(tA))$ 273
Exercises 277

8. HEAT CONDUCTION IN RIGID BODIES AND SIMILAR PROBLEMS
8.1. Introduction 280
8.2. A linear heat-conduction problem 281
8.3. A semilinear heat-conduction problem 284
8.4. Positive solutions 288
Exercises 291

9. NEUTRON TRANSPORT
9.1. Introduction 295
9.2. Linear neutron transport in L^2 297
9.3. Spectral properties of the transport operator 304
9.4. A semilinear neutron transport problem 308
Exercises 316

10. A SEMILINEAR PROBLEM FROM KINETIC THEORY OF VEHICULAR TRAFFIC
10.1. Introduction 319
10.2. Preliminary lemmas 321
10.3. The operators F, K_1, and K_2 326
10.4. The operators J and K_3 328
10.5. Global solution of the abstract problem (10.11) 331
Exercises 335

11. THE TELEGRAPHIC EQUATION AND THE WAVE EQUATION
11.1. Introduction 338
11.2. Preliminary lemmas 340
11.3. The abstract version of the telegraphic system (11.4) 348
11.4. The telegraphic equation and the wave equation 350
Exercises 355

12. A PROBLEM FROM QUANTUM MECHANICS
12.1. Introduction 357
12.2. Spectral properties of iA 358
12.3. Bounded perturbations 360
Exercises 363

13. A PROBLEM FROM STOCHASTIC POPULATION THEORY
13.1. Introduction 365
13.2. The abstract problem 366
13.3. Preliminary lemmas 367
13.4. Strict solution of the approximating problem (13.7) 369
13.5. A property of the strict solution of the approximating problem 371
13.6. Strict solution of problem (13.6) 375
13.7. The equation for the first moment $\langle n \rangle(t)$ of the bacteria population 377
Exercises 381

BIBLIOGRAPHY 383

Subject index 385

INTRODUCTION

Let S be a 'physical' system (for instance, a rigid conductor of heat, a set of interacting populations, a neutron-multiplying medium, an atom of hydrogen, etc.) characterized by a suitable 'state vector' $\Psi = \Psi(x;t)$, where x is a set of space-like variables and t is a time-like parameter. It is usually assumed that Ψ belongs to a family of functions X and that the physical properties of S, which are being investigated, can be summarized by a set of operations A which transform Ψ into another element of X. In other words, A is the mathematical model of the physical phenomena which make Ψ undergo a change as time goes on:

$$\frac{\partial \Psi}{\partial t} = A(\Psi) \ , \qquad t > 0. \tag{0.1}$$

We remark that, in most cases, (0.1) is a balance equation, i.e. it may be the mathematical formulation of some conservation principle.

If the initial state of S is known, we also have

$$\lim_{t \to 0+} \Psi(x;t) = \Psi_0(x) \tag{0.2}$$

where Ψ_0 is a given element of X. Finally, $\Psi(x;t)$ must usually satisfy some conditions if x belongs to the 'boundary' of S. However, denote by $D(A)$, the domain of A, the following subset of X:

$$D(A) = \{f : f \in X;\ A(f) \in X;\ f \text{ satisfies the prescribed boundary conditions}\} \tag{0.3}$$

where the right-hand side of (0.3) indicates the subset of X composed of *all* $f \in X$ satisfying the prescribed boundary conditions and such that $A(f)$ is still an element of X. Then, the boundary conditions are implicitly taken into account provided that we state that the solution of (0.1) + (0.2) must be sought in $D(A)$.

System (0.1) + (0.2) together with the condition $\Psi \in D(A)$ $t \geq 0$ is an *initial-value problem*, (Kato 1966, p.478).

The following example illustrates a simple and familiar initial-value problem. Let

$$A = \begin{pmatrix} a_{11} & a_{12} \\ a_{21} & a_{22} \end{pmatrix}$$

$$D(A) = X = R^2$$

$$y(t) = \begin{pmatrix} y_1(t) \\ y_2(t) \end{pmatrix}$$

$$y_0 = \begin{pmatrix} y_{01} \\ y_{02} \end{pmatrix}$$

where the a_{ij}'s are given real numbers. Moreover, let the state vector $y(t)$ satisfy the (linear) system

$$\frac{\mathrm{d}y}{\mathrm{d}t} = Ay(t) \qquad t > 0 \tag{0.4}$$

$$\lim_{t \to 0+} y(t) = y_0 \, . \tag{0.5}$$

We point out that the space-like variable in (0.4) and in (0.5) is discrete and only takes the values 1 and 2. Furthermore, the symbol $\mathrm{d}y/\mathrm{d}t$ means

$$\begin{pmatrix} \mathrm{d}y_1/\mathrm{d}t \\ \mathrm{d}y_2/\mathrm{d}t \end{pmatrix}$$

and the limit (0.5) means

$$\lim_{t \to 0+} y_i(t) = y_{0i} \qquad i = 1,2 \, .$$

If we integrate both sides of (0.4), then we obtain

$$y(t) = y_0 + \int_0^t Ay(s)\,ds \qquad t \geq 0 \tag{0.6}$$

where we took into account condition (0.5). Equation (0.6) can be solved by the standard method of successive approximations

$$y^{(j+1)}(t) = y_0 + \int_0^t Ay^{(j)}(s)\,ds \qquad j = 0,1,2,\ldots$$
$$y^{(0)}(t) = y_0 \tag{0.7}$$

and hence,

$$y^{(1)}(t) = y_0 + t\,Ay_0$$
$$y^{(2)}(t) = y_0 + \int_0^t A(y_0 + sAy_0)\,ds = y_0 + tAy_0 + \frac{t^2}{2}A^2y_0$$
$$\vdots$$

and (0.7) leads to the following expression

$$y(t) = \exp(tA)y_0 \qquad t \geq 0 \tag{0.8}$$

where

$$\exp(tA) = \sum_{j=0}^{\infty} \frac{1}{j!}(tA)^j . \tag{0.9}$$

The preceding formal result can be made rigorous as follows. Given any f belonging to the family $X = R^2$, we introduce the 'norm'

$$\| f \| = \left\| \begin{pmatrix} f_1 \\ f_2 \end{pmatrix} \right\| = (f_1^2 + f_2^2)^{1/2} \tag{0.10}$$

which is such that

$$\| f \| \geq 0 ; \qquad \| f \| = 0 \quad \text{if and only if } f = \begin{pmatrix} 0 \\ 0 \end{pmatrix} ;$$

$$\| f+g \| \leq \| f \| + \| g \| ; \qquad \| \alpha f \| = |\alpha| \, \| f \|$$

where α is any real number and where

$$f+g = \begin{pmatrix} f_1 \\ \\ f_2 \end{pmatrix} + \begin{pmatrix} g_1 \\ \\ g_2 \end{pmatrix} = \begin{pmatrix} f_1+g_1 \\ \\ f_2+g_2 \end{pmatrix}$$

$$\alpha f = \begin{pmatrix} \alpha f_1 \\ \\ \alpha f_2 \end{pmatrix}$$

$\forall f,\ g \in R^2$, (i.e., *for any* f and g belonging to R^2).

Using (0.10), it is easy to verify that dy/dt can be defined as follows

$$\lim_{h\to 0} \| h^{-1}\{y(t+h)-y(t)\} - \frac{dy}{dt} \| = 0$$

and that (0.5) is equivalent to

$$\lim_{t\to 0+} \| y(t) - y_0 \| = 0 \ .$$

Accordingly, we say that the derivative and the limit are in the strong sense or in the sense of the norm in R^2.

Given any $f \in R^2$, we also have

$$\| Af \|^2 = \left\| \begin{pmatrix} a_{11}f_1+a_{12}f_2 \\ \\ a_{21}f_1+a_{22}f_2 \end{pmatrix} \right\|^2$$

$$= (a_{11}f_1+a_{12}f_2)^2 + (a_{21}f_1+a_{22}f_2)^2$$

$$\le (a_{11}^2+a_{12}^2)(f_1^2+f_2^2) + (a_{21}^2+a_{22}^2)(f_1^2+f_2^2) = [\sum_{i,j=1}^{2} a_{i,j}^2] \| f \|^2 \ ;$$

hence

$$\| Af \| \le b \| f \| , \quad \forall f \in R^2, \quad b = [\sum_{i,j=1}^{2} a_{i,j}^2]^{1/2} \tag{0.11}$$

and in an analogous way

$$\|A^2 f\| \leq b\|Af\| \leq b^2\|f\| , \quad \forall f \in R^2$$

$$\|A^n f\| \leq b^n\|f\| , \quad n = 1,2,3,\ldots, \quad \forall f \in R^2 . \qquad (0.12)$$

Because of (0.11), we say that the (additive and homogeneous) operator A is bounded and we write $A \in \mathscr{B}(X)$.

We can now give a rigorous definition of the symbol on the right-hand side of (0.9) using the inequalities (0.12). In fact, we have

$$\left\| \sum_{j=n}^{n+p} \frac{1}{j!} (tA)^j f \right\| \leq \sum_{j=n}^{n+p} \frac{1}{j!} |t|^j \|A^j f\|$$

$$\leq \left[\sum_{j=n}^{n+p} \frac{1}{j!} |t|^j b^j \right] \|f\| , \quad \forall f \in R^2 , \quad p = 1,2,3,\ldots$$

and so

$$\left\| \sum_{j=n}^{n+p} \frac{1}{j!} (tA)^j f \right\| \to 0 \quad \text{as} \quad n \to \infty, \quad \forall p = 1,2,3,\ldots$$

because the sum within square brackets is a partial sum of the (convergent) series defining $\exp(b|t|)$. Since R^2 is a complete space (i.e. Cauchy's theorem is valid, see section 1.1), the sequence

$$\{v_n, n=1,2,\ldots\} \subset R^2 , \quad v_n = \sum_{j=0}^{n} \frac{1}{j!} (tA)^j f$$

is convergent in (the sense of the norm of) R^2. Hence, a unique $v \in R^2$ exists, such that

$$\lim_{n\to\infty} \|v_n - v\| = 0 . \qquad (0.13)$$

The element v is usually denoted by the symbol $\exp(ta)f$, for convenience.

Relation (0.13) proves that, given any $f \in R^2$, $v = \exp(tA)f$ is another element of R^2. By a similar procedure, it can be shown that

(i) $\exp(tA) \in \mathscr{B}(R^2)$ with $\|\exp(tA)f\| \leq \exp(b|t|)\|f\|$, $\forall f \in R^2$, $\forall t \in (-\infty, +\infty)$;

(ii) (0.8) is indeed the unique continuous solution of system (0.4) + (0.5);

(iii) the one-parameter family of bounded operators

$\{\exp(tA), -\infty < t < +\infty\}$ is a group, i.e.

$\exp(tA)\exp(sA) = \exp\{(t+s)A\}$ $\forall s, t$, (see section 4.2).

1

BANACH AND HILBERT SPACES

1.1. BANACH AND HILBERT SPACES

Let X be a given set and let $\underline{K}$ be either the field of real numbers or the field of complex numbers. We say that X is a *vector space* (or a linear space) over $\underline{K}$ if, given any $f \in X$, $g \in X$, and $\alpha \in \underline{K}$, the operations $f+g$ and αf are uniquely defined and $f+g \in X$, $\alpha f \in X$. Moreover, the two operations $+$ and $\alpha\cdot$ must satisfy the following axioms:

$$f+g = g+f \tag{1.1a}$$

$$(f+g)+h = f+(g+h) \tag{1.1b}$$

$$\alpha(f+g) = \alpha f+\alpha g \tag{1.1c}$$

$$(\alpha+\beta)f = \alpha f+\beta f \tag{1.1d}$$

$$\alpha(\beta f) = (\alpha\beta)f \tag{1.1e}$$

$$1\cdot f = f \tag{1.1f}$$

$$\text{an element } \theta \in X \text{ exists such that } f+\theta = f \quad \forall f \in X \tag{1.1g}$$

$$0\cdot f = \theta \tag{1.1h}$$

where f, g, $h \in X$ and $\alpha, \beta \in \underline{K}$.

If $\underline{K}$ is the field of real numbers, then X is a real vector space, whereas X is a complex vector space if $\underline{K}$ is the field of complex numbers.

It follows from axioms (1.1) that the element θ, called the *zero element* of X, is unique. Indeed, assume that $\theta_1 \in X$ also has the property (1.1g)

$$f+\theta_1 = f, \quad \forall f \in X \ ;$$

if in particular $f = \theta$, then we have

$$\theta+\theta_1 = \theta$$

and using (1.1a)

$$\theta_1+\theta = \theta.$$

Thus $\theta_1 = \theta$ because of (1.1g) and the zero element of X is unique.

In what follows, the zero vector will often be indicated by 0 and it will not be distinguished in symbol from the real number zero. Furthermore, the element $(-1)f$ will be written $-f$; thus,

$$f-f = \theta, \quad \forall f\in X \tag{1.2}$$

because $f-f = f+(-1)f = 1\cdot f+(-1)f = (1-1)f = 0\cdot f = \theta$.

The vector space X is a *normed space* if a non-negative number $\|f\|$, the *norm* of f, can be attached to each $f\in X$ such that

$$\|f\| \geq 0, \quad \|f\| = 0 \quad \text{if and only if } f = \theta \tag{1.3a}$$

$$\|f+g\| \leq \|f\|+\|g\| \quad \text{(triangle inequality)} \tag{1.3b}$$

$$\|\alpha f\| = |\alpha| \; \|f\| \tag{1.3c}$$

where $f,g\in X$ and $\alpha\in \underline{K}$.

The notation $\|f;X\|$ will also be used for the norm of $f\in X$ whenever it is advisable or necessary to distinguish between the norms in two different spaces or between two norms in the same space.

It follows from (1.3) that

$$\big|\|f\|-\|g\|\big| \leq \|f-g\| \quad \forall f,g\in X . \tag{1.4}$$

To see this, we write f and g as

$$f = (f-g)+g \qquad g = (g-f)+f$$

and then

$$\|f\| \le \|f-g\|+\|g\| \qquad \|g\| \le \|g-f\|+\|f\|$$

where we used (1.3b) and where

$$\|g-f\| = \|-1(f-g)\| = |-1|\|f-g\| = \|f-g\| \ .$$

Hence,

$$\|f\|-\|g\| \le \|f-g\|, \quad \|g\|-\|f\| \le \|f-g\|$$

and (1.4) is proved.

We remark that, according to (1.4), it may happen that $\|f\| = \|g\|$ with $f \neq g$, i.e. with $\|f-g\| \neq 0$. However, if $\|f-g\| = 0$, i.e. if f and g are the same element of X, then $\|f\| = \|g\|$.

Now let $\{f_n, n=1,2,\ldots\}$ be a sequence in the normed (linear) space X; we say that $\{f_n\}$ is a Cauchy sequence if, given any $\varepsilon > 0$, an integer n_ε can be found such that

$$\|f_n-f_m\| < \varepsilon, \quad \forall n,m > n_\varepsilon \ . \tag{1.5}$$

The normed space X is *complete* if any Cauchy sequence $\{f_n\} \subset X$ is convergent, i.e. if, given any sequence satisfying (1.5), an element $f \in X$ exists such that

$$\lim_{n\to\infty} \|f-f_n\| = 0 \ . \tag{1.6}$$

The element f is uniquely defined because, if g also satisfies (1.6), then we have

$$\|f-g\| = \|(f-f_n)+(f_n-g)\| \le \|f-f_n\|+\|g-g_n\| \to 0$$

as $n\to\infty$, where we used the triangle inequality. In the following, we shall often write

$$f = X - \lim_{n\to\infty} f_n$$

to point out that f is the limit of $\{f_n\}$ in the sense of the norm in X.

A complete normed space is a *Banach space* (B-space). Thus, Cauchy's theorem, which is usually proved for sequences of real numbers and which is a fundamental property of the set of real numbers, is assumed to be true in B-spaces, *by definition*.

Let us now assume that, given any ordered pair of elements f and g belonging to the vector space X, a complex number, denoted by (f,g) or by $(f,g)_X$ and called the *inner product* of f and g, can be defined such that

$$(f,g) = \overline{(g,f)} \tag{1.7a}$$

$$(\alpha f+\beta g,h) = \alpha(f,h)+\beta(g,h) \tag{1.7b}$$

$$(f,f) \geq 0, \quad (f,f) = 0 \quad \text{if and only if } f = \theta \tag{1.7c}$$

where $f,g,h \in X$, $\alpha,\beta \in \underline{K}$, and where $\overline{(g,f)}$ denotes the complex conjugate of (g,f). Of course, if X is a real vector space, then (f,g) is real valued and $(f,g) = (g,f)$. We have from (1.7)

$$\begin{aligned} 0 &\leq (\alpha f-\beta g,\alpha f-\beta g) \\ &= (\alpha f,\alpha f)+(\beta g,\beta g)-(\beta g,\alpha f)-(\alpha f,\beta g) \\ &= |\alpha|^2(f,f)+|\beta|^2(g,g)-\overline{\alpha}\beta(g,f)-\alpha\overline{\beta}(f,g) \end{aligned}$$

because

$$(f,\gamma g) = \overline{(\gamma g,f)} = \overline{\gamma(g,f)} = \overline{\gamma}\,\overline{(g,f)} = \overline{\gamma}(f,g)$$

$\forall\, \gamma \in \underline{K}$. If we choose $\alpha = \overline{\alpha} = (g,g)$ and $\beta = (f,g)$, then we obtain

$$0 \leq (g,g)\{(g,g)(f,f)+|(f,g)|^2-|(f,g)|^2-|(f,g)|^2\} .$$

Thus,

$$0 \leq (g,g)(f,f) - |(f,g)|^2$$

if $g \neq \theta$, whereas

$$(g,g)(f,f) - |(f,g)|^2 = 0$$

if $g = \theta$. We conclude that the following (Schwarz) inequality is valid

$$|(f,g)|^2 \leq (f,f)(g,g) \quad , \quad \forall f,g \in X \quad . \tag{1.8}$$

Let an inner product (f,g) be defined in the vector space X; if we put

$$\| f \| = \{(f,f)\}^{1/2} \quad , \quad \forall f \in X \tag{1.9}$$

then $\| f \| \geq 0$ and $\| f \| = 0$ if and only if $f = \theta$ because of (1.7c). Moreover, we have from (1.9)

$$\| \alpha f \| = [(\alpha f, \alpha f)]^{1/2} = [|\alpha|^2 (f,f)]^{1/2} = |\alpha| \; \| f \|$$

and also

$$\| f+g \|^2 = (f+g, f+g) = \| f \|^2 + \| g \|^2 + (f,g) + (g,f)$$

$$= \| f \|^2 + \| g \|^2 + (f,g) + \overline{(g,f)}$$

$$= \| f \|^2 + \| g \|^2 + 2\mathrm{Re}(f,g)$$

where $\mathrm{Re}(f,g)$ denotes the real part of (f,g) and where $\mathrm{Re}(f,g) \leq |(f,g)| \leq \| f \| \; \| g \|$ as a result of (1.8) and (1.9). Hence,

$$\| f+g \|^2 \leq \| f \|^2 + \| g \|^2 + 2\| f \| \; \| g \| = (\| f \| + \| g \|)^2$$

and (1.9) defines a norm in the vector space X because it satisfies axioms (1.7).

We also remark that, if

$$X - \lim_{n\to\infty} f_n = f \qquad X - \lim_{n\to\infty} g_n = g$$

then

$$\lim_{n\to\infty} (f_n, g_n) = (f, g)$$

because

$$\begin{aligned} |(f_n,g_n)-(f,g)| &= |(f_n-f,g_n)+(f,g_n-g)| \\ &\le |(f_n-f,g_n)|+|(f,g_n-g)| \\ &\le \|f_n-f\|\,\|g_n\|+\|f\|\,\|g_n-g\| \to 0 \end{aligned}$$

as $n\to\infty$, where we used (1.8) and where we took into account that $\|g_n\| \to \|g\|$. In other words, the inner product is 'continuous' with respect to each of the two factors.

By definition, the space X with the norm (1.9) is a *Hilbert space* if it is complete with respect to the norm (1.9) itself.

Finally, let us assume that two norms, $\|\cdot\|_1$ and $\|\cdot\|_2$, can be defined in a given vector space X. We say that $\|\cdot\|_1$ and $\|\cdot\|_2$ are *equivalent* if two positive constants c_1 and c_2 exist, such that

$$c_1\|f\|_1 \le \|f\|_2 \le c_2\|f\|_1 , \quad \forall f\in X \tag{1.10}$$

where c_1 and c_2 are independent of $f\in X$. The two norms are equivalent in the sense that $\|f-f_n\| \to 0$ if and only if $\|f-f_n\|_2 \to 0$. Hence, if X is complete with respect to $\|\cdot\|_1$, then it is also complete with respect to $\|\cdot\|_2$, and conversely.

1.2. EXAMPLES OF BANACH AND HILBERT SPACES

Example 1.1. R^n.

The set of all ordered n-tuples of real numbers is the real vector space R^n, with

$$f+g = \begin{pmatrix} f_1 \\ \cdots \\ f_n \end{pmatrix} + \begin{pmatrix} g_1 \\ \cdots \\ g_n \end{pmatrix} = \begin{pmatrix} f_1+g_1 \\ \cdots \\ f_n+g_n \end{pmatrix} \qquad \alpha f = \begin{pmatrix} \alpha f_1 \\ \cdots \\ \alpha f_n \end{pmatrix}, \quad \theta = \begin{pmatrix} 0 \\ \cdots \\ 0 \end{pmatrix}$$

where α is any real number. R^n is a (real) Hilbert space with inner product and norm

$$(f,g) = \sum_{i=1}^{n} f_i g_i , \quad \|f\| = [(f,f)]^{1/2} = [\sum_{i=1}^{n} f_i^2]^{1/2} . \tag{1.11}$$

It is easy to prove that the inner product (1.11) satisfies axioms (1.7) (see Exercise 1.1). The following norms are equivalent to $\|\cdot\|$ given by (1.11)

$$\|f\|_1 = \max\{|f_i|,\ i=1,2,\ldots,n\} \quad \|f\|_2 = \sum_{i=1}^{n} |f_i| .$$

For instance, we have

$$\|f\|^2 = \sum_{i=1}^{n} |f_i|^2 \le \sum_{i=1}^{n} \|f\|_1^2 = n\|f\|_1^2$$

$$\|f\|_1^2 = \max\{|f_i|^2,\ i=1,2,\ldots,n\} \le \sum_{i=1}^{n} |f_i|^2 = \|f\|^2$$

and so $$\|f\|_1 \le \|f\| \le \sqrt{n}\, \|f\|_1 .$$

We finally remark that the set of all ordered n-tuples of complex numbers is a complex Hilbert space with inner product and norm

$$(f,g) = \sum_{i=1}^{n} f_i \bar{g}_i , \quad \|f\| = \{(f,f)\}^{1/2} . \tag{1.12}$$

Example 1.2. l^p, $p \ge 1$

Let l^p be the set of all sequences of complex numbers $f = (f_1, f_2, \ldots, f_i, \ldots)$, such that

$$\sum_{i=1}^{\infty} |f_i|^p < \infty$$

where p, $1 \le p < +\infty$, is a given real number. The set l^p is a complex vector space with the operations

$$f+g = (f_1, f_2, \ldots)+(g_1, g_2, \ldots) = (f_1+g_1, f_2+g_2, \ldots)$$

$$\alpha f = (\alpha f_1, \alpha f_2, \ldots)$$

and with $\theta = (0,0,\ldots)$. Further, l^p is a (complex) B-space with norm

$$\|f\|_p = [\sum_{i=1}^{\infty} |f_i|^p]^{1/p} . \tag{1.13}$$

To show that l^p is a complete normed space, assume that $\{f^{(n)}, n=1,2,\ldots\}$ is a Cauchy sequence in l^p:

$$\|f^{(n)}-f^{(m)}\|_p^p = \sum_{i=1}^{\infty} |f_i^{(n)}-f_i^{(m)}|^p < \varepsilon^p \ \forall\, n,m > n_\varepsilon \tag{1.14}$$

where $f^{(n)} = (f_1^{(n)}, f_2^{(n)}, \ldots, f_i^{(n)}, \ldots)$. It follows that

$$|f_i^{(n)}-f_i^{(m)}| < \varepsilon, \ \forall\, n,m > n_\varepsilon, \ \forall i = 1,2,\ldots$$

where n_ε does not depend on i. Hence, for any fixed index i, the sequence of complex numbers $\{f_i^{(n)}, n=1,2,\ldots\}$ is convergent to some complex number f_i

$$f_i = \lim_{n\to\infty} f_i^{(n)}, \quad i = 1,2,\ldots,$$

uniformly with respect to i. On the other hand, we have from (1.14)

$$\sum_{i=1}^{N} |f_i^{(n)}-f_i^{(m)}|^p < \varepsilon^p, \ \forall\, n,m > n_\varepsilon, \ \forall N = 1,2,\ldots$$

and passing to the limit as $n\to\infty$

$$\sum_{i=1}^{N} |f_i-f_i^{(m)}|^p < \varepsilon^p, \ \forall m > n_\varepsilon, \forall N = 1,2,\ldots$$

Since the preceding inequality is valid for any $N = 1,2,\ldots,$ we obtain as $N\to\infty$

$$\|f-f^{(m)}\|_p^p \le \varepsilon^p \ \forall\, m > n_\varepsilon$$

where $f = (f_1, f_2, \ldots)$. Hence, $\forall m > n_\varepsilon$,

$$\|f-f^{(m)}\|_p \le \varepsilon$$

$$\|f\|_p = \|f-f^{(m)}+f^{(m)}\|_p \le \varepsilon + \|f^{(m)}\|_p < \infty.$$

Thus, $f \in l^p$,

$$\lim_{m\to\infty} \| f - f^{(m)} \|_p = 0$$

and the space l^p is complete.

We remark that, in particular, l^2 is a Hilbert space with inner product and norm

$$(f,g) = \sum_{i=1}^{\infty} f_i \bar{g}_i, \quad \|f\|_2 = [\sum_{i=1}^{\infty} |f_i|^2]^{1/2}$$

and that the B-space l^1 is quite important in connection with stochastic models of physical systems.

Finally, l^∞ is the set of all sequences $f = (f_1, f_2, \ldots)$ such that $\sup\{|f_i|, \, i = 1,2,\ldots\} < \infty$

and it is a B-space with norm

$$\|f\|_\infty = \sup\{|f_i|, \, i = 1,2,\ldots\} . \tag{1.15}$$

Example 1.3. $C([a,b])$, $-\infty < a < b < +\infty$.

The set $C([a,b])$ of all complex-valued continuous functions $f = f(x)$ on the closed interval $[a,b]$ is a vector space with the operations

$$(f+g)(x) = f(x)+g(x) , \quad (\alpha f)(x) = \alpha f(x) , \quad \forall x \in [a,b];$$

the zero element is the function $\theta = \theta(x) \equiv 0$, $\forall x \in [a,b]$. Note that a generic element of $C([a,b])$ is a continuous function and not a single value taken by such a function. $C([a,b])$ is a B-space with norm

$$\| f \| = \max\{ |f(x)| , \; x \in [a,b]\} . \tag{1.16}$$

Since it is easy to show that (1.16) satisfies axioms (1.3), we shall only prove that $C([a,b])$ is a complete space. To this aim, let $\{f_n, \, n = 1,2,\ldots\}$ be a Cauchy sequence in $C([a,b])$:

$$\| f_n - f_m \| = \max\{ |f_n(x) - f_m(x)|, \, x \in [a,b]\} < \varepsilon \quad \forall n,m > n_\varepsilon$$

where n_ε does not depend on x. Hence,

$$|f_n(x)-f_m(x)| < \varepsilon, \ \forall x\in[a,b], \ \forall n,m > n_\varepsilon \tag{1.17}$$

and the sequence $\{f_n(x),\ n = 1,2,\dots\}$ of continuous functions is uniformly convergent over $[a,b]$. It follows that a continuous function $f = f(x)$, $x\in[a,b]$, exists, such that

$$\lim_{n\to\infty} f_n(x) = f(x)$$

at each $x\in[a,b]$. We then obtain from (1.17) as $n\to\infty$

$$|f(x)-f_m(x)| \le \varepsilon \ , \ \forall x\in[a,b], \ \forall m > n_\varepsilon$$

and so

$$\|f-f_m\| = \max\{|f(x)-f_m(x)|, \ x\in[a,b]\} \le \varepsilon, \ \forall m > n_\varepsilon .$$

The preceding inequality shows that $\lim\|f-f_m\| = 0$ as $m\to\infty$; thus, $C([a,b])$ is a complete space.

It is also clear from the proof of the completeness of $C([a,b])$ that convergence in $C([a,b])$ means *uniform convergence* over $[a,b]$.

Example 1.4. $L^p(a,b)$, $p \ge 1, -\infty \le a < b \le +\infty$.

To give a self-contained definition of the B-space $L^p(a,b)$, we first outline briefly some concepts from the theory of Lebesgue measure and of Lebesgue integrals.

Let (a,b) be an open interval where we assume for simplicity that $b-a < \infty$ and let Ω be a subset of (a,b) : $\Omega \subseteq (a,b)$. If $\{I_n,\ n = 1,2,\dots\}$ is a countable family of open intervals, such that

(i) $I_n = (a_n,b_n) \subset (a,b)$

(ii) $\Omega \subset \bigcup_n I_n$

then the Lebesgue *outer measure* $m^*\Omega$ of the set Ω is defined by

$$m^*\Omega = \inf\{ \sum_n (b_n-a_n)\} \tag{1.18}$$

where the infimum is taken over any countable family $\{I_n\}$ satisfying (i) and (ii). Since the family composed only of $I = (a,b)$ satisfies (i) and (ii), then $m^*\Omega \le b-a$. Now let us denote by Ω^c the *complement* of Ω with respect to (a,b):

$$\Omega^c = \{x : x\in(a,b) \ ; \ x\notin\Omega\}$$

and let $m^*\Omega^c$ be the outer measure of Ω^c. We say that Ω is Lebesgue measurable if

$$m^*\Omega + m^*\Omega^c = b-a \tag{1.19}$$

and its Lebesgue measure $m\Omega$ is given by $m^*\Omega$: $m\Omega = m^*\Omega$ (for further details on Lebesgue measure, see Royden 1963, Chapter 3). Obviously, if $\Omega = I_1 = (a_1,b_1) \subset (a,b)$, then $m^*\Omega = m\Omega = b_1-a_1$.

Let $f = f(x)$, $x\in(a,b)$, be a real-valued function; we say that f is measurable if, for each real number α, the set $\{x : x\in(a,b);\ f(x) > \alpha\}$ is measurable. If $f(x)$ is complex valued, we say that f is measurable if both $f_r(x) = \mathrm{Re}[f(x)]$ and $f_i(x) = \mathrm{Im}[f(x)]$ are measurable. The so-called *simple functions* provide an example of measurable functions: $\phi = \phi(x)$ is said to be a simple function on (a,b) if

$$\phi(x) = \sum_{i=1}^{n} a_i\ \chi_{E_i}(x) \ . \tag{1.20}$$

In definition (1.20), E_i is a measurable subset of (a,b), $\chi_{E_i}(x)$ is the *characteristic function* of E_i:

$$\chi_{E_i}(x) = 1 \ \text{ if } \ x\in E_i, \qquad \chi_{E_i}(x) = 0 \ \text{ if } \ x\notin E_i$$

and a_i is the value taken by $\phi(x)$ over E_i. In what follows, we shall assume that $E_i \cap E_j$ = the empty set if $i \ne j$ and that the a_i's are distinct and non-zero. Note that $\phi = \phi(x)$ is called simple because it only takes a finite number of values and it is constant over each E_i.

We are now in a position to define the Lebesgue integral of a bounded function $f = f(x)$ over the finite open interval (a,b), where $f(x)$ is bounded over (a,b) if two simple functions $\phi(x)$ and $\psi(x)$ exist, such that $\phi(x) \le f(x) \le \psi(x), \forall\, x\in(a,b)$. Given a simple function $\phi(x)$, its Lebesgue integral over (a,b) is defined by the quite natural relation

$$\int_a^b \phi(x)\,dx = \sum_{i=1}^{n} a_i \, m\,E_i \tag{1.21}$$

where $m\,E_i$ is the Lebesgue measure of E_i. On the other hand, if $f = f(x)$, $x\in(a,b)$, is a real-valued bounded function, then we consider the two numbers

$$\sup\{\int_a^b \phi(x)\,dx\}\;, \quad \inf\{\int_a^b \psi(x)\,dx\}\;, \tag{1.22}$$

where the supremum is taken over the family of all simple functions ϕ such that $\phi(x) \le f(x), \forall\, x\in(a,b)$ and where the infimum is taken over the family of all simple functions ψ such that $f(x) \le \psi(x)$ $\forall\, x\in(a,b)$. If the two numbers (1.22) are equal, then we say that f is *Lebesgue integrable* over (a,b) and we write

$$\begin{aligned}\int_a^b f(x)\,dx &= \sup\{\int_a^b \phi(x)\,dx\} \\ &= \inf\{\int_a^b \psi(x)\,dx\}.\end{aligned} \tag{1.23}$$

It can be shown that the bounded function $f(x)$ is Lebesgue integrable over (a,b) if and only if it is measurable over (a,b) (Royden 1963, p.65). If f is a measurable complex-valued function on (a,b), with $f_r(x) = \mathrm{Re}[f(x)]$ and $f_i(x) = \mathrm{Im}[f(x)]$ bounded, then the Lebesgue integral of f is defined by

$$\int_a^b f(x)\,dx = \int_a^b f_r(x)\,dx + i\int_a^b f_i(x)\,dx\;. \tag{1.24}$$

We also mention that, if $f\in C([a,b])$, then the Riemann integral of f over $[a,b]$ is equal to its Lebesgue integral.

For further details on the Lebesgue integral of bounded functions and on the Lebesgue integral of measurable functions (non-necessarily bounded) over finite or infinite intervals, we refer to Chapter 4 of Royden (1963).

Now we can define the vector space $L^p(a,b)$ as the set of all complex-valued Lebesgue-measurable functions $f = f(x)$ on (a,b), such that

$$\int_a^b |f(x)|^p \, dx < \infty \qquad 1 \leq p < \infty$$

where, for the time being, p is a given real number not smaller than unity. The operations + and $\alpha\cdot$ are defined in $L^p(a,b)$ as in $C([a,b])$; see Example 1.3. $L^p(a,b)$ is a B-space with norm

$$\|f\|_p = [\int_a^b |f(x)|^p \, dx]^{1/p} . \tag{1.25}$$

Definition (1.25) obviously satisfies axiom (1.3c), whereas the triangle inequality (1.3b) follows from the Minkowski inequality (Royden 1963, p.95). Moreover, $\|f\|_p$ is clearly a non-negative number. Finally, we have from the theory of Lebesgue integrals that $\|f\|_p = 0$ if and only if $f(x) = 0$ *almost everywhere* on (a,b) (which we abbreviate to a.e. on (a,b)), i.e. if and only if

$$m\{x : x \in (a,b); \; f(x) \neq 0\} = 0.$$

Hence, definition (1.25) also satisfies axiom (1.3a) provided that we consider any two measurable functions $f(x)$ and $g(x)$ to be the *same* element of $L^p(a,b)$ if $f(x)-g(x) = 0$ a.e. on (a,b), i.e. if

$$m\{x : x \in (a,b); \; f(x)-g(x) \neq 0\} = 0.$$

In an analogous way, the zero element of $L^p(a,b)$ is any function $\theta = \theta(x)$ such that $\theta(x) = 0$ a.e. on (a,b).

We also remark that $L^p(a,b)$ is a complete normed space because of the Riesz-Fisher theorem (Royden 1963, p.99).

Among L^p spaces, L^1 and L^2 are the most commonly used in applied mathematics. In fact, $\|f\|_1$ may represent for instance the total number of particles or the total thermal energy contained in a physical system S. On the other hand, L^2 is a Hilbert space with inner product and norm

$$(f,g) = \int_a^b f(x)\,\overline{g(x)}\, dx, \qquad \|f\|_2 = [\int_a^b |f(x)|^2 dx]^{1/2} \tag{1.26}$$

and it is the 'space of quantum mechanics'.

Finally, $L^\infty(a,b)$ is the set of all measurable functions

$f = f(x)$ which are bounded a.e. on (a,b) and it is a B-space with norm

$$\|f\|_\infty = \text{ess sup}\{|f(x)|, x\epsilon(a,b)\}$$

where

$$\text{ess sup } \{f(x)|, x\epsilon(a,b)\} = \inf[M: m\{x:x\epsilon(a,b);|f(x)| > M\} = 0]$$

i.e. $\|f\|_\infty$ is the 'smallest' constant M, such that the Lebesgue measure of the set $\{x : x\epsilon(a,b);\ |f(x)| > M\} = 0$ is zero. In other words, $|f(x)| \leq \|f\|_\infty$ a.e. on (a,b). We remark that (1.27) is in a sense a generalization of the norm (1.16).

By a similar procedure, the spaces $L^p(\Omega)$, $1 \leq p \leq \infty$, can be defined, where Ω is an open set of R^n (Adams 1975, Chapter 2).

Example 1.5. Closed linear subsets of a B-space X.

Let X be a B-space with norm $\|\cdot\|$ and let $f_0\epsilon X$ and $r > 0$ be given. By definition, the subset of X

$$S(f_0;r) = \{f : f\epsilon X;\ \|f-f_0\| < r\}$$

is an open ball (sphere) with centre f_0 and radius r. Now let Y be a subset of X: $Y \subset X$, where the word 'subset' and the symbol $\subset$ do not exclude the possibility that $Y = X$. An element (point) $f_0\epsilon Y$ is an *interior point* of Y if $S(f_0;r) \subset Y$ for some r. A subset $Y \subset X$ is *open* (in X) if each $f_0\epsilon Y$ is an interior point. A subset $X_0 \subset X$ is *closed* (in X) if its complement with respect to X

$$X_0^c = \{f : f\epsilon X;\ f\notin X_0\}$$

is open in X.

Given any $X_1 \subset X$, the *closure* of X_1, $\text{cl}(X_1)$, is the intersection of all closed sets containing X_1. Of course, $\text{cl}(X_1)$ is a closed subset of X and $X_1 \subset \text{cl}(X_1)$; moreover, X_1 is closed if and only if $X_1 = \text{cl}(X_1)$.

If X_0 is a closed subset of X, then any Cauchy sequence

$\{f_n\} \subset X_0$ is convergent to $f_0 \in X_0$. In fact, since X is a B-space by assumption, a unique $f_0 \in X$ exists such that $\lim \|f_0 - f_n\| = 0$ as $n \to \infty$. Now if $f_0 \in X_0^c$, then f_0 is an interior point of the open set X_0^c and consequently $S(f_0;\varepsilon) \subset X_0^c$ provided that $\varepsilon > 0$ is suitably chosen. Since $\|f_0 - f_n\| < \varepsilon, \forall n > n_\varepsilon$, then $f_n \in S(f_0;\varepsilon) \subset X_c^0$, $\forall n > n_\varepsilon$, and this leads to a contradiction because $f_n \in X_0$ and $X_0 \cap X_0^c$ = the empty set. We conclude that f_0 must belong to X_0.

A subset $X_0 \subset X$ is called *linear* if $\alpha f + \beta g \in X_0$, $\forall f, g \in X_0$ and $\forall \alpha, \beta \in \underline{K}$.

Finally, let X_0 be a linear and closed subset of the B-space X. Then, X_0 is itself a B-space with norm $\|\cdot\|$. In fact, X_0 is a vector space because it is a linear subset of the vector space X; moreover, X_0 is complete as we have proved above.

Example 1.6. $X = X_1 \times X_2$.

Let X_1 and X_2 be B-spaces over the same field $\underline{K}$, with norms $\|\cdot; X_1\|$ and $\|\cdot; X_2\|$. Then, $X = X_1 \times X_2$ is the space of all ordered pairs

$$f = \begin{pmatrix} f_1 \\ f_2 \end{pmatrix}, \quad f_1 \in X_1, \quad f_2 \in X_2$$

and it is a vector space with the operations

$$f+g = \begin{pmatrix} f_1 \\ f_2 \end{pmatrix} + \begin{pmatrix} g_1 \\ g_2 \end{pmatrix} = \begin{pmatrix} f_1+g_1 \\ f_2+g_2 \end{pmatrix}, \quad \alpha f = \begin{pmatrix} \alpha f_1 \\ \alpha f_2 \end{pmatrix}$$

$\forall f, g \in X$, $\forall \alpha \in \underline{K}$. We point out that the operation + in X is defined in terms of the operation + in X_1 and in X_2; the same remark applies to the operation $\alpha\cdot$. X is a B-space with norm

$$\|f\| = [\|f_1;X_1\|^2 + \|f_2;X_2\|^2]^{1/2} \quad . \tag{1.28}$$

1.3. GENERALIZED DERIVATIVES

Let Ω be an open subset of R^n and let $f(x)$ be a function defined a.e. on Ω. We say that $f(x)$ is *locally integrable* on Ω and we write $f \in L^1_{loc}(\Omega)$ if $f \in L^1(\Omega_0)$ for *every* $\Omega_0 \subset \Omega$, such that (i) Ω_0 is Lebesgue measurable, (ii) Ω_0 is bounded, i.e. $\Omega_0 \subset S(0;r)$ for some $r > 0$, and (iii) the closure $\text{cl}(\Omega_0)$ of Ω_0 is contained in Ω, where we recall that $\text{cl}(\Omega_0)$ is the intersection of all closed sets containing Ω_0. Note that Ω is Lebesgue measurable because it is an open subset of R^n (Royden 1963, p.50) and that Ω may be bounded or unbounded (for instance, Ω may coincide with the whole R^n).

Further, let $\text{supp}(f)$ denote the *support* of the function $f(x)$, $x \in \Omega$, where

$$\text{supp}(f) = \text{cl}\{x : x \in \Omega;\ f(x) \neq 0\}$$

i.e. $\text{supp}(f)$ is the closure of the subset of Ω on which $f(x)$ takes non-zero values. Thus, if $x \notin \text{supp}(f)$, then certainly $f(x) = 0$.

Finally, we indicate by $C_0^\infty(\Omega)$ the vector space of all functions $\Phi(x)$ which are continuous on Ω together with their partial derivatives of any order and which have support bounded and contained in Ω (i.e. $\text{supp}(\Phi) \subset S(0;r)$ for some $r > 0$ and $\text{supp}(\Phi) \subset \Omega$).

Example 1.7. A function belonging to $C_0^\infty(-2,+2)$

The function

$$\Phi(x) = \exp[-(1-x^2)^{-1}] \quad \text{if} \quad |x| < 1$$

$$\Phi(x) = 0 \quad \text{if} \quad |x| \geq 1$$

belongs to $C_0^\infty(-2,+2) \subset C_0^\infty(-\infty,+\infty)$.

Instead, the function $\Phi_1(x) = \exp(-x^2)$ does not belong to $C_0^\infty(-\infty,+\infty)$ because $\text{supp}(\Phi_1) = (-\infty,+\infty)$. □

Given a function $f \in L^1_{\text{loc}}(\Omega)$, we now define its *generalized* (or *distributional*) derivative

$$D^k f = \frac{\partial^{|k|}}{\partial x_1^{k_1} \ldots \partial x_n^{k_n}} f, \quad k = (k_1, k_2, \ldots, k_n), \quad |k| = \sum_{i=1}^{n} k_i$$

as follows. If a $g_k \in L^1_{\text{loc}}(\Omega)$ exists such that

$$\int_\Omega f(x)\ D^k\ \Phi(x)dx = (-1)^{|k|} \int_\Omega g_k(x)\ \Phi(x)dx, \quad \forall\, \Phi\in C_0^\infty(\Omega) \quad (1.29)$$

where $dx = dx_1 \ldots dx_n$, then $g_k = D^k f$ by definition. Note that, in (1.29), $\Phi(x) \equiv 0$ if $x \notin \mathrm{supp}(\Phi)$ where $\mathrm{supp}(\Phi)$ is a bounded subset of Ω. Relation (1.29) defines a unique element of $L^1_{loc}(\Omega)$. In fact, if both g_k and h_k satisfy (1.29), then we have

$$\int_\Omega [g_k(x) - h_k(x)]\Phi(x)dx = 0 \ , \quad \forall\, \Phi\in C_0^\infty(\Omega)$$

and consequently $g_k(x) = h_k(x)$ a.e. on Ω (Adams 1975, p.59). Hence, g_k and h_k are the same element of $L^1_{loc}(\Omega)$.

Remark 1.1.

Assume for simplicity that Ω is the bounded open rectangle $\{x : x = (x_1, x_2);\ a_i < x_i < b_i,\ i = 1,2\} \subset R^2$ and that the 'classical' derivative $\partial f/\partial x_1$ exists and is continuous on Ω. Then, we obtain on integrating by parts

$$\int_\Omega f(x_1,x_2)\ \frac{\partial \Phi}{\partial x_1}\ (x_1,x_2)\,dx_1 dx_2 = - \int_\Omega \frac{\partial f}{\partial x_1}\ (x_1,x_2)\,\Phi(x_1,x_2)\,dx_1 dx_2$$

$\forall\, \Phi\in C_0^\infty(\Omega)$. The preceding relation motivates the definition (1.29).

□

Remark 1.2.

More generally, if the 'classical' partial derivative $D^k f$ exists and is continuous on $\Omega \subset R^n$, then it coincides with the corresponding generalized derivative. This justifies the adjective 'generalized'.

Example 1.8. The generalized derivative of $f(x) = |x|$

Let $\Omega = (-1,1) \subset R^1$ and let $f(x) = |x| \in L^1(-1,1) \subset L^1_{loc}(-1,1)$; then the generalized derivative of f exists and is given by

$$\begin{aligned} Df = \mathrm{sign}(x) &= -1 \quad \text{if} \quad -1 < x < 0 \\ &= +1 \quad \text{if} \quad 0 < x < 1 \ . \end{aligned}$$

In fact, given any $\Phi\in C_0^\infty(-1,1)$, we have

$$\int_{-1}^{1} |x| \frac{d\Phi}{dx} dx = -\int_{-1}^{0} x \frac{d\Phi}{dx} + \int_{0}^{1} x \frac{d\Phi}{dx} dx$$

$$= \int_{-1}^{0} \Phi(x)dx - \int_{0}^{1} \Phi(x)dx = -\int_{-1}^{1} \text{sign}(x)\ \Phi(x)dx .$$

□

Now let f and g be locally integrable over Ω and assume that the generalized derivatives $D^k f$ and $D^k g$ both exist. Then we have directly from (1.29) (see Exercise 1.8)

$$D^k(c_1 f + c_2 g) = c_1\ D^k f + c_2\ D^k g \tag{1.30}$$

i.e. generalized derivatives are *additive* and *homogeneous* operations.

Finally, we briefly examine how generalized differentiability can be related to absolute continuity (for simplicity we assume that $\Omega = (a,b) \subset R^1$). We recall that the continuous function $f = f(x)$, $x\in[a,b]$, is absolutely continuous if, given any $\varepsilon > 0$, a $\delta = \delta(\varepsilon) > 0$ can be found such that

$$\sum_{i=1}^{m} |f(b_i)-f(a_i)| < \varepsilon$$

for every finite family of subintervals $\{[a_i,b_i],\ i = 1,2,\ldots,m\}$ with $[a_i,b_i] \subset [a,b]$ and with

$$\sum_{i=1}^{m} (b_i-a_i) < \delta .$$

Now, if $f(x)$ is absolutely continuous on $[a,b]$, then it is differentiable (in the 'classical' sense) a.e. on $[a,b]$ and $f' = df/dx\in L^1(a,b)$. Moreover

$$f(x) = f(a) + \int_a^x f'(y)dy , \quad x\in[a,b] \tag{1.31}$$

where the integral is a Lebesgue integral (Royden 1963, Chapter 5). On the other hand, given any $\Phi\in C_0^\infty(a,b)$, we have

$$\int_a^b f(y)\Phi'(y)\,dy = [f(y)\Phi(y)]_{y=a}^{y=b} - \int_a^b f'(y)\Phi(y)\,dy$$
$$= -\int_a^b f'(y)\Phi(y)\,dy$$

because $\Phi(a) = \Phi(b) = 0$ (supp(Φ) is a closed set contained in the open set (a,b)). We conclude that f' is also the generalized derivative of f. Conversely, it can be shown that, if $f \in L^1_{loc}(a,b)$ is differentiable in the generalized sense, then f is absolutely continuous and formula (1.31) is valid.

1.4. SOBOLEV SPACES OF INTEGER ORDER

Generalized derivatives allow us to define in a straightforward way the Sobolev spaces of integer order $W^{m,p}(\Omega)$, $m = 0,1,\ldots$, $p \geq 1$, $\Omega \subset R^n$, which, together with the spaces of fractional order, play a fundamental role in the theory of partial differential equations.

Let $W^{m,p}(\Omega)$ ($m = 0,1,\ldots$, $p \geq 1$, $\Omega \subset R^n$) be the vector space of all complex-valued functions $f = f(x)$ which belong to $L^p(\Omega)$ together with all their generalized partial derivatives $D^k f$ up to order m:

$$W^{m,p}(\Omega) = \{f\colon\ D^k f = \frac{\partial^{|k|}}{\partial x_1^{k_1} \cdots \partial x_n^{k_n}} f \in \ L^p(\Omega),$$
$$0 \leq |k| = \sum_{i=1}^{n} k_i \leq m\}$$

where $D^0 f = f$. The operations + and $\alpha\cdot$ and the zero element are defined in $W^{m,p}(\Omega)$ in the usual way (see Example 1.3). $W^{m,p}(\Omega)$ can be normed as follows

$$\|f\|_{m,p} = [\sum_{0 \leq |k| \leq m} \|D^k f\|_p^p]^{1/p}, \quad 1 \leq p < \infty \qquad (1.32)$$

$$\|f\|_{m,\infty} = \max\{\|D^k f\|_\infty, \quad 0 \leq |k| \leq m\} \qquad (1.33)$$

where m is a given non-negative integer and p is either a given real number not smaller than unity or $p = \infty$, and where we recall that

$$\|g\|_p = [\int_\Omega |g(x)|^p \, dx]^{1/p}$$

$$\|g\|_\infty = \text{ess sup}\{|g(x)|, x\in\Omega\}$$

(see Example 1.4). Note that the sum on the right-hand side of (1.32) is extended to all the n-tuples of non-negative integers $k = (k_1,\ldots,k_n)$ such that $0 \le \Sigma_{i=1}^n k_i \le m$. If $W^{m,p}(\Omega)$ is normed according to (1.32) or to (1.33), then it is complete, i.e. it is a B-space.

Theorem 1.1. $W^{m,p}(\Omega)$ is a Banach space

Proof. We first observe that $\|\cdot\|_{m,p}$ $(1 \le p \le \infty)$ satisfies axioms (1.3) because it is defined in terms of the norm $\|\cdot\|_p$. Of course, as in $L^p(\Omega)$, f and g are the *same* element of $W^{m,p}(\Omega)$ if $f(x) = g(x)$ a.e. on Ω (see Example 1.4). To prove that $W^{m,p}(\Omega)$ is a complete space, we consider a Cauchy sequence $\{f^{(i)}, i = 1,2,\ldots\} \subset W^{m,p}(\Omega)$:

$$[\|f^{(i)}-f^{(j)}\|_{m,p}]^p = \sum_{0\le|k|\le m} [\|D^k f^{(i)}-D^k f^{(j)}\|_p]^p < \varepsilon^p$$

$\forall i,j > j_0 = j_0(\varepsilon)$. Then

$$\|D^k f^{(i)}-D^k f^{(j)}\|_p < \varepsilon, \quad \forall i,j > j_0, \quad \forall\, 0 \le |k| \le m$$

and $\{D^k f^{(i)}, i = 1,2,\ldots\}$ is a Cauchy sequence in $L^p(\Omega)$ for each $k = (k_1,\ldots,k_n)$ with $0 \le |k| \le m$. Since $L^p(\Omega)$ is a B-space, an element $g_k \in L^p(\Omega)$ exists such that

$$\lim_{i\to\infty} \|D^k f^{(i)}-g_k\|_p = 0 , \quad 0 \le |k| \le m ; \qquad (1.34)$$

if in particular $|k| = 0$, i.e. $k = (0,\ldots,0)$, then we have from (1.34)

$$\lim_{i\to\infty} \|f^{(i)}-g\|_p = 0$$

where $g = g_{(0,\ldots,0)} \in L^p(\Omega)$.

To prove that in fact $g_k = D^k g$, we first observe that

$$\int_\Omega \Phi\, D^k f^{(i)} dx = (-1)^{|k|} \int_\Omega f^{(i)}\, D^k\Phi\, dx, \qquad \forall\, \Phi \in C_0^\infty(\Omega)$$

because $f^{(i)} \in W^{m,p}(\Omega)$ and consequently its generalized derivative $D^k f^{(i)}$ satisfies (1.29). It follows that

$$\int_\Omega g_k\, \Phi\, dx - (-1)^{|k|} \int_\Omega g\, D^k\Phi dx = \int_\Omega [g_k - D^k f^{(i)}]\Phi\, dx +$$

$$(-1)^{|k|} \int_\Omega [f^{(i)} - g] D^k\, \Phi\, dx$$

$$\left|\int_\Omega g_k\, \Phi dx - (-1)^{|k|} \int_\Omega g\, D^k\Phi\, dx\right|$$

$$\le \int_\Omega |g_k - D^k f^{(i)}|\, |\Phi|\, dx + \int_\Omega |f^{(i)} - g|\, |D^k\Phi|\, dx \tag{1.35}$$

$$\le \|\Phi\|_q\, \|g_k - D^k f^{(i)}\|_p + \|D^k\Phi\|_q\, \|f^{(i)} - g\|_p$$

where we used the Hölder inequality:

$$\int_\Omega |F(x)|\, |G(x)| dx \le \|F\|_p\, \|G\|_q, \qquad \forall\, F \in L^p(\Omega),\ G \in L^q(\Omega) \tag{1.36}$$

with $q = p/(p-1)$ if $p > 1$ and with $q = \infty$ if $p = 1$ (Adams 1975, p.23; Royden 1963, p.95). Passing to the limit as $i \to \infty$ in (1.35), we obtain

$$\int_\Omega g_k\, \Phi\, dx = (-1)^{|k|} \int_\Omega g\, D^k\, \Phi\, dx$$

because of (1.34). Thus, $g_k = D^k g$ and $g \in W^{m,p}(\Omega)$ with $\|g - f^{(i)}\|_{m,p} \to 0$ as $i \to \infty$ again because of (1.34). We conclude that $W^{m,p}(\Omega)$ is a complete space. □

Note that $W^{0,p}(\Omega) = L^p(\Omega)$, $\|\cdot\|_{0,p} = \|\cdot\|_p$ and that $W^{m,2}(\Omega)$ is a Hilbert space with inner product

$$(f, g)_m = \sum_{0 \le |k| \le m} (D^k f,\ D^k g)$$

when $(\cdot,\cdot) = (\cdot,\cdot)_0$ is the inner product in $L^2(\Omega)$.

We also remark that $C_0^\infty(\Omega) \subset W^{m,p}(\Omega)$ because each $\Phi \in C_0^\infty(\Omega)$ is certainly continuous on Ω together with its partial derivatives up to order m (see Remark 1.2). Moreover $\Lambda = \text{supp}(\Phi)$ is a closed and bounded set contained in Ω (see § 1.3) and $\Phi(x) \equiv 0$ if $x \notin \Lambda$. Then,

$$\|\Phi\|_{m,p} = [\sum_{0\leq|k|\leq m} \int_{\Omega} |D^k\Phi(x)|^p dx]^{1/p}$$

$$= [\sum_{0\leq|k|\leq m} \int_{\Lambda} |D^k\Phi(x)|^p dx]^{1/p} < \infty$$

and $\Phi \in W^{m,p}(\Omega)$. Thus $C_0^\infty(\Omega)$ is a linear subset of $W^{m,p}(\Omega)$.

The closure of $C_0^\infty(\Omega)$ in the sense of $\|\cdot\|_{m,p}$, i.e. the intersection of all sets which are closed in $W^{m,p}(\Omega)$ and which contain $C_0^\infty(\Omega)$ will be denoted by $W_0^{m,p}(\Omega)$.

Remark 1.3.

$W_0^{m,p}(\Omega)$ is made up of all the elements of $C_0^\infty(\Omega)$ and of all $f \in W^{m,p}(\Omega)$ such that $\|f-f^{(i)}\|_{m,p} \to 0$ as $i \to \infty$ with $\{f^{(i)}, i = 1,2,\ldots\} \subset C_0^\infty(\Omega)$. Basically, $W_0^{m,p}(\Omega)$ consists of all the $f \in W^{m,p}(\Omega)$ which 'vanish on the boundary of Ω'. □

$W_0^{m,p}(\Omega)$ is itself a B-space with norm (1.32), or (1.33), because it is a closed and linear subset of $W^{m,p}(\Omega)$ (see Example 1.5).

Let Y be a B-space with norm $\|\cdot;Y\|$ and let $X_0 \subset Y$. We say that X_0 is *dense* in Y if the closure $\text{cl}(X_0)$ of X_0 coincides with Y. Then, given any $f \in Y = \text{cl}(X_0)$ and any $\varepsilon > 0$, a suitable $g \in X_0$ can be found such that $\|f-g;Y\| < \varepsilon$. In other words, any elements of Y can be approximated by a suitable element of X_0 with an error smaller than ε. To see this, assume that $f_0 \in Y = \text{cl}(X_0)$ and $\varepsilon_0 > 0$ exist such that

$$\|g-f_0;Y\| \geq \varepsilon_0 \quad \forall g \in X_0 .$$

It follows that the open ball

$$S(f_0;\varepsilon_0) = \{\phi : \phi \in Y;\ \|\phi - f_0;Y\| < \varepsilon_0\}$$

does not contain any $g \in X_0$; consequently, the complement of $S(f_0;\varepsilon_0)$

$$S^c = \{f\colon f\in Y;\ f\notin S(f_0;\varepsilon_0)\} \subset Y$$

is a closed set and it contains X_0. However, $\mathrm{cl}(X_0)$ is the intersection of all closed sets containing X_0 and so $Y = \mathrm{cl}(X_0) \subset S^c$, which is a contradiction because S^c is contained in Y and it does not coincide with it.

Now, according to Remark 1.3, $C_0^\infty(\Omega)$ is dense in the space $W_0^{m,p}(\Omega)$. Similarly, let $C^m(\Omega)$ be the set of all functions $f(x)$ continuous on Ω together with their partial derivatives up to order m. Then, the set

$$\{f\colon f\in C^m(\Omega);\ \|f\|_{m,p} < \infty\}$$

is a dense subset of $W^{m,p}(\Omega)$ (Adams 1975, p.52).

Finally, we list without proof some 'imbedding' characteristics of Sobolev spaces (Adams 1975, chapter 5).

Let X and Y be B-spaces with norms $\|\cdot;X\|$ and $\|\cdot;Y\|$, where X is also a linear subset of Y (see Examples 1.9 and 1.10 below). We say that X is *imbedded* in Y and we write $X\to Y$ if

$$\|f;Y\| \leq \mu\ \|f;X\|, \quad \forall f\in X \tag{1.37}$$

where the constant μ does not depend on $f\in X$. Hence, $X\to Y$ means that any element f of the B-space X can also be considered as an element of the B-space Y and that inequality (1.37) holds.

Example 1.9. $W_0^{m,p}(\Omega) \to W^{m,p}(\Omega) \to L^p(\Omega)$

In fact, since the norm in both $W_0^{m,p}(\Omega)$ and $W^{m,p}(\Omega)$ is given by (1.32), or by (1.33), we have

$$\|f\|_p = [\int_\Omega |f(x)|^p \mathrm{d}x]^{1/p}$$

$$\leq [\sum_{0\leq|k|\leq m} \int_\Omega |\mathrm{D}^k f(x)|^p \mathrm{d}x]^{1/p} = \|f\|_{m,p}$$

$$\|f\|_\infty \leq \max\{\|\mathrm{D}^k f\|_\infty,\quad 0\leq|k|\leq m\} = \|f\|_{m,\infty}$$

and (1.37) is valid with $\mu = 1$. □

Example 1.10. $L^{p''}(\Omega) \to L^{p'}$, $1 \le p' \le p'' \le \infty$, $m\Omega < \infty$

If the Lebesgue measure of Ω is finite, then we have (see Exercise 1.9)

$$\|f\|_{p'} \le [m\Omega]^{1/p'-1/p''} \|f\|_{p''}, \ \forall f \in L^{p''}(\Omega)$$

and (1.37) is valid with $\mu = [m\Omega]^{1/p'-1/p''}$. □

Theorem 1.2.

(a) If $mp < n$, *then*

$$W_0^{j+m,p}(\Omega) \to W_0^{m,q}(\Omega), \quad p \le q \le np/(n-mp), \ j = 0,1,2,\ldots \,. \tag{1.38}$$

(b) If $mp = n$, *then*

$$W_0^{j+m,p}(\Omega) \to W_0^{j,q}(\Omega), \quad p \le q < \infty, \ j = 0,1,2,\ldots \,. \tag{1.39}$$

(c) If $p = 1$, $m = n$, *then*

$$W_0^{j+m,1}(\Omega) \to C_B^j(\Omega), \quad j = 0,1,2,\ldots \,. \tag{1.40}$$

(d) If $mp > n$, *then*

$$W_0^{j+m,p}(\Omega) \to C_B^j(\Omega), \quad j = 0,1,2,\ldots \,. \quad \square \tag{1.41}$$

In (1.40) and in (1.41), $C_B^j(\Omega)$ is the set of all $f(x)$ which are continuous and bounded over Ω together with their partial derivatives up to order j. $C_B^j(\Omega)$ is a B-space with norm

$$\|f; C_B^j(\Omega)\| = \max_{0 \le |k| \le j} \{\sup |D^k f(x)|, \ x \in \Omega\} \tag{1.42}$$

(see also Exercise 1.11).

If we want to extend the results of Theorem 1.2 to the spaces $W^{m,p}(\Omega)$, then we have to restrict ourselves to open sets $\Omega \subset R^n$ which have the so-called *cone property* (Adams 1975, p.66). The set $C_x \subset R^n$ is a *finite cone* with vertex $x \in R^n$ if

$$C_x = S_1 \cap \{z : z \in R^n;\ z = x+\alpha(y-x),\quad \forall \alpha > 0, \forall y \in S_2\}$$

where $S_1 = S(x;r_1)$ and $S_2 = S(x_2;r_2)$ are open balls such that $x \notin S_2$ (i.e. $0 < r_2 < |x_2-x|$). Hence, C_x is the set of all $z \in R^n$ which belong to S_1 and to the infinite cone with vertex x and tangent to the ball S_2. A set $\Omega \subset R^n$ has the *cone property* if each $x \in \Omega$ can be taken as the vertex of a finite cone C_x contained in Ω and congruent to a suitable finite cone C.

Theorem 1.3. If Ω is an open set of R^n which has the cone property, then the imbeddings (1.38) - (1.41) are valid for $W^{m,p}(\Omega)$. □

Example 1.11. $W^{1,2}(R^1) \to C_B(R^1)$

If in particular $m = n = 1$, $p = 2$, $\Omega \equiv R^1$, then we have from Theorems 1.2 and 1.3 that

$$W^{1,2}(R^1) \to C_B(R^1) .$$

Such an imbedding property can be proved *directly* as follows. Given any $f \in W^{1,2}(R^1)$, $f' = df/dx$ exists in the generalized sense and it belongs to $L^2(R^1)$, i.e. $|f'|^2 \in L^1(R^1)$. Hence,

$$\begin{aligned} |f(x)|^2 &= f(x)\,\overline{f(x)} = \int_{-\infty}^{x} \frac{d}{dy}\,[f(y)\overline{f(y)}]\,dy \\ &= \int_{-\infty}^{x} [f'(y)\overline{f(y)}+f(y)\overline{f'(y)}]\,dy \le 2\int_{-\infty}^{x} |f'(y)|\ |f(y)|\,dy \\ &\le 2\|f'\|_2\,\|f\|_2 \le \|f\|_2^2+\|f'\|_2^2 = \|f\|_{1,2}^2 \end{aligned}$$

where we used the Schwarz inequality (1.8) and where

$$\|f\|_2 = \|f\|_{0,2} = [\int_{-\infty}^{+\infty} |f(y)|^2 dy]^{1/2}$$

$$\|f'\|_2 = \|f'\|_{0,2} = [\int_{-\infty}^{+\infty} |f'(y)|^2 dy]^{1/2}$$

Thus

$$|f(x)|^2 \leq \|f\|^2_{1,2} , \quad \forall\, x \in R^1$$

and so

$$\|f;\, C_B(R^1)\| = \sup\{|f(x)|,\, x \in R^1\} \leq \|f\|_{1,2}$$

and (1.37) holds with $\mu = 1$. □

1.5. SOBOLEV SPACES OF FRACTIONAL ORDER

We shall restrict here to define the fractional-order spaces $L^{s,p}(\Omega)$ with $s \geq 0$, $p = 2$, and $\Omega = R^n$ by using the Fourier-Plancherel transform (Kato 1966, p.259; Adams 1975, p.219). We first recall that the Fourier transform is defined for any $g \in L^1(R^n)$ as follows

$$\mathscr{F}g = (\mathscr{F}g)(y) = (2\pi)^{-n/2} \int \exp(-iy \cdot x) g(x)\, dx \tag{1.43}$$

where $i = \sqrt{-1}$ and where

$$y \cdot x = \sum_{j=1}^{n} y_j x_j \,, \quad \int \ldots\, dx = \int_{R^n} \ldots\, dx_1 \ldots dx_n \,.$$

To define the Fourier-Plancherel transform of any $f \in L^2(R^n)$, we introduce the sequence $\{f_N,\, N = 1,2,\ldots\}$ with

$$\left.\begin{aligned} f_N(x) &= f(x) \quad \text{if } |x| = [\sum_{j=1}^{n} x_j^2]^{1/2} \leq N \\ f_N(x) &= 0 \qquad \text{if } |x| > N. \end{aligned}\right\} \tag{1.44}$$

We have

Lemma 1.1. (i) $\{f_N\} \subset L^1(R^n) \cap L^2(R^n)$; (ii) $\lim \|f - f_N\|_2 = 0$ *as* $N \to \infty$; (iii) $\mathscr{F} f_N$ *exists, belongs to* $L^2(R^n)$, *and* $\{\mathscr{F} f_N\}$ *is a*

Cauchy sequence in $L^2(R^n)$.

Proof. We have from definition (1.44)

$$\|f_N\|_1 = \int_{|x|\le N} 1\cdot|f(x)|\,dx \le [\int_{|x|\le N} 1^2 dx \int_{|x|\le N} |f(x)|^2 dx]^{1/2}$$

$$\le [\int_{|x_j|\le N} dx]^{1/2}\|f\|_2 = (2N)^{n/2}\|f\|_2 < \infty,$$

$$\|f_N\|_2 = [\int_{|x|\le N} |f(x)|^2 dx]^{1/2} \le \|f\|_2 < \infty.$$

Thus, f_N belongs both to $L^1(R^n)$ and to $L^2(R^n)$ and (i) is proved. As far as (ii) is concerned, we obtain from (1.44)

$$\|f-f_N\|_2^2 = \int|f(x)-f_N(x)|^2 dx = \int_{|x|>N} |f(x)|^2 dx \to 0$$

as $N\to\infty$ because $f\in L^2(R^n)$ and so $|f|^2\in L^1(R^n)$. Finally, to prove (iii) we first remark that $\mathscr{F}f_N$ exists since $f_N\in L^1(R^n)$ because of (i). On the other hand, the Parseval relation (Titchmarsh 1948, Chapter 2)

$$\int|(\mathscr{F}F)(y)|^2 dy = \int |F(x)|^2 dx \quad , \quad \forall F\in L^1(R^n) \cap L^2(R^n) \quad (1.45)$$

shows that

$$\|\mathscr{F}f_N\|_2 = \|f_N\|_2 < \infty. \quad (1.46)$$

Moreover, using (1.45) again we have

$$\|\mathscr{F}f_N-\mathscr{F}f_M\|_2 = \|f_N-f_M\|_2 \le \|f_N-f\|_2+\|f-f_M\|_2$$

because $f_N-f_M\in L^1(R^n) \cap L^2(R^m)$ according to (i), and so

$$\|\mathscr{F}f_N-\mathscr{F}f_M\|_2 < \varepsilon, \ \forall N,M > N_\varepsilon$$

because of (ii). Thus $\{\mathscr{F}f_N\}$ is a Cauchy sequence in $L^2(R^n)$ and Lemma 1.1 is completely proved. □

Since $L^2(R^n)$ is a B-space, a unique $\hat{f}\in L^2(R^n)$ exists such that

$$\lim_{N\to\infty} \|\mathscr{F}f_N - \hat{f}\|_2 = 0 \ .$$

By definition, $\hat{f}$ is the *Fourier-Plancherel transform* of $f \in L^2(R^n)$ and it will be still denoted by the symbol $\mathscr{F}f$ (see Remark 1.4). We have the following theorem.

Theorem 1.4. The Fourier-Plancherel transform of any $f \in L^2(R^n)$ is defined by

$$\hat{f} = \hat{f}(y) = \lim_{N\to\infty} \mathscr{F}f_N = \lim_{N\to\infty}[(2\pi)^{-n/2} \int_{|x|\le N} \exp(-iy\cdot x)f(x)\,dx] \tag{1.47}$$

where the limit is in the sense of $\|\cdot\|^2$ and where

$$\|\hat{f}\|_2 = \|f\|_2 \quad \forall f \in L^2(R^n) \tag{1.48}$$

Proof. The existence of the limit (1.47) was proved in Lemma 1.1, and (1.48) follows directly from (1.46) and from (ii) of Lemma 1.1. □

Remark 1.4.

If $g \in L^1(R^n) \cap L^2(R^n)$, then the sequence $\{g_N\}$, defined by (1.44) with g instead of f, converges to g in the sense of both $\|\cdot\|_2$ (see (ii) of Lemma 1.1) and $\|\cdot\|_1$ (as it is easy to verify). Similarly, the sequence $\{\mathscr{F}g_N\}$ converges to the Fourier Transform $\mathscr{F}g$ of g in the sense of $\|\cdot\|_1$ and to the Fourier-Plancherel $\hat{g}$ in the sense of $\|\cdot\|_2$. However, the Fourier transform $\mathscr{F}g$ belongs to $L^2(R^n)$ because of (1.45) and it can be shown that $\mathscr{F}g$ and the Fourier-Plancherel transform $\hat{g}$ are the same element of $L^2(R^n)$. This justifies using the symbol $\mathscr{F}g$ for the Fourier-Plancherel tranform of g. □

We are now in position to define the spaces $L^{s,2}(R^n)$, $s \ge 0$, as follows

$$L^{s,2}(R^n) = \{f\colon f \in L^2(R^n);\ [(1+|y|^2)^{s/2}\hat{f}(y)] \in L^2(R^n)\} \tag{1.49}$$

where $\hat{f}$ is the Fourier-Plancherel transform of f. $L^{s,2}(R^n)$

(as a vector space) is a linear subset of $L^2(R^n)$; moreover, $L^{s,2}(R^n)$ is a Hilbert space with inner product and norm

$$\left.\begin{aligned} (f,g)_{s,2} &= \int_{R^n} (1+|y|^2)^s \, \hat{f}(y) \, \overline{\hat{g}(y)} \, dy \\ \|f\|_{s,2} &= [(f,f)_{s,2}]^{1/2} = \Big[\int_{R^n} (1+|y|^2)^s \, |\hat{f}(y)|^2 \, dy\Big]^{1/2} \end{aligned}\right\} \tag{1.50}$$

We finally point out that the spaces $L^{s,2}(R^n)$ may be considered as a generalization of the spaces $W^{m,2}(R^n)$, because it can be proved that $L^{s,2}(R^n)$ coincides with $W^{s,2}(R^n)$ if $s = 0,1,2,\ldots$ (Adams 1975). Furthermore, we have

Theorem 1.5.

(a) If $0 < s < n/2$, *then*

$$L^{s,2}(R^n) \to L^r(R^n), \qquad 2 \le r \le 2n/(n-2s) \; . \tag{1.51}$$

(b) If $s = n/2$, *then*

$$L^{s,2}(R^n) \to L^r(R^n), \qquad 2 \le r < \infty \; . \tag{1.52}$$

(c) If $s > n/2$, *then*

$$L^{s+j,2}(R^n) \to C^j_{\mathrm{B}}(R^n), \quad j = 0,1,2,\ldots \; . \tag{1.53}$$

EXERCISES

1.1. Show that the inner product (1.11) satisfies axioms (1.7).

1.2. Prove that the norm $\|\cdot\|_2$ of Example 1.1 is equivalent to $\|\cdot\|$ given by (1.11). Hint: $\|f\|_2^2 = [\sum_{i=1}^{n} |f_i|]^2 \le n \sum_{i=1}^{n} |f_i|^2$.

1.3. Prove that the norm (1.16) satisfies axioms (1.3).

1.4. Let $C^j([a,b])$ be the vector space of all complex-valued j-times continuously differentiable functions $f = f(x)$ on the closed interval $[a,b]$. Show that $C^j([a,b])$ is a B-space with norm

$$\| f;C^j([a,b]) \| = \max\{\| f;C \|;\| \mathrm{d}f/\mathrm{d}x;C \|; \ \ldots \ ;\| \mathrm{d}^j f/\mathrm{d}x^j;C \|\}$$

where

$$\| g;C \| = \max\{ |g(x)|, \ x \in [a,b]\}$$

is the norm in $C([a,b])$ (see Example 1.3).

1.5. Prove directly that the norm (1.26) satisfies the triangle inequality. Hint: $\| f+g \|^2 = (f+g,f+g) = \| f \|^2 + \| g \|^2 + 2\mathrm{Re}(f,g)$.

1.6. Show that the set $\{f: f \in C([a,b]); f(a) = f(b) = 0\}$ is a B-space with norm given (1.16).

1.7. Show that the norm $\| f \|^* = \| f_1 \|_1 + \| f_2 \|_2$ is equivalent to the norm (1.28).

1.8. Prove formula (1.30). Hint: multiply both sides of (1.29) by $c_1 \ldots$.

1.9. Prove that, if $m\Omega = \int_\Omega 1\mathrm{d}x < \infty$ and if $1 \le p' \le p'' \le \infty$ then each $f \in L^{p''}(\Omega)$ also belongs to $L^{p'}(\Omega)$ with $\| f \|_{p'} \le [m\Omega]^{1/p' - 1/p''} \| f \|_{p''}$. Hint: write $(\| f \|_{p'})^{p'} = \int_\Omega [1][|f(x)|^{p'}]\mathrm{d}x$ and use (1.36) with $p = p''/p'$.

1.10. Prove inequality (1.36) in the particular case $p = q = 2$. Hint: (1.36) with $p = q = 2$ is equivalent to

$$\int_\Omega \frac{|F(x)|}{\| F \|_2} \frac{|G(x)|}{\| G \|_2} \mathrm{d}x \le \frac{1}{2} + \frac{1}{2} = \frac{1}{2} \int_\Omega \frac{|F(x)|^2}{\| F \|_2^2} \mathrm{d}x + \frac{1}{2} \int_\Omega \frac{|G(x)|^2}{\| G \|_2^2} \mathrm{d}x$$

where we assume that $\| F \|_2 \| G \|_2 \ne 0$. Hence, the following inequality must be proved.

$$0 \le \int_\Omega \{|F(x)| \, \| G \|_2^2 + |G(x)| \, \| F \|_2^2 - 2|F(x)| \, \| G \|_2 \, |G(x)| \, \| F \|_2\}\mathrm{d}x \, .$$

1.11. Prove that the set $C^j(\overline{\Omega})$ of all $f(x)$, which are uniformly continuous and bounded on Ω together with their partial derivatives up to order j, is a B-space with norm

$$\| f; C^j(\overline{\Omega}) \| = \max_{0\le|k|\le j} \{\sup|D^k f(x)|,\ x\in\Omega\} .$$

Furthermore, show that $C^j(\overline{\Omega}) \subset C_B^j(\Omega)$.

1.12. Show that the finite rectangle $\Omega = \{x\colon x = (x_1,x_2),\ a_i < x_i < b_i,\ i = 1,2\}$ has the cone property.

1.13. Prove that $L^{s,2}(R^n)$ is a Hilbert space with inner product given by (1.50) and show that $L^{s,2}(R^n) \to L^2(R^n)$.

2

OPERATORS IN BANACH SPACES

2.1. NOTATION AND BASIC DEFINITIONS

Let X and Y be two given B-spaces over the same field $\underline{K}$ (see section 1.1), and let D be a subset of X. A mapping A that sends each $f \in D$ into a unique $g \in Y$ is called an *operator* with *domain* $D(A) = D$ and *range*

$$R(A) = \{g\colon g \in Y;\ g = A(f),\ f \in D(A)\}$$

contained in Y. The element $g = A(f)$ is the *image* of $f \in D(A)$ under the mapping A. In the following, we shall generally use the word 'operator' for mappings between X and Y with $X \neq R^1$ and reserve the word 'function' for mappings from a subset of R^1 into a B-space Y.

If B is another operator such that $D(B) \subset D(A)$ and $B(f) = A(f)\, \forall f \in D(B)$, then we say that B is a *restriction* of A and that A is an *extension* of B.

If the mapping A is one-to-one, i.e. if $A(f) \neq A(f_1)$ $\forall f, f_1 \in D(A)$ with $f \neq f_1$, then an *inverse* operator A^{-1} can be defined, with domain $D(A^{-1}) = R(A)$ and with range $R(A^{-1}) = D(A)$, such that

$$A^{-1}(A(f)) = f, \quad \forall f \in D(A)\ . \tag{2.1}$$

Note that, given the B-spaces X and Y, an operator A is properly defined if both its formal expression and its domain $D(A)$ are specified. The following simple example demonstrates how some basic features of an operator may depend on the definition of its domain.

Example 2.1. The operator $B(f) = f^2$ and its inverse

Let X and Y both coincide with the (real) B-space $C([0,1])$ of all real-valued continuous functions $f = f(x)$ on $[0,1]$. If we put

$$A(f) = f^2\ , \quad D(A) = C([0,1]) \tag{2.2}$$

$$B(f) = f^2 , \quad D(B) = \{f\colon f\in C([0,1]);\ f(x) \geq 0\ \forall x\in[0,1]\} \tag{2.3}$$

then A and B have the same formal expression but different domains. In fact, B is a restriction of A. Moreover, the inverse of A does not exist because f^2 is the image of both $-f$ and $+f$, whereas we have

$$B^{-1}(g) = \surd g, \quad \forall g\in D(B^{-1}) = R(B) \tag{2.4}$$

where $\surd g$ indicates the continuous function $(\surd g)(x) = \surd\{g(x)\}$ $\forall x\in[0,1]$. □

Of course, the properties of an operator A also depend on the nature of the spaces $X \supset D(A)$ and $Y \supset R(A)$, which are usually chosen in such a way that the norms $\|\cdot;X\|$ and $\|\cdot;Y\|$ have some simple 'physical' meaning.

2.2. BOUNDED LINEAR OPERATORS

An operator A with domain $D(A) \subset X$ and with range $R(A) \subset Y$ is *linear* if

(i) $D(A)$ is a linear subset of X (see Example 1.5);
(ii) $A(\alpha f+\beta f_1) = \alpha A(f)+\beta A(f_1)$, $\forall f,f_1\in D(A)$, $\forall \alpha,\beta\in \underline{K}$.

We point out that $D(A)$ has to be a linear subset of X so that $\alpha f+\beta f_1\in D(A)$ $\forall f,f_1\in D(A)$, $\forall\alpha,\beta\in\underline{K}$.

In the following, we shall write Af, rather than $A(f)$, when A is a linear operator.

We have the following lemma.

Lemma 2.1. If A is linear, then the inverse operator A^{-1} exists if and only if the equation $Af = \theta_Y$ has only the trivial solution $f = \theta_X$, where θ_X and θ_Y denote the zero elements of X and of Y.

Proof. Let $Af = \theta_Y$ imply $f = \theta_X$ and let f_1 and f_2 belong to $D(A)$, with $f \neq f_1$. Then, $Af_1-Af_2 = A[f_1-f_2] \neq \theta_Y$ and $Af_1 \neq Af_2$, i.e. A is one-to-one. Note that $\theta_X\in D(A)$ because

$D(A)$ is a linear subset of X and so $\theta_X = 0g \in D(A)$ with $g \in D(A)$. Conversely, if A^{-1} exists, then A is one-to-one and $Af_1 \neq Af_2$ $\forall f_1, f_2 \in D(A)$ with $f_1 \neq f_2$. However, A is linear and so $A\theta_X = \theta_Y$ (see Exercise 2.1); thus if we also had $Af = \theta_Y$, then $f = \theta_X$, i.e. $Af = \theta_Y$ implies $f = \theta_X$. □

The linear operator A is *bounded* (over $D(A)$) if a constant M, independent of $f \in D(A)$, can be found such that

$$\| Af; Y \| \leq M \| f; X \| , \quad \forall f \in D(A) \tag{2.5}$$

where $\| Af; Y \|$ is the norm in Y of $Af \in Y$ and $\| f; X \|$ is the norm in X of $f \in D(A) \subset X$. Of course, the constant M may depend in general on the particular A under consideration. The *norm* of the linear bounded operator A, denoted by $\| A \|$, is defined by

$$\| A \| = \sup \left\{ \frac{\| Af; Y \|}{\| f; X \|} , \; f \in D(A), \; \| f; X \| \neq 0 \right\} . \tag{2.6}$$

We shall also use the symbol $\| A; X, Y \|$, rather than $\| A \|$, whenever it is necessary to point out that $D(A) \subset X$ and $R(A) \subset Y$. We have

Lemma 2.2. The norm of the linear bounded operator A is the least constant M that can be used in (2.5).

Proof. Definition (2.6) gives

$$\| Af; Y \| \leq \| A \| \, \| f; X \| \quad \forall f \in D(A) \tag{2.7}$$

and so $\| A \|$ is one of the constants M. On the other hand, we have from (2.5)

$$\frac{\| Af; Y \|}{\| f; X \|} \leq M, \quad \forall f \in D(A), \quad \| f; X \| \neq 0$$

and by taking the supremum we obtain $\| A \| \leq M$. Thus $\| A \|$ is the least of all constants M. □

Remark 2.1.

Since $\|A\|$ is the least of all constants M, (2.7) is the 'best' inequality of the type (2.5). Note that, if a particular M_0 is known for which (2.5) holds, then $\|A\| \leq M_0$ (see Examples 2.2-2.7). □

The family of all bounded linear operators with domain $D \subset X$ and with range contained in Y will be denoted by $\mathscr{B}(D,Y)$ or by $\mathscr{B}(X)$ if $D = X = Y$. Thus, any $B \in \mathscr{B}(D,Y)$ is a linear bounded operator with domain $D(B) = D$ and inequality (2.7) holds with a suitable $\|B\|$ and $\forall f \in D$. We have the following theorem

Theorem 2.1. $\mathscr{B}(D,Y)$ is a B-space with norm $\|A;\mathscr{B}(D,Y)\| = \|A\|$.

Proof. $\mathscr{B}(D,Y)$ is a vector space over the field $\underline{K}$ with the operations + and $\alpha\cdot$ defined as follows:

$$(A+B)f = Af+Bf \ , \quad (\alpha A)f = \alpha Af \ , \quad \forall f \in D, \ \forall \alpha \in \underline{K}$$

i.e. the operations + and $\alpha\cdot$ in $\mathscr{B}(D,Y)$ are defined in terms of the corresponding ones in Y. The zero element 0_{XY} of $\mathscr{B}(D,Y)$ is the operator that sends each element of D into θ_Y, the zero element of Y.

Furthermore, $\|A\|$ satisfies axioms (1.3). In fact, according to definition (2.6), $\|A\|$ is certainly non-negative and $\|0_{XY}\| = 0$ because $\|0_{XY}f;Y\| = \|\theta_Y;Y\| = 0 \ \forall f \in D$. On the other hand, if $\|A\| = 0$, then $\|Af;Y\| = 0 \ \forall f \in D$, i.e. $Af = \theta_Y$ $\forall f \in D$ and $A = 0_{XY}$. Thus $\|A\|$ satisfies (1.3a). Moreover, we have from (2.6)

$$\|\alpha A\| = \sup\left\{\frac{|\alpha|\,\|Af;Y\|}{\|f;X\|}\ , \ f \in D, \ \|f;X\| \neq 0\right\} = |\alpha| \ \|A\|$$

$$\|A+B\| = \sup\left\{\frac{\|(A+B)f;Y\|}{\|f;X\|}\ , \ f \in D, \ \|f;X\| \neq 0\right\}$$

$$\leq \sup\left\{\frac{\|Af;Y\|}{\|f;X\|} + \frac{\|Bf;Y\|}{\|f;X\|}\ , \ f \in D, \ \|f;X\| \neq 0\right\} \leq \|A\| + \|B\|.$$

and $\|A\|$ also satisfies axioms (1.3b) and (1.3c). We conclude that $\|A\|$ is indeed a norm.

We finally show that $\mathscr{B}(D,Y)$ is a B-space, i.e. that it is a complete normed space. To see this, let $\{A_n, n=1,2,\ldots\}$ be a Cauchy sequence in $\mathscr{B}(D,Y)$:

$$\|A_n - A_m\| < \varepsilon, \quad \forall n,m > n_\varepsilon. \tag{2.8}$$

where n_ε depends only on ε. Given any $f \in D$, put $g_n = A_n f$, $n = 1,2,\ldots$; then we obtain from (2.7) and from (2.8)

$$\|g_n - g_m; Y\| = \|(A_n - A_m)f; Y\| \le \|A_n - A_m\| \|f; X\| \le \varepsilon \|f; X\| \quad \forall n,m > n_\varepsilon$$

and $\{g_n = A_n f, n = 1,2,\ldots\}$ is a Cauchy sequence in the B-space Y. Hence, a $g \in Y$ exists such that $\|g - g_n; Y\| = \|g - A_n f; Y\| \to 0$ as $n \to \infty$. Of course, such a $g \in Y$ depends on the element $f \in D$ from which we started: $g = G(f)$. The mapping G, which is defined over the whole D, is linear because

$$\|G(\alpha f + \beta f_1) - \alpha G(f) - \beta G(f_1); Y\| = \|G(\alpha f + \beta f_1) - A_n[\alpha f + \beta f_1] + \alpha A_n f +$$

$$+ \beta A_n f_1 - \alpha G(f) - \beta G(f_1); Y\|$$

$$\le \|G(\alpha f + \beta f_1) - A_n[\alpha f + \beta f_1]; Y\| + |\alpha| \|A_n f - G(f); Y\| +$$

$$+ |\beta| \|A_n f_1 - G(f_1); Y\| \to 0$$

as $n \to \infty$, $\forall f, f_1 \in D$, $\forall \alpha, \beta \in \underline{K}$, because $\|G(\phi) - A_n \phi; Y\| \to 0$ as $n \to \infty$, $\forall \phi \in D$, as we have proved above. Thus, $G(\alpha f + \beta f_1) = \alpha G(f) + \beta G(f_1)$. The linear operator G also satisfies (2.5), i.e. $G \in \mathscr{B}(D,Y)$. To see this, write $\alpha_n = \|A_n\| \ge 0$, $n = 1,2,\ldots$; then $\{\alpha_n, n = 1,2,\ldots\}$ is a Cauchy sequence in R^1 because

$$|\alpha_n - \alpha_m| = |\|A_n\| - \|A_m\|| \le \|A_n - A_m\| < \varepsilon \quad \forall n,m > n_\varepsilon$$

owing to (1.4). Hence, a non-negative $\alpha_0 \in R^1$ exists such that $|\alpha_m - \alpha_0| < \varepsilon \; \forall n > n_\varepsilon$, i.e.

$$\alpha_0 - \varepsilon < \alpha_n < \alpha_0 + \varepsilon, \quad \forall n \ge n_0 = n_\varepsilon + 1$$

and also

$0 \le \alpha_n \le M = \max\{\alpha_0+\varepsilon; \alpha_1; \alpha_2; \ldots; \alpha_{n_0}\}, \quad \forall n = 1,2,\ldots$.

In other words, the sequence $\{A_n, n = 1,2,\ldots\} \subset \mathscr{B}(D,Y)$ is *bounded*:

$$0 \le \| A_n \| \le M, \quad \forall n = 1,2,\ldots .$$

Thus,

$$\| A_n f; Y \| \le \| A_n \| \, \| f; X \| \le M \| f; X \|, \quad \forall f \in D, \ \forall n = 1,2,\ldots$$

and passing to the limit as $n\to\infty$ we obtain

$$\| Gf; Y \| \le M \| f; X \| , \quad \forall f \in D .$$

We conclude that G is bounded over D and so $G \in \mathscr{B}(D,Y)$.

To complete the proof of Theorem 2.1, we have to show that $\| G-A_n; \mathscr{B}(D,Y) \| = \| G-A_n \| \to 0$ as $n\to\infty$, i.e. that the sequence $\{A_n\}$ converges to G in the sense of the norm of $\mathscr{B}(D,Y)$. We have from (2.7) and from (2.8)

$$\| A_n f - Gf; Y \| \le \| A_n f - A_m f; Y \| + \| A_m f - Gf; Y \|$$

$$\le \| A_n - A_m \| \, \| f; X \| + \| A_m f - Gf; Y \| \le \varepsilon \| f; X \| + \| A_m f - Gf; Y \|$$

$$\forall n,m > n_\varepsilon, \quad \text{and} \quad \text{as } m\to\infty$$

$$\| A_n f - Gf; Y \| \le \varepsilon \| f; X \| , \quad \forall n > n_\varepsilon, \ \forall f \in D$$

where we recall that n_ε depends *only* on ε. Hence

$$\| A_n - G \| = \sup\left\{\frac{\| A_n f - Gf; Y \|}{\| f; X \|}, \ f \in D, \ \| f; X \| \ne 0\right\} < \varepsilon, \ n > n_\varepsilon$$

and $\| An - G \| \to 0$ as $n\to\infty$. Thus $A_n \to G$ in $\mathscr{B}(D,Y)$ and $\mathscr{B}(D,Y)$ is a complete space. □

Remark 2.2.

The elements of the B-space $\mathscr{B}(D,Y)$ are bounded linear operators.

Thus $\mathscr{B}(D,Y)$ is an example of a B-space whose elements are not functions in the elementary sense. □

Remark 2.3.

If in particular $D = X$ and Y is the B-space $\mathbb{C}$ of all complex numbers with norm $\|\cdot;\mathbb{C}\| = |\cdot|$, then the elements of $\mathscr{B}(X,\mathbb{C})$ are called complex-valued bounded linear *functionals* on X. □

2.3. EXAMPLES OF LINEAR BOUNDED OPERATORS

Example 2.2. Operators on R^n

The linear operator

$$Af = \begin{pmatrix} a_{11},\ldots,a_{1n} \\ a_{21},\ldots,a_{2n} \\ \ldots\ldots\ldots \\ a_{m1},\ldots,a_{mn} \end{pmatrix} \begin{pmatrix} f_1 \\ f_2 \\ \ldots \\ f_n \end{pmatrix} , \quad D = R^n , \quad R(A) \subset R^m$$

where the a_{ij}'s are given real numbers belongs to the real B-space $\mathscr{B}(R^n,R^m)$. In fact if we put $g = Af$, then we have

$$g = \begin{pmatrix} g_1 \\ \ldots \\ g_m \end{pmatrix} , \quad g_i = \sum_{j=1}^{n} a_{ij} f_j , \quad i = 1,2,\ldots,m$$

$$\|Af;R^m\|^2 = \|g;R^m\|^2 = \sum_{i=1}^{m} |g_i|^2$$

$$\leq \sum_{i=1}^{m} [\sum_{j=1}^{n} |a_{ij}| \, |f_j|]^2$$

$$= \sum_{i=1}^{m} [(\gamma,\delta)_{R^n}]^2$$

where $(\cdot,\cdot)_{R^n}$ is the inner product in R^n (see Example 1.1) and where

$$\gamma = \begin{pmatrix} |a_{i1}| \\ \cdots \\ |a_{in}| \end{pmatrix}, \quad \delta = \begin{pmatrix} |f_1| \\ \cdots \\ |f_n| \end{pmatrix} .$$

Using Schwarz inequality (1.8), we obtain

$$\|Af;R^m\|^2 \le \sum_{i=1}^{m} [\|\gamma;R^n\|^2 \ \|\delta;R^n\|^2]$$

$$= \sum_{i=1}^{m} [\sum_{j=1}^{n} |a_{ij}|^2 \ \sum_{j=1}^{n} |f_j|^2]$$

$$= [\sum_{i=1}^{m} \sum_{j=1}^{n} |a_{ij}|^2] \ \|f;R^n\|^2 .$$

Hence,

$$\|Af;R^m\| \le M_0 \ \|f;R^n\|, \quad \forall f \in R^n \tag{2.9}$$

$$M_0 = [\sum_{i=1}^{m} \sum_{j=1}^{n} |a_{ij}|^2]^{1/2}$$

and consequently (see Remark 2.1)

$$\|A\| = \|A;R^n,R^m\| \le [\sum_{i=1}^{m} \sum_{j=1}^{n} |a_{ij}|^2]^{1/2} . \quad \square \tag{2.10}$$

Example 2.3. Operators on l^1.

Consider the linear operator

$$Af = \begin{pmatrix} a_{11}, a_{12}, \ldots, a_{1n}, \ldots \\ a_{21}, a_{22}, \ldots, a_{2n}, \ldots \\ \cdots\cdots\cdots\cdots\cdots\cdots \end{pmatrix} \begin{pmatrix} f_1 \\ f_2 \\ \cdots \end{pmatrix} , \quad f = \begin{pmatrix} f_1 \\ f_2 \\ \cdots \end{pmatrix} \in l^1$$

where the a_{ij}'s are given complex numbers and where the B-space l^1 was defined in Example 1.2. If we put $g = Af$, then we have

$$g = \begin{pmatrix} g_1 \\ g_2 \\ \cdots \end{pmatrix} , \quad g_i = \sum_{j=1}^{\infty} a_{ij} f_j , \quad i = 1,2,\ldots$$

$$\|g\|_1 = \|Af\|_1 = \sum_{i=1}^{\infty} |g_i| = \sum_{i=1}^{\infty} |\sum_{j=1}^{\infty} a_{ij} f_j|$$

$$\le \sum_{i=1}^{\infty} \sum_{j=1}^{\infty} |a_{ij}| |f_j| = \sum_{j=1}^{\infty} [\sum_{i=1}^{\infty} |a_{ij}|] |f_j|$$

where $\|\cdot\|_1 = \|\cdot;l^1\|$ is the norm in l^1. Thus if we assume that the infinite matrix which defines A is such that

$$\sum_{i=1}^{\infty} |a_{ij}| \le M_0 < \infty, \quad \forall j = 1,2,... \tag{2.11}$$

then we obtain

$$\|Af\|_1 \le M_0 \|f\|_1, \quad \forall f \in l^1 \tag{2.12}$$

and so

$$\|A\| = \|A;\, l^1, l^1\| \le M_0 . \tag{2.13}$$

Inequality (2.12) shows that A is bounded and that it is defined over the whole l^1. Hence, $A \in \mathscr{B}(l^1, l^1) = \mathscr{B}(l^1)$, provided that condition (2.11) is satisfied. □

Example 2.4. The Fredholm integral operator on $L^2(a,b)$.

The following (Fredholm) integral operator

$$Af = \int_a^b k(x,y) f(y)\,dy , \quad f \in L^2(a,b)$$

belongs to $\mathscr{B}(L^2(a,b))$ provided that the *kernel* $k = k(x,y)$ is an element of $L^2((a,b)\times(a,b))$, i.e.

$$|||\, k \,||| = [\int_a^b dx \int_a^b |k(x,y)|^2 dy]^{1/2} < \infty . \tag{2.14}$$

In fact, if $g = g(x) = Af$, then we have

$$|g(x)|^2 \le [\int_a^b |k(x,y)|\ |f(y)|\,dy]^2$$

$$\le \int_a^b |k(x,y)|^2 dy \int_a^b |f(y)|^2 dy$$

$$= \int_a^b |k(x,y)|^2 dy \; \|f\|_2^2$$

where $\|\cdot\|_2 = \|\cdot;L^2(a,b)\|$ is the norm in $L^2(a,b)$ and where we use the Schwarz inequality. Thus,

$$\|Af\|_2^2 = \|g\|_2^2 = \int_a^b |g(x)|^2 dx \le |||k|||^2 \; \|f\|_2^2$$

$$\|Af\|_2 \le |||k||| \; \|f\|_2, \quad \forall f \in L^2(a,b) \tag{2.15}$$

and so

$$\|A\| = \|A;L^2(a,b),L^2(a,b)\| \le |||k||| \; . \tag{2.16}$$

Inequality (2.15) shows that A is defined over the whole $L^2(a,b)$ and that $A \in \mathscr{B}(L^2(a,b))$, provided that $|||k||| < \infty$.

Note that A can also be considered as an operator from $C([a,b])$ into itself or into $L^2(a,b)$, if $k = k(x,y) \in C([a,b]\times[a,b])$, with $-\infty < a < b < +\infty$ (see Exercise 2.3). □

Example 2.5. The operator $A = \mathrm{d}/\mathrm{d}x$, *with* $D(A) = W^{1,2}(R^1)$ *and* $R(A) \subset L^2(R^1)$

The linear operator

$$Af = \frac{\mathrm{d}f}{\mathrm{d}x}, \quad f \in W^{1,2}(R^1) = D(A), \quad R(A) \subset L^2(R^1)$$

where $\mathrm{d}f/\mathrm{d}x$ is a generalized derivative, belongs to $\mathscr{B}(W^{1,2}(R^1),L^2(R^1))$ because

$$\|Af\|_2^2 = \left\|\frac{\mathrm{d}f}{\mathrm{d}x}\right\|_2^2 \le \|f\|_2^2 + \left\|\frac{\mathrm{d}f}{\mathrm{d}x}\right\|_2^2 = \|f\|_{1,2}^2$$

where $\|\cdot\|_{1,2}$ is the norm in $W^{1,2}(R^1)$ (see section 1.4), and so

$$\|A\| = \|A;W^{1,2}(R^1),L^2(R^1)\| \le 1.$$

We remark that the operator

$$Af = \frac{\mathrm{d}f}{\mathrm{d}x}, \quad D(A) = C^1([0,1]) \subset C([0,1]), \quad R(A) \subset C([0,1])$$

where df/dx is now a classical derivative, does *not* belong to $\mathscr{B}(C([0,1,]))$. In fact, consider the sequence

$$\{f_n = \sin(nx),\ n = 1,2,\ldots\} \subset D(A);$$

then

$$0 < \|f_n\| = \max\{|\sin(nx)|\ ,\ \ x\in[0,1]\} \le 1,\ \ \forall n = 1,2,\ldots$$

where $\|\cdot\|$ is the norm in $C([0,1])$, and also

$$\|Af_n\| = \|n\ \cos(nx)\| = n\ \max\{|\cos(nx)|\ ,\ x\in[0,1]\} = n.$$

Thus, we have

$$\|Af_n\| \ge n\|j_n\|, \quad n = 1,2,\ldots\ ;$$

consequently $\|Af_n\|/\|f_n\|$ is not bounded as $n\to\infty$ and $\|A\|$ cannot be finite (see (2.6)). □

Example 2.6. Extension of a densely defined bounded operator

Let the bounded linear operator B have domain $D(B) \subset X$ and range $R(B) \subset Y$. If $D(B)$ is a dense subset of X, then we say that B is *densely defined*. Under the assumption that B is densely defined, given any $f\in X$ and $\varepsilon > 0$, a suitable $f_0\in D(B)$ can be found such that $\|f-f_0;X\| < \varepsilon$ (see section 1.4, after Example 1.10). Hence, if we choose $\varepsilon = 1/n$, $n = 1,2,\ldots$, a sequence $\{f_n\} \subset D(B)$ can be determined such that $\|f-f_n;X\| < 1/n$, i.e. such that $X\text{-}\lim f_n = f$ as $n\to\infty$. Consider then the sequence $\{g_n = Bf_n,\ n = 1,2,\ldots\} \subset Y$, for which we have

$$\|g_n-g_m;Y\| = \|B[f_n-f_m];Y\| \le \|B\|\ \|f_n-f_m;X\|$$

$$\le \|B\|\{\|f_n-f;X\| + \|f-f_m;X\|\} \le \left(\frac{1}{n}+\frac{1}{m}\right)\|B\|\ .$$

Thus, $\{Bf_n\}$ is a Cauchy sequence in the B-space Y and a $g\in Y$ exists such that $g = Y\text{-}\lim Bf_n$ as $n\to\infty$. Note that the element $g\in Y$ is *uniquely* determined by the element $f\in X$ from which we started.

In fact, if $f = X\text{-lim } \phi_n$, i.e. if f is also the X-limit of another sequence $\{\phi_n\} \subset D(B)$, then we obtain

$$\|Bf_n - B\phi_n; Y\| \le \|B\| \, \|f_n - \phi_n; X\|$$

$$\le \|B\| \{\|f_n - f; X\| + \|f - \phi_n; X\|\} \to 0 \quad \text{as } n\to\infty$$

and $Y\text{-lim } Bf_n = Y\text{-lim } B\phi_n$. Hence, we can write that $g = A(f)$ $\forall f \in X$ and the operator A with domain $D(A) = X$ is properly defined by the relation

$$A(f) = Y\text{-} \lim_{n\to\infty} Bf_n \tag{2.17}$$

where $\{f_n\}$ is *any* sequence contained in $D(B)$ and convergent to f in the sense of the norm of X.

It is easy to verify that A is linear since B has such a property (see the proof of Theorem 2.1); moreover, we have from (2.17)

$$\big| \|Af; Y\| - \|Bf_n; Y\| \big| \le \|Af - Bf_n; Y\| \to 0$$

as $n\to\infty$ and so

$$\|Af; Y\| = \lim_{n\to\infty} \|Bf_n; Y\|$$

$$\|Af; Y\| \le \lim_{n\to\infty} \|B\| \, \|f_n; Y\| = \|B\| \, \|f; Y\| \, , \quad \forall f \in X$$

because $\|Bf_n; Y\| \le \|B\| \, \|f_n; X\|$ $\forall n = 1,2,\ldots$. Hence, $\|A\| = \|A; X, Y\| \le \|B\|$ and $A \in \mathscr{B}(X,Y)$.

The operator A is an extension of B (see section 2.1), since $Bf = Af$ $\forall f \in D(B) \subset D(A) = X$. In fact, if $f \in D(B)$, then the sequence $\{f_n = f, \ n = 1,2,\ldots\}$ belongs to $D(B)$ and it is obviously such that $X\text{-lim } f_n = f$ and $Af = Y\text{-lim } Bf_n = Bf$. Finally, since

$$\|A\| = \sup\left\{\frac{\|Af; Y\|}{\|f; X\|} \, , \quad f \in D(A) = X, \quad \|f; X\| \ne 0\right\}$$

$$\ge \sup\left\{\frac{\|Af; Y\|}{\|f; X\|} \, , \quad f \in D(B), \quad \|f; X\| \ne 0\right\} = \|B\|$$

we conclude that $\|A\| = \|B\|$.

Thus, a linear bounded operator B, which is densely defined in X, can be extended to the whole space X by the above procedure, without changing its norm. □

Example 2.7. Linear functionals on $L^2(a,b)$.

The linear operator

$$Hf = \int_a^b \overline{h(y)} f(y)\,dy = (f,h), \quad f \in L^2(a,b) ,$$

where h is a *given* element of $L^2(a,b)$ is a bounded linear functional on $L^2(a,b)$, i.e. $H \in \mathscr{B}(L^2(a,b),\mathbb{C})$ (see Remark 2.3), and $\|H\| = \|h\|_2$. In fact,

$$\|Hf;\mathbb{C}\| = |Hf| = |(f,h)| \le \|h\|_2 \|f\|_2 \ , \ \forall f \in L^2(a,b)$$

and so

$$\|H\| = \|H;L^2(a,b),\mathbb{C}\| \le \|h\|_2$$

(see Remark 2.1). Since $h \in L^2(a,b)$, we also have

$$Hh = (h,h) = \|h\|_2^2$$

$$\|Hh;\mathbb{C}\| = |Hh| = \|h\|_2^2 \ .$$

However, according to (2.7)

$$\|Hh;\mathbb{C}\| \le \|H\| \, \|h\|_2$$

and consequently

$$\|h\|_2^2 \le \|H\| \, \|h\|_2, \quad \|h\|_2 \le \|H\| \quad (\|h\|_2 \ne 0) .$$

Since we have already proved that $\|H\| \le \|h\|_2$, we conclude that

$$\|H\| = \|h\|_2 \ . \quad □$$

2.4. LIPSCHITZ OPERATORS

An operator A, with domain $D(A) \subset X$ and with range $R(A) \subset Y$, is a *Lipschitz* operator if

$$\|A(f)-A(f_1);Y\| \leq M\|f-f_1;X\|, \quad \forall f,f_1 \in D(A) \tag{2.18}$$

where the constant M does not depend on f and f_1, but it does depend in general on the particular A under consideration. If A is a Lipschitz operator, its *Lipschitz constant* is defined by the relation

$$\nu(A) = \sup\left\{\frac{\|A(f)-A(f_1);Y\|}{\|f-f_1;X\|}, \ f,f_1 \in D(A), \ f \neq f_1\right\}. \tag{2.19}$$

It follows from (2.18) and from definition (2.19) that $0 \leq \nu(A) < \infty$ and that

$$\|A(f)-A(f_1);Y\| \leq \nu(A)\|f-f_1;X\|, \quad \forall f,f_1 \in D(A) \tag{2.20}$$

provided that $f \neq f_1$. However, (2.20) is valid even if $f = f_1 \in D(A)$ because then $\|A(f)-A(f_1);Y\| = 0$ owing to (2.18). Hence, $\nu(A)$ is a particular constant M for which (2.18) holds. On the other hand, we have from (2.18)

$$\frac{\|A(f)-A(f_1);Y\|}{\|f-f_1;X\|} \leq M, \quad \forall f,f_1 \in D(A), \quad f \neq f_1$$

and taking the supremum we obtain that $\nu(A) \leq M$. We conclude that $\nu(A)$ is the *least* constant M which can be used in (2.18).

In what follows, we shall use the symbol $\mathrm{Lip}(D,Y)$ to denote the family of all Lipschitz operators with domain $D \subset X$ and with range contained in Y (Martin 1976, p.63). Note that

$$\mathscr{B}(D,Y) \subset \mathrm{Lip}(D,Y) \tag{2.21}$$

i.e. $\mathscr{B}(D,Y)$ is a (linear) subset of $\mathrm{Lip}(D,Y)$. To see this, assume that A is linear and that it satisfies (2.5) with $D(A) = D$:

$$\| A\phi ; Y \| \le M \| \phi ; X \| \ , \quad \forall\ \phi \in D.$$

However, D is now a linear subset of X and so $\phi = f - f_1 \in D$, $\forall f, f_1 \in D$. Thus (2.18) follows from the preceding inequality and $A \in \mathscr{B}(D, Y)$ implies $A \in \mathrm{Lip}(D, Y)$. Moreover, $\nu(A) = \| A \|$, as it is easy to verify.

Example 2.8. The operator $A(f) = f^2$ with $D(A) \subset C([a,b])$

The operator

$$\left.\begin{aligned} &A(f) = f^2 \\ &D(A) = D = \{f : f \in C([a,b]) ; \| f \| \le r\}, \quad R(A) \subset C([a,b]) \end{aligned}\right\} \quad (2.22)$$

belongs to $\mathrm{Lip}(D, C([a,b]))$ because we have from (2.22)

$$A(f) - A(f_1) = [f + f_1][f - f_1]$$

$$\| A(f) - A(f_1) \| = \max\{[|f(x) + f_1(x)| \, |f(x) - f_1(x)|], x \in [a,b]\}$$

$$\le \max\{|f(x) + f_1(x)|, x \in [a,b]\} \max\{|f(x) - f_1(x)|, x \in [a,b]\}$$

$$= \| f + f_1 \| \, \| f - f_1 \| \le (\| f \| + \| f_1 \|) \| f - f_1 \| \ , \ \forall f, f_1 \in C([a,b])$$

where $\| \cdot \|$ is the norm in $C([a,b])$. Hence,

$$\| A(f) - A(f_1) \| \le 2r \| f - f_1 \|, \ \forall f, f_1 \in D$$

and $A \in \mathrm{Lip}(D, C([a,b]))$ with $\nu(A) \le 2r$. Note that $A \notin \mathrm{Lip}(C([a,b], C([a,b]))$ because, if $D = C([a,b])$, then $(\| f \| + \| f_1 \|)$ is not bounded by any constant M independent of f and of f_1. □

Example 2.9. The operator $A(f) = \mu f^2/(1+f^2)$ with $D(A) = L^2(a,b)$

The operator

$$A(f) = \mu \frac{f^2}{1+f^2} \ , \quad D(A) = L^2(a,b) \ , \quad R(A) \subset L^2(a,b) \quad (2.23)$$

where μ is a given real constant, belongs to $\mathrm{Lip}(L^2(a,b), L^2(a,b))$ (here $L^2(a,b)$ is the real B-space of real-valued square-summable functions on (a,b)). Note first that A is indeed defined over the whole $L^2(a,b)$ because

$$[\|A(f)\|_2]^2 = \int_a^b \left| \frac{\mu f^2(x)}{1+f^2(x)} \right|^2 dx$$

$$= \mu^2 \int_a^b \left[\frac{|f(x)|}{1+|f(x)|^2}\right]^2 |f(x)|^2\, dx$$

$$\leq \frac{\mu^2}{4} \int_a^b |f(x)|^2\, dx$$

$$= \frac{\mu^2}{4} [\|f\|_2]^2 < \infty, \quad \forall f \in L^2(a,b)$$

where we took into account that $|f(x)|/[1+|f(x)|^2] \leq 1/2$, as can easily be verified. Further, given any $f, f_1 \in D(A)$, we have from (2.23)

$$|A(f)-A(f_1)| = |\mu| \frac{|f^2(x)-f_1^2(x)|}{[1+f^2(x)][1+f_1^2(x)]}$$

$$\leq |\mu| \frac{|f(x)|+|f_1(x)|}{[1+f^2(x)][1+f_1^2(x)]} |f(x)-f_1(x)|$$

$$\leq |\mu|(1/2+1/2)\,|f(x)-f_1(x)|$$

because

$$\frac{|f|}{\{[1+f^2][1+f_1^2]\}} \leq \frac{|f|}{[1+f^2]} \leq \frac{1}{2} \ .$$

Hence

$$[\|A(f)-A(f_1)\|_2]^2 \leq \mu^2 \int_a^b |f(x)-f_1(x)|^2 dx$$

$$= \mu^2 [\|f-f_1\|_2]^2$$

and A is a Lipschitz operator with $\nu(A) \leq |\mu|$. □

Example 2.10. The operator $A(f) = f^2$ with $D(A) \subset W^{1,2}(R^1)$

Let $X = W^{1,2}(R^1)$ be the (real) Sobolev space

$$W^{1,2}(R^1) = \{f: f\in L^2(R^1); f'\in L^2(R^1)\}$$

$$\|f\|_{1,2} = [\|f\|^2+\|f'\|^2]^{1/2}$$

$$\|f\| = [\int_{-\infty}^{+\infty} |f(x)|^2 dx]^{1/2}$$

where $f = f(x)$ is real valued, $f' = df/dx$ is a generalized derivative (see sections 1.3, 1.4) and where $\|\cdot\| = \|\cdot\|_2$ is the norm in the (real) space $L^2(a,b)$ (see Example 1.4).

The operator

$$\left.\begin{aligned} &A(f) = f^2,\ D(A) = D = \{f: f\in X;\ \|f-f_0\|_{1,2} \le r\} \subset X = W^{1,2}(R^1) \\ &R(A) \subset X \end{aligned}\right\} \tag{2.24}$$

where f_0 is a given element of X, belongs to $\text{Lip}(D,X)$.

To verify that A is properly defined on D, we have to show that $f^2\in X\ \forall f\in D$. More generally, we shall prove the following inequality

$$\|fg\|_{1,2} \le 2\ \|f\|_{1,2}\ \|g\|_{1,2}\quad \forall f,g\in X\ ; \tag{2.25}$$

hence, if $f = g\in X$, then $\|f^2\|_{1,2} \le 2\|f\|^2_{1,2} < \infty$ and $f^2\in X$. Given any $f\in X$, we have

$$f^2(x) = \int_{-\infty}^{x} \frac{df^2}{dy}(y)\,dy = 2\int_{-\infty}^{x} f(y)f'(y)\,dy$$

since both f and f' belong to $L^2(R^1)$ and so $(ff')\in L^1(R^1)$. Thus,

$$f^2(x) \le 2[\int_{-\infty}^{x} |f(y)|^2 dy \int_{-\infty}^{x} |f'(y)|^2 dy]^{1/2} \le 2\ \|f\|\ \|f'\|$$

$$\le \|f\|^2 + \|f'\|^2 = \|f\|^2_{1,2} \quad \text{for almost all } x\in R^1$$

and in an analogous way

$$g^2(x) \le \|g\|^2_{1,2} \quad \text{for almost all } x\in R^1$$

because both f and g belong to X. It follows that

$$\left.\begin{aligned}\|fg\|^2 &= \int_{-\infty}^{+\infty} |f(x)|^2|g(x)|^2dx \le \|f\|_{1,2}^2 \int_{-\infty}^{+\infty} |g(x)|^2dx = \|f\|_{1,2}^2 \|g\|^2 \\ \|fg'\|^2 &= \int_{-\infty}^{+\infty} |f(x)|^2|g'(x)|^2dx \le \|f\|_{1,2}^2\|g'\|^2 \\ \|f'g\|^2 &= \int_{-\infty}^{+\infty} |f'(x)|^2|g(x)|^2dx \le \|g\|_{1,2}^2 \|f'\|^2 .\end{aligned}\right\} \quad (2.26)$$

On the other hand,

$$\|fg\|_{1,2}^2 = \|fg\|^2+\|[fg]'\|^2 = \|fg\|^2+\|fg'+f'g\|^2$$

$$\le \|fg\|^2+[\|fg'\|+\|f'g\|]^2 \le \|fg\|^2+2[\|fg'\|^2+\|f'g\|^2]$$

because $(|\alpha|+|\beta|)^2 \le 2(|\alpha|^2+|\beta|^2)$. Hence, using (2.26)

$$\|fg\|_{1,2}^2 \le \|f\|_{1,2}^2\|g\|^2+2\|f\|_{1,2}^2\|g'\|^2+2\|g\|_{1,2}^2\|f'\|^2$$

$$\le 2\|f\|_{1,2}^2[\|g\|^2+\|g'\|^2]+2\|g\|_{1,2}^2[\|f\|^2+\|f'\|^2]$$

$$= 4\|f\|_{1,2}^2\|g\|_{1,2}^2$$

and inequality (2.25) is proved.

Now we have from (2.24) and from (2.25)

$$\|A(f)-A(f_1)\|_{1,2} = \|(f+f_1)(f-f_1)\|_{1,2} \le 2\|f+f_1\|_{1,2}\|f-f_1\|_{1,2}$$

$$\le 2[\|f\|_{1,2}+\|f_1\|_{1,2}]\|f-f_1\|_{1,2}$$

$$\le 4(r+\|f_0\|_{1,2})\|f-f_1\|_{1,2}$$

because f and f_1 belong to D and so

$$\|f\|_{1,2} = \|f-f_0+f_0\|_{1,2} \le \|f-f_0\|_{1,2} + \|f_0\|_{1,2} \le r+\|f_0\|_{1,2}$$

$$\|f_1\|_{1,2} \le r+\|f_0\|_{1,2} .$$

We conclude that $A\in\text{Lip}(D,X)$ with $\nu(A) \le 4(r+\|f_0\|_{1,2})$. We finally mention that (2.25) is a particular case of the following inequality:

$$\|fg\|_{m,p} \le k\|f\|_{m,p}\,\|g\|_{m,p} \quad \forall f,g\in W^{m,p}(\Omega) \tag{2.27}$$

where Ω is an open set of R^n with $n < mp$ and where k does not depend on f and on g (Adams 1975, p.115). □

Example 2.11. The operator $F(f) = (Af)(Bf)$ with A and B belonging to $\mathscr{B}(X,C([a,b]))$

Assume that A and B are bounded linear operators with domains $D(A) = D(B) = X$ and with ranges contained in $Y = C([a,b])$: $A,B\in\mathscr{B}(X,C([a,b]))$; thus, $\phi = \phi(x) = Af$ and $\gamma = \gamma(x) = Bf$ are continuous functions of $x\in[a,b], \forall f\in X$. If we put

$$F(f) = (Af)(Bf) \ , \quad f\in X \tag{2.28}$$

then F is a non-linear operator defined over the whole X and with range contained in $C([a,b])$, because the product of the two continuous functions $\phi(x) = Af$ and $\gamma(x) = Bf$ is still a continuous function, i.e. $F(f)\in C([a,b])\ \forall f\in X$. We have from (2.28)

$$\begin{aligned}\|F(f)-F(f_1);Y\| &= \|(Af)(Bf)-(Af_1)(Bf_1);Y\| \\ &= \|[Bf-Bf_1](Af)+[Af-Af_1](Bf_1);Y\| \\ &\le \|Bf-Bf_1;Y\|\,\|Af;Y\|+\|Af-Af_1;Y\|\,\|Bf_1;Y\|\end{aligned}$$

where $\|\cdot;Y\|$ is the norm in $Y = C([a,b])$. Hence

$$\begin{aligned}\|F(f)-F(f_1);Y\| \le\ & [\|B\|\,\|f-f_1;X\|][\|A\|\,\|f;X\|] \\ & + [\|A\|\,\|f-f_1;X\|][\|B\|\,\|f_1;X\|] = \|A\|\,\|B\|\,[\|f;X\| \\ & + \|f_1;X\|]\,\|f-f_1;X\|\end{aligned}$$

where we used (2.7). Thus, if we define D as

$$D = \{g:\ g\in X;\ \|g-f_0;X\| \le r\}$$

where f_0 is a given element of X, then we obtain

$$\| F(f)-F(f_1);Y\| \leq \|A\| \|B\| [2(r+\|f_0;X\|)]\|f-f_1;X\|$$

$\forall f,f_1\in D$ because any $g\in D$ is such that $\|g;X\| = \|g-f_0+f_0;X\|$ $\leq \|g-f_0;X\|+\|f_0;X\| \leq r+\|f_0;X\|$.

We conclude that $F\in \text{Lip}(D,C([a,b]))$ with $\nu(A) \leq \|A\| \|B\| \cdot$ $\cdot [2(r+\|f_0;X\|)]$. □

Example 2.12. Contraction mapping theorem

Assume that A is an operator with domain $D \subset X$, with range $R(A) \subset Y = X$ and such that

(i) $A\in \text{Lip}(D,X)$ with $q = \nu(A) < 1$ (then, A is called *strictly contractive*);
(ii) A maps the *closed* subset $D_0 \subset D$ into itself, i.e. $A(f)\in D_0$ $\forall f\in D_0$.

Then the equation

$$f = A(f) \tag{2.29}$$

has one and only one solution belonging to D_0 (*contraction mapping theorem*). To see this, consider the 'successive approximations' method

$$f_n = A(f_{n-1}) \quad n = 1,2,... \tag{2.30}$$

with $f_0\in D_0$. Then, $f_n\in D_0$ $\forall n = 0,1,2,...$ because of assumption (ii); moreover, since $A\in \text{Lip}(D,X)$, we have

$$\|f_{n+1}-f_n\| = \|A(f_n)-A(f_{n-1})\| \leq q\|f_n-f_{n-1}\| = q\|A(f_{n-1})-A(f_{n-2})\|$$
$$\leq q^2\|f_{n-1}-f_{n-2}\| \leq ... \leq q^n\|f_1-f_0\|$$

where we used (2.20) and where $\|\cdot\| = \|\cdot;X\|$ is the norm in X. Thus, for any integer $m \geq n+1$, we have

$$\|f_m-f_n\| = \|f_m-f_{m-1}+f_{m-1}-f_{m-2}+ \dots +f_{n+1}-f_n\|$$
$$\leq \sum_{j=n}^{m-1} \|f_{j+1}-f_j\| \leq [\sum_{j=n}^{m-1} q^j] \|f_1-f_0\|$$

and also

$$\|f_m-f_n\| \leq \frac{q^n}{1-q} \|f_1-f_0\|, \quad m \geq n+1 \tag{2.31}$$

because

$$\sum_{j=n}^{m-1} q^j = q^n[1+q+\dots+q^{m-n-1}] \leq q^n \sum_{j=0}^{\infty} q^j = \frac{q^n}{1-q}$$

since $0 \leq q < 1$ by assumption. Inequality (2.31) shows that $\{f_n, n = 1,2,\dots\}$ is a Cauchy sequence in X because $0 \leq q < 1$ and so $q^n \to 0$ as $n\to\infty$. Hence, $f\in X$ exists such that $\|f-f_n\| \to 0$ as $n\to\infty$ and $f\in D_0$ since $\{f_n\} \subset D_0$ and D_0 is a closed subset of X (see Example 1.5). Further, we have from (2.30)

$$f-A(f) = [f-f_n]+[f_n-A(f)] = [f-f_n]+[A(f_{n-1})-A(f)]$$

$$\|f-A(f)\| \leq \|f-f_n\|+\|A(f_{n-1})-A(f)\| \leq \|f-f_n\|+q\|f_{n-1}-f\| \to 0$$

as $n\to\infty$. Thus $\|f-A(f)\| = 0$ and f satisfies (2.29). Finally, f is the unique solution of (2.29) belonging to D_0 since, if $g\in D_0$ also satisfies (2.29), then we have

$$f-g = A(f)-A(g), \quad \|f-g\| = \|A(f)-A(g)\| \leq q\|f-g\|$$

$$0 \leq (1-q)\|f-g\| \leq 0$$

and $\|f-g\| = 0$, i.e. $f = g$. Note that, if $\theta_x \in D_0$ and if $A(\theta_x) = \theta_x$, then $f = \theta_x$ is the only solution of (2.29) in D_0.

For example, let

$$B(f) = f_0+\lambda Af, \quad f\in L^2(a,b) = D_0 = X$$

where A is the integral operator of Example 2.4, λ is a real constant, and where f_0 is a given element of $L^2(a,b)$. Note that

B is non-linear if $\|f_0\|_2 \neq 0$. We have from (2.15)

$$\|B(f)-B(f_1)\|_2 = |\lambda| \ \|Af-Af_1\|_2 \leq |\lambda| \ |||k||| \ \|f-f_1\|_2$$

and $B \in \mathrm{Lip}(L^2(a,b), L^2(a,b))$, with $q = \nu(A) = |\lambda| \ |||k|||$. Hence, if $|\lambda| < 1/|||k|||$, then B is strictly contractive on $D_0 = L^2(a,b)$. Since B obviously maps $L^2(a,b)$ into itself, the equation

$$f(x) = f_0(x) + \lambda \int_a^b k(x,y) f(y) \, dy$$

has one and only one solution belonging to $L^2(a,b)$.

As a second example, consider the real-valued function $\phi = \phi(x), x \in [0,1] = D_0 \subset R^1$ and assume that (a) $0 \leq \phi(x) \leq 1$, $\forall x \in D_0$ and (b) $\phi(x)$ is continuously differentiable over D_0 with $|\phi'(x)| \leq q < 1 \ \forall x \in D_0$. Then, $\phi(x)$ maps the closed interval D_0 into itself because of (a); moreover, using the Mean Value Theorem, we have

$$|\phi(x)-\phi(x_1)| = |(x-x_1)\phi'(\hat{x})| \leq q|x-x_1| \quad \forall x, x_1 \in D_0$$

where $\hat{x}$ is suitably chosen. Hence, $\phi \in \mathrm{Lip}(D_0, R^1)$, it satisfies conditions (i) and (ii), and so the equation $x = \phi(x)$ has one and only one solution $x \in D_0$.

We finally point out that *both* conditions (i) and (ii) must always be checked. Thus, if we know that $A \in \mathrm{Lip}(D_0, X)$ with $D_0 = \{g\colon g \in X;\ \|g-f_0\| \leq r\}$ and with

$$\|A(f)-A(f_1)\| \leq q\|f-f_1\|, \quad \forall f, f_1 \in D_0$$

where usually $q = q(r)$, then we first choose the *radius* r so that $0 \leq q(r) < 1$ (if it is possible, of course). However, we still have to verify that A maps D_0 into itself and this can be done as follows:

$$\|A(f)-f_0\| = \|A(f)-A(f_0)+A(f_0)-f_0\| \leq \|A(f)-A(f_0)\| +$$

$$\|A(f_0)-f_0\| \leq q(r)\|f-f_0\| + \|A(f_0)-f_0\| \leq rq(r) + \|A(f_0)-f_0\|$$

$\forall f \in D_0$. Hence, $\|A(f)-f_0\| \leq r, \forall f \in D_0$, i.e. $A(f) \in D_0 \ \forall f \in D_0$, provided that

$$rq(r) + \|A(f_0)-f_0\| \le r, \quad q(r) + \frac{\|A(f_0)-f_0\|}{r} \le 1$$

which is a second condition to be imposed on $q(r)$. □

2.5. CLOSED OPERATORS

Let A be an operator with domain $D(A) \subset X$ and with range $R(A) \subset Y$; the *graph* $G(A)$ of A is defined by

$$G(A) = \left\{\phi\colon \phi = \begin{pmatrix} f \\ g \end{pmatrix} \quad \forall f \in D(A), \quad g = A(f)\right\} \tag{2.32}$$

and it is a subset of the B-space $X \times Y$ (see Example 1.6). Note that if $D(A) = (a,b) \subset R^1$ and $R(A) \subset R^1$, then $G(A) \subset R^2$ and it is the graph of a real function of a real variable in the elementary sense.

We say that A is a *closed* operator if $G(A)$ is a closed subset of $X \times Y$, i.e. if given any sequence $\{\phi_n, n = 1,2,\ldots\}$ $\subset G(A)$ such that $\|\phi_n - \phi\| \to 0$ as $n \to \infty$, then $\phi \in G(A)$, where $\|\cdot\|$ is the norm in $X \times Y$: $\|\cdot\| = [\|\cdot;X\|^2 + \|\cdot;Y\|^2]^{1/2}$. We have the following theorem

Theorem 2.2. The operator A is closed if and only if given any sequence $\{f_n, n = 1,2,\ldots\}$ such that (i) $\{f_n\} \subset D(A)$, (ii) X-$\lim f_n = f$ as $n \to \infty$, (iii) Y-$\lim A(f_n) = g$ as $n \to \infty$, then (iv) $f \in D(A)$ and $A(f) = g$.

Proof. Assume that A is closed and that the sequence $\{f_n\}$ satisfies (i), (ii), and (iii). Then

$$\phi_n = \begin{pmatrix} f_n \\ \\ A(f_n) \end{pmatrix} = \begin{pmatrix} f_n \\ \\ g_n \end{pmatrix} \subset G(A)$$

and

$$\|\phi_n - \phi\|^2 = \|f_n - f; X\|^2 + \|g_n - g; Y\|^2 \to 0 \quad \text{as} \quad n \to \infty$$

where $\phi = \binom{f}{g}$. However, $G(A)$ is a closed subset of $X \times Y$ because A is closed and so $\phi \in G(a)$, i.e. $f \in D(A)$ and $g = A(f)$

and $\{f_n\}$ satisfies (iv).

Conversely, let any sequence satisfying (i), (ii), and (iii) be such that (iv) holds. Consider then the sequence

$$\{\phi_n,\ n = 1,2,\ldots\} = \left\{\begin{pmatrix} f_n \\ A(f_n) \end{pmatrix},\ n = 1,2,\ldots\right\} \subset G(A)$$

and assume that $\{\phi_n\}$ is convergent to some $\phi = \begin{pmatrix} f \\ g \end{pmatrix} \in X\times Y$:

$$\|\phi_n - \phi; X\|^2 = \|f_n - f; X\|^2 + \|A(f_n) - g; Y\|^2 \to 0 \quad \text{as} \quad n\to\infty .$$

Then, $\{f_n\}$ satisfies (i), (ii), and (iii) and so (iv) holds, i.e. $f\in D(A)$ and $g = A(f)$. It follows that

$$\phi = \begin{pmatrix} f \\ g \end{pmatrix} = \begin{pmatrix} f \\ A(f) \end{pmatrix} \in G(A)$$

and $G(A)$ is a closed subset of $X\times Y$. □

The family of all *linear* and *closed* operators with domain contained in X and with range contained in Y will be denoted by $\mathscr{C}(X,Y)$ or by $\mathscr{C}(X)$ if $X = Y$ (Kato 1966, p.164).

Remark 2.4.

The elements of $\mathscr{C}(X,Y)$ are linear operators, not necessarily bounded (see the Examples of section 2.7). We also point out that, if $\mathscr{C}(D,Y)$ is the subset of $\mathscr{C}(X,Y)$ composed of all closed operators with domain D, then $\mathscr{C}(D,Y)$ is *not* a linear subset of $\mathscr{C}(X,Y)$. In fact, (A_1+A_2) with $A_1, A_2\in\mathscr{C}(D,Y)$ is not necessarily a closed operator. □

Remark 2.5.

Closed operators are important and 'handy' because they possess the following commutation property with the symbol X-lim

$$A(X\text{-}\lim_{n\to\infty} f_n) = A(f) = g = Y\text{-}\lim_{n\to\infty} A(f_n) \tag{2.33}$$

provided that $\{f_n\}$ satisfied (i), (ii), and (iii) of Theorem 2.2. □

Some of the basic properties of *linear* closed operators are summarized by the following.

Theorem 2.3. (a) A bounded linear operator B is closed if and only if $D(B)$ is a closed subset of X; (b) if $A \in \mathscr{C}(X,Y)$ and B is a bounded operator with $D(B) \supset D(A)$, then $A+B \in \mathscr{C}(X,Y)$; (c) if A^{-1} exists, then $A \in \mathscr{C}(X,Y)$ if and only if $A^{-1} \in (Y,X)$. □

Proof.

(*a*) See Exercise 2.7.

(*b*) We first note that $(A+B)f$ means $Af+Bf$; hence, $D(A+B) = D(A) \cap D(B) = D(A)$ because $D(A) \subset D(B)$ by assumption. Now let $\{f_n\} \subset D(A+B)$ be such that $X\text{-}\lim f_n = f$ and $Y\text{-}\lim\, (A+B)f_n = g$ as $n \to \infty$. Since B is bounded and $\{f_n\} \subset D(B)$ is convergent in X, we have because of (2.7)

$$\|Bf_m - Bf_n;Y\| \leq \|B\|\,\|f_m - f_n;X\| \quad \text{as} \quad m,n \to \infty$$

i.e. $\{Bf_n\}$ is a Cauchy sequence in the B-space Y and so a $\hat{g} \in Y$ exists such that

$$\hat{g} = Y\text{-}\lim_{n\to\infty} Bf_n \;.$$

It follows that

$$g = Y\text{-}\lim_{n\to\infty} (A+B)f_n = Y\text{-}\lim_{n\to\infty} Af_n + \hat{g}$$

$$g - \hat{g} = Y\text{-}\lim_{n\to\infty} Af_n$$

and the sequence $\{f_n\}$ satisfies conditions (i), (ii), and (iii) of Theorem 2.2 (i.e. $\{f_n\} \subset D(A)$; $X\text{-}\lim f_n = f$; $Y\text{-}\lim Af_n = g-\hat{g}$). Since $A \in \mathscr{C}(X,Y)$, we conclude that $f \in D(A) = D(A+B)$ and that $g - \hat{g} = Af$. On the other hand, f also belongs to $D(B) \supset D(A)$ and so

$$\|Bf - Bf_n;Y\| \leq \|B\|\,\|f - f_n;X\| \to 0 \quad \text{as} \quad n \to \infty$$

i.e. $Bf = Y\text{-}\lim Bf_n$. Then

$$g = (g-\hat{g})+\hat{g} = Af+Y\text{-}\lim_{n\to\infty} Bf_n = Af+Bf = (A+B)f$$

and $(A+B)\in\mathscr{C}(X,Y)$ because of Theorem 2.2.
(*c*) Finally, assume that $A\in\mathscr{C}(X,Y)$ and that A^{-1} exists. To prove that $A^{-1}\in\mathscr{C}(X,Y)$, let $\{k_n\}\subset Y$ satisfy (i), (ii), and (iii) of Theorem 2.2 (with respect to A^{-1}): $\{k_n\}\subset D(A^{-1}) = R(A)$; $Y\text{-}\lim k_n = k$; $X\text{-}\lim A^{-1}k_n = h$. If we put $h_n = A^{-1}k_n$, then we have that $\{h_n\}\subset R(A^{-1}) = D(A)$, $X\text{-}\lim h_n = h$, $Y\text{-}\lim Ah_n = k$ and so $h\in D(A)$, $k = Ah$ because $A\in\mathscr{C}(X,Y)$ (see Theorem 2.2). It follows that $h\in R(A^{-1})$, $k = Ah\in R(A) = D(A^{-1})$; consequently, $h = A^{-1}k$ and $A^{-1}\in\mathscr{C}(Y,X)$ because of (the converse part of) Theorem 2.2. Since it can be shown by a similar procedure that $A^{-1}\in\mathscr{C}(Y,X)$ implies $A\in\mathscr{C}(X,Y)$, Theorem 2.3 is completely proved. □

It follows from (*a*) of Theorem 2.3 that

$$\mathscr{B}(X,Y)\subset\mathscr{C}(X,Y) \tag{2.34}$$

because, if $B\in\mathscr{B}(X,Y)$, then $D(B)$ coincides with X and so it is a closed set.

Finally, let A be a linear operator with domain and range contained in the B-space X (over the field K). If a $z\in$ K exists such that

$$-(zI-A)^{-1}\in\mathscr{B}(X) \tag{2.35}$$

then $A\in\mathscr{C}(X)$ (in (2.35) I is the *identity* operator: $If = f\ \forall f\in X$). In fact, $-(zI-A)^{-1}\in\mathscr{C}(X)$ owing to (2.34) and so $-(zI-A)\in\mathscr{C}(X)$ because of (*c*) of Theorem 2.3. Hence, $A = -(zI-A)+zI\in\mathscr{C}(X)$ in view of (*b*) of Theorem 2.3 because $zI\in\mathscr{B}(X)$, as is easy to verify. Relation (2.35) will be used in section 2.8 where several examples of closed operators are discussed.

2.6. SELF-ADJOINT OPERATORS

Let X be a *Hilbert space* with inner product (f,g) and with norm $\|f\| = [(f,f)]^{1/2}$ and let A be a linear operator with domain $D(A)$ *dense* in X and with range $R(A)\subset X$. Then $Af\in X$

$\forall f \in D(A)$ and the inner product (Af,g) is defined for any $g \in X$. Now assume that g is such that a suitable $g^* \in X$ can be found with the property

$$(Af,g) = (f,g^*) \quad \textit{for all } f \in D(A) \ . \tag{2.36}$$

The subset of X composed of all elements g which possess the property (2.36) will be denoted by D^*. Thus, if $g \in D^*$, then a $g^* \in X$ can be found such that (2.36) holds. Note that g^* is uniquely determined by $g \in D^*$ for if a $g_1^* \in X$ also exists such that

$$(Af,g) = (f,g_1^*) \quad \forall f \in D(A)$$

then $(f,g^*) = (f,g_1^*)$ $\forall f \in D(A)$, i.e.

$$0 = (f,g_1^*-g^*) \quad \forall f \in D(A) \ . \tag{2.37}$$

However, $D(A)$ is dense in X by assumption and so, given any $\phi \in X$ and $\varepsilon < 0$, a suitable $f \in D(A)$ can be found such that $\| f-\phi \| < \varepsilon$. It then follows from (2.37)

$$(\phi,g_1^*-g^*) = (\phi-f,g_1^*-g^*)+(f,g_1^*-g^*) = (\phi-f,g_1^*-g^*)$$

$$|(\phi,g_1^*-g^*)| = |(\phi-f,g_1^*-g^*)| \le \|\phi-f\| \, \|g_1^*-g^*\| < \varepsilon \|g_1^*-g^*\| \ .$$

Since the positive number ε can be chosen arbitrarily, then

$$(\phi,g_1^*-g^*) = 0 \qquad \forall \phi \in X;$$

in other words, (2.37) can be *extended* from $D(A)$ (which is dense in X) to the whole space X. If in particular we take $\phi = g_1^*-g^*$, then we obtain

$$(g_1^*-g^*,g_1^*-g^*) = \|g_1^*-g^*\|^2 = 0$$

and $g_1^* = g^*$, i.e. g_1^* and g^* are the same element of the Hilbert space X. Thus g^* is uniquely determined by $g \in D^*$ and so the operator $g^* = A^*(g)$ is properly defined for any

$g \in D^*$. The operator A^* is linear for, if g and $\hat{g}$ belong to D^*, then

$$(Af,g) = (f,A^*(g)), \quad (Af,\hat{g}) = (f,A^*(\hat{g})), \quad \forall f \in D(A)$$

$$\begin{aligned}(Af,\alpha g+\beta\hat{g}) &= \overline{\alpha}(Af,g)+\overline{\beta}(Af,\hat{g}) \\ &= \overline{\alpha}(f,A^*(g))+\overline{\beta}(f,A^*(g)) \\ &= (f,\alpha A^*(g)+\beta A^*(\hat{g})), \quad \forall f \in D(A).\end{aligned}$$

Hence, the element $h^* = \alpha A^*(g)+\beta A^*(\hat{g}) \in X$ is such that

$$(Af,\alpha g+\beta\hat{g}) = (f,h^*), \quad \forall f \in D(A)$$

and so $\alpha g+\beta\hat{g} \in D^*$ with $h^* = A^*(\alpha g+\beta\hat{g})$, i.e. $\alpha A^*(g)+\beta A^*(\hat{g}) = A^*(\alpha g+\beta\hat{g})$ and A^* is a linear operator with domain $D(A^*) = D^*$. A^* is called the *adjoint* of A and it satisfies the relation

$$(Af,g) = (f,A^*g), \quad \forall f \in D(A), \forall g \in D(A^*). \tag{2.38}$$

We have

$$A^* \in \mathscr{C}(X); \tag{2.39}$$

to see this, let $\{h_n\} \subset D(A^*)$ be chosen so that $X\text{-}\lim h_n = h$, $X\text{-}\lim A^*h_n = k^*$ (see Theorem 2.2). Then, $\forall f \in D(A)$

$$|(Af,h_n)-(Af,h)| = |(Af,h_n-h)| \leq \|Af\| \, \|h_n-h\| \to 0$$

as $n\to\infty$, and in an analogous way

$$|(f,A^*h_n)-(f,k^*)| \to 0 \quad \text{as} \quad n\to\infty$$

i.e.

$$\lim_{n\to\infty} (Af,h_n) = (Af,h)$$

$$\lim_{n\to\infty} (f,A^*h_n) = (f,k^*) \quad \forall f \in D(A).$$

On the other hand,

$$(Af,h_n) = (f,A^*h_n), \quad \forall f \in D(A)$$

because of (2.38) and so

$$\lim_{n\to\infty} (Af,h_n) = \lim_{n\to\infty} (f,A^*h_n), \quad \forall f \in D(A).$$

Hence, we obtain

$$(Af,h) = (f,k^*) \ , \quad \forall f \in D(A)$$

which implies that $h \in D^* = D(A^*)$ and that $k^* = A^*h$. Thus, $A^* \in \mathscr{C}(X)$ because of Theorem 2.2 and (2.39) is proved.

A linear operator A, with domain $D(A)$ dense in X and with range $R(A) \subset X$, is called *symmetric* if

$$(Af,g) = (f,Ag) \ , \quad \forall f \in D(A), \forall g \in D(A). \tag{2.40}$$

If we compare (2.40) with (2.36), then we see that each $g \in D(A)$ also belongs to $D^* = D(A^*)$ and that $g^* = A^*g = Ag$ $\forall g \in D(A)$. Hence, $D(A) \subset D(A^*)$ and A^* is an extension of A (see section 2.1).

A densely defined symmetric operator is called *self-adjoint* if $A = A^*$, i.e. if $D(A) = D(A^*)$. Thus, a linear operator A is self-adjoint if it is densely defined and

$$(Af,g) = (g,Af) \quad \forall f,g \in D(A) = D(A^*). \tag{2.41}$$

Note that each self-adjoint operator is closed because of (2.39).

Now assume that a densely defined linear operator A is given; then relation (2.41) can be used to prove that A is self-adjoint. However, since it is sometimes difficult to verify that $D(A) = D(A^*)$, the following sufficient condition is quite useful in practice.

Theorem 2.4. Let A be a symmetric operator with domain dense in the (complex) Hilbert space X. Then A is self-adjoint *if a complex number z can be found such that the ranges of the operators $(zI-A)$ and $(\bar{z}I-A)$ coincide with the whole X, i.e. $(zI-A)[D(A)] = R(zI-A) = X$ and $(\bar{z}I-A)\ [D(A)] = R(\bar{z}I-A) = X$.*

Proof. Since A is a densely defined and symmetric operator, A^* is an extension of A: $D(A^*) \supset D(A)$, $Af = A^*f\ \forall f \in D(A)$. Hence if we show that $D(A^*) \subset D(A)$, then $D(A) = D(A^*)$ and Theorem 2.4 is proved because of (2.41). Now let z be such that $R(zI-A) = R(\bar{z}I-A) = X$ and let $f_0 \in D(A^*)$ exist for which $(zI-A^*)f_0 = 0$ with $\|f_0\| \neq 0$. Then

$$0 = (f,(zI-A^*)f_0) = \bar{z}(f,f_0)-(f,A^*f_0)$$

$$= \bar{z}(f,f_0)-(Af,f_0) = ((\bar{z}I-A)f,f_0),\ \ \forall f \in D(A)$$

because of (2.38). However, given any $g \in X$, a suitable $f \in D(A)$ can be determined such that $(\bar{z}I-A)f = g$ because $R(\bar{z}I-A) = X$ by assumption. Hence,

$$0 = (g,f_0),\ \forall g \in X$$

and if in particular $g = f_0$

$$0 = (f_0,f_0) = \|f_0\|^2 .$$

Thus $\|f_0\| = 0$ which contradicts the assumption $\|f_0\| \neq 0$. We conclude that

$$(zI-A^*)f_0 \neq 0, \qquad \forall f_0 \in D(A^*) \quad \text{with} \quad \|f_0\| \neq 0.$$

Further, let ϕ be any given element of $D(A^*)$; then, $(zI-A^*)\phi = \phi_1 \in X$ and a suitable $f_1 \in D(A)$ exists for which

$$(zI-A)f_1 = (zI-A^*)\phi$$

because the range of $(zI-A)$ is the whole space X. However, A^* is an extension of A and so $Af_1 = A^*f_1$. Hence,

$$(zI-A^*)[f_1-\phi] = 0.$$

It follows from what we have proved above that $\| f_1-\phi \| = 0$ and so $\phi = f_1 \in D(A)$. We then conclude that $D(A^*) \subset D(A)$ because any $\phi \in D(A^*)$ also belongs to $D(A)$ and Theorem 2.4 is proved. □

Remark 2.6.

Under the assumptions of Theorem 2.4, A is self-adjoint and so $A = A^* \in \mathscr{C}(X)$ because of (2.39). Since $zI \in \mathscr{B}(X)$, $(zI-A)\in \mathscr{C}(X)$ in view of (b) of Theorem 2.3. Moreover, $(zI-A)^{-1}$ exists with domain $D((zI-A)^{-1}) = R(zI-A) = X$, because, as we have shown in the proof of Theorem 2.4, $(zI-A)\phi = (zI-A^*)\phi \neq 0 \ \forall \phi \in D(A) = D(A^*)$ such that $\|\phi\| \neq 0$ (see Lemma 2.1 with $zI-A$ instead of A). The inverse operator $(zI-A)^{-1} \in \mathscr{C}(X)$ because of (c) of Theorem 2.3 and so $(zI-A)^{-1}$ is a closed operator defined over the whole space X. It follows that $(zI-A)^{-1} \in \mathscr{B}(X)$ (see Kato 1966, Theorem 5.20, p.166). The above results apply if in particular z can be taken equal to zero.

Example 2.13. The Fredholm integral operator

Let the Fredholm integral operator A of Example 2.4 be such that $k(x,y) = \overline{k(y,x)}$, where as usual the bar denotes the complex conjugate. Given any f and g belonging to $D(A) = L^2(a,b)$, we have

$$(Af,g) = \int_a^b \left[\int_a^b k(x,y)f(y)\,\mathrm{d}y\right]\overline{g(x)}\,\mathrm{d}x =$$

$$\int_a^b f(y)\left[\int_a^b k(x,y)\overline{g(x)}\,\mathrm{d}x\right]\mathrm{d}y = \int_a^b f(y)\left[\int_a^b \overline{k(y,x)}\,\overline{g(x)}\,\mathrm{d}x\right]\mathrm{d}y = (f,Ag)$$

where we used Fubini's theorem to interchange the order of integration (Royden 1963, p.233). Hence, A is symmetric and densely defined because $D(A) = L^2(a,b) = X$. On the other hand, since in general $D(A) \subset D(A^*) \subset X$, we have that $D(A) = D(A^*) = X$ and so A is self-adjoint. □

Example 2.14. Consider the heat-diffusion operator

$Af = kf''(x)$, $D(A) = \{f : f \in L^2(a,b); f'' \in L^2(a,b); f(a) = f(b) = 0\} \subset$

$$L^2(a,b),\ R(A) \subset L^2(a,b) \tag{2.42}$$

where k is a positive constant, *the heat-diffusion coefficient*, and $f'' = d^2f/dx^2$ is a generalized derivative. Note that $f'' \in L^1_{loc}(a,b)\ \forall f \in D(A)$ by the definition of a generalized derivative, and so $f'(x)$ is an absolutely continuous function on $[a,b]$. Hence, $f(x)$ is certainly continuous on $[a,b]$ and it makes sense to impose the *boundary conditions* $f(a) = f(b) = 0$. In other words, since $D(A)$ is a linear subset of $W^{2,2}(a,b)$ and since $W^{2,2}(a,b) \to C_B(a,b)$ (see Theorem 1.3 with $\Omega = (a,b)$, $j = 0$, $m = p = 2$, $n = 1$), then each element of $D(A)$ can be regarded as an element of $C_B(a,b)$.

Now A is densely defined because $D(A)$ obviously contains $C_0^\infty(a,b)$ (see section 1.3), which is dense in $L^2(a,b)$ (Kato 1966, p.130). Hence,

$$C_0^\infty(a,b) \subset D(A) \subset L^2(a,b)$$

$$L^2(a,b) = \text{cl}(C_0^\infty(a,b)) \subset \text{cl}(D(A) \subset L^2(a,b)$$

and so $\text{cl}(D(A)) = L^2(a,b)$, where $\text{cl}(D(A))$ denotes the closure of $D(A)$ (see Example 1.5). On the other hand, we have $\forall f,g \in D(A)$

$$(Af,g) = k\int_a^b f''(x)\overline{g(x)}\,dx = -k\int_a^b f'(x)\overline{g'(x)}\,dx$$

$$(f,Ag) = k\int_a^b f(x)\overline{g''(x)}\,dx = -k\int_a^b f'(x)\overline{g'(x)}\,dx$$

where we took into account the boundary condition $f(a) = f(b) = 0$, $g(a) = g(b) = 0$ (see the definition of $D(A)$). Hence, A is symmetric and densely defined. Further, let g be any given element of $L^2(a,b)$; if we put

$$f = f(x) = \frac{1}{k}\left[\int_a^x dy \int_a^y g(s)\,ds - \frac{x-a}{b-a}\int_a^b dy \int_a^y g(s)\,ds\right]$$

then it is easy to verify that $f(a) = f(b) = 0$ and that $kf''(x) = g(x)$. Hence, $f \in D(A)$, $Af = g$. Thus, given any $g \in L^2(a,b)$, a suitable $f \in D(A)$ exists such that $Af = g$; as a consequence,

$R(A) = L^2(a,b)$ and A is self-adjoint because of Theorem 2.4 with $z = 0$.

We finally remark that the self-adjoint operator A has the property

$$(Af,f) = -k \int_a^b |f'(x)|^2 dx \le 0 \quad \forall f \in D(A);$$

correspondingly, we say that $-A$ is a *non-negative* operator. We also say that A is *dissipative* because the property $(Af,f) \le 0$ $\forall f \in D(A)$ is usually related to some form of dissipation of energy.

□

Example 2.15.

Let the linear operator A be defined as follows

$$Af = \mathscr{F}^{-1}[-ky^2 \mathscr{F} f], \quad D(A) = L^{s+2,2}(R^1) \subset L^{s,2}(R^1)$$

$$R(A) \subset L^{s,2}(R^1) \tag{2.43}$$

where $L^{s,2}(R^1)$ is a Sobolev space of fractional order $s \ge 0$ (see section 1.5). In (2.43), $k > 0$ and $\mathscr{F} f = \hat{f} = \hat{f}(y)$ and $\mathscr{F}^{-1}\phi$ are respectively the Fourier-Plancherel transform of $f \in L^{s,2}(R^1) \subset L^2(R^1)$ and the inverse Fourier-Plancherel transform of ϕ (see Theorem 1.4 and Exercise 2.8). In other words, if $g = Af$, then $\hat{g} = \hat{g}(y) = -ky^2 \hat{f}(y)$. Now, according to (1.50), any $f \in D(A)$ is such that

$$[\|f\|_{s+2,2}]^2 = \int_{-\infty}^{+\infty} (1+y^2)^{s+2} |\hat{f}(y)|^2 dy < \infty$$

and so

$$[\|Af\|_{s,2}]^2 = \int_{-\infty}^{+\infty} (1+y^2)^s |(Af)^\wedge|^2 dy = k^2 \int_{-\infty}^{+\infty} (1+y^2)^s |y^2 \hat{f}(y)|^2 dy$$

$$= k^2 \int_{-\infty}^{+\infty} (1+y^2)^s (y^2)^2 |\hat{f}(y)|^2 dy$$

$$\le k^2 \int_{-\infty}^{+\infty} (1+y^2)^{s+2} |\hat{f}(y)|^2 dy$$

$$= k^2 [\| f \|_{s+2,2}]^2 < \infty$$

and A is correctly defined on $D(A)$ (which is dense in the Hilbert space $L^{s,2}(R^1)$ because it contains $C_0^\infty(R^1)$ (Adams 1975, p.221)). Furthermore, if f and g belong to $D(A)$, then we have from definition (2.43) and from (1.50)

$$(Af,g)_{s,2} = -k \int_{-\infty}^{+\infty} (1+y^2)^s \, y^2 \, \hat{f}(y) \, \overline{\hat{g}(y)} \, dy$$

because $\mathscr{F}[Af] = -k \, y^2 \, \hat{f}(y)$; on the other hand,

$$(f,Ag)_{s,2} = -k \int_{-\infty}^{+\infty} (1+y^2)^s \, \hat{f}(y) \, \overline{y^2 \hat{g}(y)} \, dy = (Af,g)_{s,2}$$

and A is a densely defined symmetric operator.

Finally, given any $g \in L^{s,2}(R^1)$ and any $z > 0$, we consider the equation

$$(zI-A)f = g.$$

Using the definition (2.43), we obtain

$$(z+ky^2)\hat{f}(y) = \hat{g}(y)$$

and so

$$\hat{f}(y) = (z+ky^2)^{-1} \, \hat{g}(y) \tag{2.44}$$

where $z+ky^2 \geq z > 0$. It follows from (2.44) that

$$[\| f \|_{s,2}]^2 = \int_{-\infty}^{+\infty} (1+y^2)^s (z+ky^2)^{-2} |\hat{g}(y)|^2 dy$$

$$\leq z^{-2} \int_{-\infty}^{+\infty} (1+y^2)^s |\hat{g}(y)|^2 dy$$

$$= z^{-2} [\| g \|_{s,2}]^2 < \infty \tag{2.45}$$

and also

$$[\|f\|_{s+2,2}]^2 = \int_{-\infty}^{+\infty} (1+y^2)^s \frac{(1+y^2)^2}{(z+ky^2)^2} |\hat{g}(y)|^2 \, dy$$

$$\leq \eta^2 \int_{-\infty}^{+\infty} (1+y^2)^s |\hat{g}(y)|^2 dy = \eta^2 [\|g\|_{s,2}]^2 < \infty$$

where $\eta = \max\{1/z; 1/k\}$ because

$$(1+y^2)/(z+ky^2) \leq \max\{1/z; 1/k\}, \forall y \in R^1.$$

Hence, the element f, whose Fourier-Plancherel transform is given by (2.44), belongs to $D(A)$ and it is such that $(zI-A)f = g$. In other words, given any $g \in L^{s,2}(R^1)$, a suitable $f \in D(A)$ can be determined such that $(zI-A)f$ where z is any given positive number. As a consequence, $R(zI-A) = L^{s,2}(R^1)$ and A is self-adjoint because of Theorem 2.4.

It also follows from (2.45) that

$$\|f\|_{s,2} = \|(zI-A)^{-1}g\|_{s,2} \leq z^{-1}\|g\|_{s,2}, \; \forall z > 0 \; \forall g \in L^{s,2}(R^1) \tag{2.46}$$

and so

$$(zI-A)^{-1} \in \mathscr{B}(L^{s,2}(R^1)), \quad \|(zI-A)^{-1}\| \leq z^{-1}, \quad \forall z > 0 \tag{2.47}$$

which agrees with the discussion of Remark 2.6.

The preceding results obviously hold if $s = m = 0,1,2,\ldots$, i.e. if s is a non-negative integer; correspondingly, $L^{s,2}(R^1) = W^{s,2}(R^1)$ and $D(A)$ coincides (as a vector space) with $W^{s+2,2}(R^1)$ (see section 1.5). In particular, if $s = 0$, then $L^{0,2}(R^1) = L^2(R^1)$, $D(A) = W^{2,2}(R^1)$ and $Af = kf''$ where the derivative is in the generalized sense (see Example 2.14).

Similar conclusions can be reached for the heat diffusion operator in R^3

$$Af = \mathscr{F}^{-1}[-k|y|^2\hat{f}(y)], \quad D(A) = L^{s+2,2}(R^3), R(A) \subset L^{s,2}(R^3) \tag{2.48}$$

where

$$y = (y_1, y_2, y_3), \quad |y|^2 = y_1^2+y_2^2+y_3^2 \quad . \quad \square$$

2.7. SPECTRAL PROPERTIES: BASIC DEFINITIONS

Let A be the matrix operator

$$Af = \begin{pmatrix} a_{11}, & a_{22} \\ a_{21}, & a_{22} \end{pmatrix} \begin{pmatrix} f_1 \\ f_2 \end{pmatrix}$$

$$f = \begin{pmatrix} f_1 \\ f_2 \end{pmatrix} \in D(A) = X = \mathbb{C}\times\mathbb{C},\ R(A) \subset X$$

where the a_{ij}'s are given real constants and X is the B-space of all ordered pairs of complex numbers with norm

$$\| f \| = \left\| \begin{pmatrix} f_1 \\ f_2 \end{pmatrix} \right\| = [|f_1|^2+|f_2|^2]^{1/2}, \quad f_1, f_2 \in \mathbb{C}.$$

As in Example 2.2, (see also (0.11)) we have

$$A \in \mathscr{B}(X), \quad \| Af \| \le [\sum_{i,j=1}^{2} a_{i,j}^2]^{1/2} \| f \|, \quad \forall f \in X.$$

Now consider the following initial-value problem (remember (0 4), (0.5), and the discussion leading to (0.8))

$$\frac{\mathrm{d}}{\mathrm{d}t} y(t) = Ay(t), \ t > 0; \quad \lim_{t\to 0+} \| y(t) - y_0 \| = 0 \tag{2.49}$$

where

$$y_0 = \begin{pmatrix} y_{01} \\ y_{02} \end{pmatrix}$$

is a given element of X. From the elementary theory of linear differential equations with constant coefficients, we have that the solution of system (2.49) has the form

$$y(t) = \exp(z_1 t) f^{(1)} + \exp(z_2 t) f^{(2)}, \quad t \ge 0 \tag{2.50}$$

where z_1 and z_2 are the roots of the equation

$$\det\begin{pmatrix} a_{11}-z, & a_{12} \\ a_{21}, & a_{22}-z \end{pmatrix} = (a_{11}-z)(a_{22}-z)-a_{12}a_{21} = 0 \qquad (2.51)$$

and where $f^{(1)}$ and $f^{(2)}$ are suitable elements of X, uniquely determined by y_0 (here we assume for simplicity that $z_1 \neq z_2$). However, the equation

$$(zI-A)f = 0 \qquad (2.52)$$

has a non-zero solution $f \in X$ if and only if z is a root of equation (2.51). We then conclude that the solution of the initial-value problem (2.49) can be written down explicitly provided that the values of z are known for which (2.52) has non-zero solutions, i.e. for which the inverse matrix $(zI-A)^{-1}$ does *not* exist (see Lemma 2.1 with $zI-A$ instead of A).

More generally, given a linear operator A with domain $D(A)$ and range $R(A)$ contained in the B-space X, we consider the following classification of the complex numbers z.

(*a*) If $z \in \mathbb{C}$ is such that $zI-A$ is not one-to-one, i.e. $(zI-A)^{-1}$ does not exist, then we say that z is an *eigenvalue* of A. The set of all the eigenvalues of A is called the *point spectrum* of A and it is denoted by $P_\sigma(A)$. Thus, in view of Lemma 2.1, $z \in P_\sigma(A)$ if and only if the equation

$$(zI-A)f = 0 \qquad (2.53)$$

has at least a solution $f \in D(A)$ such that $\|f;X\| \neq 0$ and so

$$P_\sigma(A) = \{z : z \in \mathbb{C};\ (zI-A)^{-1} \text{ does not exist}\}$$

$$= \{z : z \in \mathbb{C};\ \text{equation (2.53) has at least a non-zero solution } f \in D(A)\}. \qquad (2.54a)$$

The non-zero elements of $D(A)$ that satisfy (2.53) are called *eigenfunctions* of A corresponding to the eigenvalue z.

Remark 2.7.

If $z = 0 \in P\sigma(A)$, then A^{-1} does not exist and A is not invertible. However, going back to the simple initial-value problem (2.49) if 0 is an eigenvalue of the matrix A and f is a corresponding eigenfunction, then $y(t) \equiv f$ is a time-independent solution of the first of (2.49). □

Remark 2.8.

If $z \in P\sigma(A)$ and f is a corresponding eigenfunction, then $g = \alpha f$ also satisfies (2.53) and it belongs to $D(A)$ $\forall \alpha \in \underline{K}$. In other words, if f is an eigenfunction corresponding to $z \in P_\sigma(A)$, then αf is still an eigenfunction $\forall \alpha \in \underline{K}$ with $\alpha \neq 0$. If in particular we choose $\alpha = 1/\|f;X\|$, then $g = \alpha f$ is such that $\|g;X\| = 1$ and it is called a *normalized* eigenfunction. □

(*b*) If $z \in \mathbb{C}$ is such that $(zI-A)^{-1}$ exists but its domain $D((zI-A)^{-1}) = R(zI-A)$ is *not* dense in X, then we say that z belongs to the *residual spectrum* $R_\sigma(A)$ of A. Hence,

$$R_\sigma(A) = \{z : z \in \mathbb{C};\ (zI-A)^{-1} \text{ exists but it is not densely defined}\} \tag{2.54b}$$

(*c*) If $z \in \mathbb{C}$ is such that $(zI-A)^{-1}$ exists and its domain $D((zI-A)^{-1}) = R(zI-A)$ is dense in X but $(zI-A)^{-1}$ is *not* a bounded operator on $D((zI-A)^{-1})$, then we say that z belongs to the *continuous spectrum* $C_\sigma(A)$ of A. Hence,

$$C_\sigma(A) = \{z : z \in \mathbb{C}; (zI-A)^{-1} \text{ exists, it is densely-defined but it is not a bounded operator}\}. \tag{2.54c}$$

The *spectrum* $\sigma(A)$ of A is then defined as follows:

$$\sigma(A) = P_\sigma(A) \cup R_\sigma(A) \cup C_\sigma(A) . \tag{2.55}$$

Roughly speaking, if $z_0 \in P_\sigma(A)$, then it is 'more singular' than $z_1 \in R_\sigma(A)$, which in turn is 'more singular' than $z_2 \in C_\sigma(A)$.

(*d*) Finally, if $z \in \mathbb{C}$ is such that $(zI-A)^{-1}$ is densely defined and bounded, then we say that z belongs to the *resolvent set* $\rho(A)$ of A. Hence,

$$\rho(A) = \{z : z \in \mathbb{C}; (zI-A)^{-1} \text{ is densely defined and bounded}\} \quad (2.54d)$$

Note that the sets $P_\sigma(A)$, $R_\sigma(A)$, $C_\sigma(A)$, and $\rho(A)$ are mutually disjoint and that

$$P_\sigma(A) \cup R_\sigma(A) \cup C_\sigma(A) \cup \rho(A) = \sigma(A) \cup \rho(A) = \mathbb{C}. \quad (2.56)$$

Remark 2.9.

If $A \in \mathscr{C}(X)$ and $z \in \rho(A)$, then $zI-A$ and $(zI-A)^{-1}$ are both closed operators because of (*b*) and (*c*) of Theorem 2.3 and so $(zI-A)^{-1}$ is densely defined, bounded, and closed. According to (*a*) of Theorem 2.3, the domain of $(zI-A)^{-1}$ is a closed subset of X, i.e.

$$D((zI-A)^{-1}) = \mathrm{cl}[D((zI-A)^{-1})] = X$$

because $D((zI-A)^{-1}$ is dense in X. Thus, $(zI-A)^{-1} \in \mathscr{B}(X)$ and

$$\rho(A) = \{z : z \in \mathbb{C}; (zI-A)^{-1} \in \mathscr{B} X)\}, \quad A \in \mathscr{C}(X). \quad (2.57)$$

Conversely, if $(zI-A)^{-1} \in \mathscr{B}(X)$, then

$$-(zI-A)^{-1} \in \mathscr{B}(X) \subset \mathscr{C}(X)$$

$$-(zI-A) \in \mathscr{C}(X)$$

$$A = -(zI-A) + zI \in \mathscr{C}(X)$$

(see the discussion after (2.35)).

If $z \in \rho(A)$, $A \in \mathscr{C}(X)$, then the inverse operator $(zI-A)^{-1}$ is usually denoted by $R(z,A)$ and is called the *resolvent* of A. □

Remark 2.10.

If $A\in\mathscr{C}(X)$, then $z\in\rho(A)$ if and only if $R(z,A) = (zI-A)^{-1}\in\mathscr{B}(X)$ (see Remark 2.9). To find out whether or not z belongs to $\rho(A)$, we consider the equation

$$(zI-A)f = g \tag{2.58}$$

where g is any given element of X and where the 'unknown' f must obviously be sought in $D(A)$. Now assume that (2.58) has a unique solution $f\in X$ such that

$$\|f;X\| \leq M\|g;X\|, \quad \forall g\in X$$

where M is a suitable constant (independent of g). Then, $f = R(z,A)g = (zI-A)^{-1}g$, $\|R(z,A)\| \leq M$ and so $R(z,A)\in\mathscr{B}(X)$ and $z\in\rho(A)$ (see Examples 2.16 - 2.21). □

2.8. SPECTRAL PROPERTIES : EXAMPLES

Example 2.16. Resolvent of $A\in\mathscr{B}(X)$

Given $A\in\mathscr{B}(X)$, consider the sequence $\{B_n\}\subset\mathscr{B}(X)$:

$$B_n f = \sum_{j=0}^{n} z^{-(j+1)} A^j f, \ \forall f\in X, \quad n = 0,1,2,\ldots \tag{2.59}$$

where $A^0 = I$ and where the complex parameter z is such that $|z| > \|A\|$. We have

$$\|B_m - B_n\| = \|\sum_{j=n+1}^{m} z^{-(j+1)} A^j\| \leq \sum_{j=n+1}^{m} |z|^{-(j+1)}\|A\|^j$$

$$= |z|^{-1} \sum_{j=n+1}^{m} [|z|^{-1}\|A\|]^j, \qquad m \geq n+1$$

where $[|z|^{-1}\|A\|]^{\cdot} < 1$ by assumption and where $\|\cdot\| = \|\cdot;\mathscr{B}(X)\|$ is the norm in $\mathscr{B}(X)$ (see Theorem 2.1 with $D = Y = X$). It follows that $\{B_n\}$ is a Cauchy sequence in the B-space $\mathscr{B}(X)$, i.e.

$$\|B_m - B_n\| < \varepsilon, \ \forall m > n > n_\varepsilon$$

because the geometric series

$$\sum_{j=0}^{\infty} [|z|^{-1}\|A\|]^j = (1-|z|^{-1}\|A\|)^{-1}$$

is convergent. Hence, a suitable $B\in\mathscr{B}(X)$ exists such that

$$B = \mathscr{B}(X) - \lim_{n\to\infty} B_n = \sum_{j=0}^{\infty} z^{-(j+1)} A^j , \quad |z| > \|A\| \tag{2.60}$$

with

$$\|B\| = \lim_{n\to\infty} \|B_n\| \le (|z|-\|A\|)^{-1} , \quad |z| > \|A\| \tag{2.61}$$

because

$$\begin{aligned}\|B_n\| &\le |z|^{-1} \sum_{j=0}^{n} [|z|^{-j}\|A\|^j] \le |z|^{-1}(1-|z|^{-1}\|A\|)^{-1}\\ &= (|z|-\|A\|)^{-1} .\end{aligned}$$

On the other hand, we have from (2.59) and from (2.60)

$$B_n[zI-A]-I = \sum_{j=0}^{n} z^{-j} A^j - \sum_{j=0}^{n} z^{-(j+1)} A^{j+1} - I = -z^{-n-1} A^{n+1}$$

$$\begin{aligned}\|B[zI-A]-I\| &\le \|B[zI-A]-B_n[zI-A]\| + \|B_n[zI-A]-I\|\\ &\le \|B-B_n\| \, \|zI-A\| + |z|^{-n-1}\|A^{n+1}\|\\ &\le (|z|+\|A\|)\|B-B_n\| + [|z|^{-1}\|A\|]^{n+1} \to 0, \quad \text{as } n\to\infty.\end{aligned}$$

Thus ,

$$B[zI-A]f = f, \quad \forall f\in X, \quad |z| > \|A\|$$

and $B = (zI-A)^{-1}$ (see also Exercise 2.9). Furthermore, (2.60) shows that

$$\left.\begin{aligned}\|(zI-A)^{-1}\| &\le (|z|-\|A\|)^{-1}\\ (zI-A)^{-1} &= \sum_{j=0}^{\infty} z^{-(j+1)} A^j \in\mathscr{B}(X)\end{aligned}\right\} \tag{2.62}$$

provided that $|z| > \|A\|$ and so

$$\begin{aligned} \{z : z\in\mathbb{C};\ |z| > \|A\|\} &\subset \rho(A) \\ \{z : z\in\mathbb{C};\ |z| \le \|A\|\} &\supset \sigma(A) \end{aligned} \tag{2.63}$$

because of (2.57) and because $\rho(A)$ and $\sigma(A)$ are disjoint subsets of $\mathbb{C}$. The preceding results can be refined as follows. If the *spectral radius* $\operatorname{spr}(A)$ of $A\in\mathscr{B}(X)$ is defined by

$$\operatorname{spr}(A) = \lim_{n\to\infty} \|A^n\|^{1/n} \tag{2.64}$$

where the limit exists for any $A\in\mathscr{B}(X)$ (Kato 1966, p.153), then we have

$$\sigma(A) \subset \{z : z\in\mathbb{C};\ |z| \le \operatorname{spr}(A)\} \tag{2.65}$$

and at least one $z_0\in\sigma(A)$ is such that $|z_0| = \operatorname{spr}(A)$ (Kato 1966, p.176). Note that (2.65) improves (2.63) because

$$\|A^n\|^{1/n} \le [\|A\|^n]^{1/n} = \|A\|$$

and so $\operatorname{spr}(A) \le \|A\|$.

Of course the above results hold if in particular A is the matrix operator of Example 2.2. □

Example 2.17. Spectral properties of an operator in l^1

Let $X = l^1$ (see Example 1.2) and let A be the operator defined by the matrix $\{a_{i,j}; i,j, = 1,2,\ldots\}$ with $a_{i,j} = 0$ if $i \ne j$ and with $a_{j,j} = -(j-1)$, $j = 1,2,\ldots$:

$$Af = \begin{pmatrix} 0, & 0, & 0, & \ldots \\ 0, & -1, & 0, & \ldots \\ 0, & 0, & -2, & \ldots \\ \ldots & \ldots & \ldots & \ldots \end{pmatrix} \begin{pmatrix} f_1 \\ f_2 \\ f_3 \\ \ldots \end{pmatrix}$$

$$f = \begin{pmatrix} f_1 \\ f_2 \\ f_3 \\ \cdots \end{pmatrix} \in D(A) = \{f : f \in l^1; \quad \sum_{j=1}^{\infty} j|f_j| < \infty\}, \qquad R(A) \subset l^1 .$$

A is densely defined because $D(A)$ contains D_0, the subset of l^1 composed of all vectors with a *finite* number of non-zero components, which is dense in l^1 as is easy to verify. Note also that $\|Af\|_1 < \infty$, $\forall f \in D(A)$.

We now consider the equation

$$(zI-A)f = 0 \tag{2.66}$$

and, according to definition (2.54a), we try to find out whether or not values of the complex parameter z exist for which (2.66) has non-zero solutions belonging to $D(A)$. We have from (2.66)

$$zf_1 = 0, \ (z+1)f_2 = 0, \ (z+2)f_3 = 0, \ldots, \ (z+n-1)f_n = 0, \ \ldots \tag{2.67}$$

since the components of Af are $0, -f_2, -2f_3, \ldots, -(n-1)f_n, \ldots$, and so if we choose $z = z_n = -(n-1)$ and

$$f = f^{(n)} = \begin{pmatrix} f_1^{(n)} \\ f_2^{(n)} \\ \cdots \end{pmatrix}, \quad f_1^{(n)} = f_2^{(n)} = \ldots = f_{n-1}^{(n)} = f_{n+1}^{(n)} = \ldots = 0, \ f_n^{(n)} = 1$$

then $f^{(n)}$ satisfies (2.67) with $z = z_n = -(n-1)$. Moreover,

$$\|f^{(n)}\|_1 = \sum_{j=1}^{\infty} |f_j^{(n)}| = 1$$

$$\|Af^{(n)}\|_1 = \sum_{j=1}^{\infty} (j-1)|f_j^{(n)}| = n-1 < \infty.$$

We conclude that the non-zero element $f^{(n)}$ belongs to $D(A)$ and that it satisfies (2.66) with $z = z_n = -(n-1)$. Hence, $z_n \in P_\sigma(A)$, $f^{(n)}$ is the corresponding normalized eigenfunction, and

$$\{z : z = z_n = -(n-1), \ n = 1,2,\ldots\} \subset P_\sigma(A)$$

i.e. $0,-1,-2,\ldots$, are eigenvalues of A.

The next step is usually to investigate the structure of $\rho(A)$; to this aim, we consider the equation

$$(zI-A)f = g \tag{2.68}$$

where g is any given element of l^1 and where $z = \alpha+i\beta \neq z_n$ $\forall\, n = 1,2,\ldots$. Equation (2.68) can be written as follows
$zf_1 = g_1,\ (z+1)f_2 = g_2,\ldots,(z+n-1)f_n = g_n,\ldots$

and so

$$f_1 = \frac{g_1}{z}\,,\quad f_2 = \frac{g_2}{z+1}\,,\ldots,\ f_n = \frac{g_n}{z+n-1}\,,\ldots\,. \tag{2.69}$$

Note that the element f, whose components are given by (2.69), is such that

$$\begin{aligned}\|f\|_1 &= \sum_{j=1}^{\infty} |f_j| = \sum_{j=1}^{\infty} \frac{1}{|z+j-1|}\,|g_j| \\ &= \sum_{h=0}^{\infty} \frac{1}{|z+h|}\,|g_{h+1}|\end{aligned} \tag{2.70a}$$

$$\begin{aligned}\|Af\|_1 &= \sum_{j=1}^{\infty} (j-1)\,|f_j| = \sum_{j=1}^{\infty} \frac{j-1}{|z+j-1|}\,|g_j| \\ &= \sum_{h=0}^{\infty} \frac{h}{|z+h|}\,|g_{h+1}|\end{aligned} \tag{2.70b}$$

where $|z+h| = [(\alpha+h)^2+\beta^2]^{1/2}$ and $z \neq z_n$ $\forall n = 1,2,\ldots$. Assume first that $z = \alpha > 0$; then $z \neq z_n$ $\forall n = 1,2,\ldots$ and we obtain from (2.70)

$$\|f\|_1 \leq \frac{1}{z} \sum_{h=0}^{\infty} |g_{h+1}| = \frac{1}{z}\,\|g\|_1\,,\quad \|Af\|_1 \leq \|g\|_1 \quad \forall\, z > 0$$

because now $|z+h| = z+h \geq z$, $h/|z+h| < 1$. Since $\|Af\|_1 < \infty$, the element f given by (2.69) belongs to $D(A)$ and it is the unique solution of equation (2.68). Thus,

$$f = (zI-A)^{-1}g$$

$$= \begin{pmatrix} z^{-1}, & 0 & , & 0 & , 0,\ldots \\ 0 & , (z+1)^{-1}, & 0 & , 0,\ldots \\ 0, & 0 & , (z+2)^{-1}, & 0,\ldots \\ \cdot & \cdot & \cdot & \cdot & \cdot \end{pmatrix} \begin{pmatrix} g_1 \\ g_2 \\ g_3 \\ \cdot\cdot \end{pmatrix} \forall z > 0$$

where $(zI-A)^{-1}$ is defined for any $g \in l^1$ and

$$\|f\|_1 = \|(zI-A)^{-1}\ g\|_1 \le \frac{\|g\|_1}{z} \quad \forall z > 0$$

and so

$$(zI-A)^{-1} = R(z,A) \in \mathscr{B}(l^1), \quad \|R(z,A)\| \le z^{-1}, \ \forall z > 0 \tag{2.71}$$

$$\{z : z = \alpha > 0\} \subset \rho(A) \ . \tag{2.72}$$

Now assume that $z = \alpha+i\beta$ with $\beta \neq 0$; then we have from (2.70)

$$\|f\|_1 = \sum_{h=0}^{\infty} [(\alpha+h)^2+\beta^2]^{-1/2} \ |g_{h+1}| \le |\beta|^{-1} \|g\|_1$$

$$\|Af\|_1 = \sum_{h=0}^{\infty} \{h[(\alpha+h)^2+\beta^2]^{-1/2}\}|g_{h+1}|$$

$$\le (\alpha^2+\beta^2)^{1/2}|\beta|^{-1}\|g\|_1$$

because $h^2[(\alpha+h)^2+\beta^2]^{-1} \le (\alpha^2+\beta^2)\beta^{-2} \ \forall\, h \in R^1$. Hence, the element f given by (2.69) still belongs to $D(A)$ and it is the unique solution of equation (2.68). Thus, $(zI-A)^{-1}$ exists, its domain is the whole space l^1, and

$$(zI-A)^{-1} = R(z,A) \in \mathscr{B}(l^1), \quad \|R(z,A)\| \le |\beta|^{-1} \ \forall z = \alpha+i\beta, \ \beta \neq 0,$$

$$\{z : z = \alpha+i\beta, \beta \neq 0\} \subset \rho(A).$$

The case $z = \alpha < 0$, $\alpha \neq z_n$, $n = 1,2,\ldots$ remains to be examined; since now $\alpha = -|\alpha|$, we have from (2.70a)

$$\|f\|_1 = \sum_{h=0}^{\infty} \frac{1}{|h-|\alpha||} |g_{h+1}|$$

$$\leq \frac{1}{|h_\alpha - |\alpha||} \sum_{h=0}^{\infty} |g_{h+1}| = \frac{1}{|h_\alpha - |\alpha||} \|g\|_1$$

where the non-negative integer h_α is such that $|h_\alpha - |\alpha|| \leq |h-|\alpha||$ $\forall h = 0,1,2,\ldots$. Moreover, we obtain from (2.70b)

$$\|Af\|_1 = \sum_{h=0}^{\infty} \frac{h-|\alpha|+|\alpha|}{|h-|\alpha||} |g_{h+1}| \leq \sum_{h=0}^{\infty} \frac{|h-|\alpha||+|\alpha|}{|h-|\alpha||} |g_{h+1}|$$

$$= \|g\|_1 + |\alpha| \sum_{h=0}^{\infty} \frac{1}{|h-|\alpha||} |g_{h+1}| \leq \|g\|_1 + \frac{|\alpha|}{|h_\alpha - |\alpha||} \|g\|_1 < \infty.$$

As before we conclude that

$$\{z : z = \alpha < 0,\ \alpha \neq z_n,\ n = 1,2,\ldots\} \subset \rho(A) .$$

We may summarize the preceding results as follows:

$$\sigma(A) = P_\sigma(A) = \{z : z = z_n = -(n-1),\ n = 1,2,\ldots\}$$
$$\rho(A) = \{z : z = \alpha + i\beta,\ z \neq z_n,\ n = 1,2,\ldots\} \tag{2.73}$$

where we point out that relation (2.71) holds only if $z = \alpha > 0$. Note also that $A \in \mathscr{C}(l^1)$ because $-(zI-A)^{-1} \in \mathscr{B}(l^1)$ $\forall z > 0$ (see (2.35)).

We finally mention that the matrix operator A of this example arises from the stochastic theory of neutron multiplication in low-power nuclear reactors. □

Example 2.18. The convection operator

Let $X = L^2(a,b)$ with $-\infty < a < b < +\infty$ and with norm $\|\cdot\|_2$ (see Example 1.4) and denote by A the *convection* operator

$$Af = -v \frac{\mathrm{d}f}{\mathrm{d}x} \qquad D(A) = \{f : f \in L^2(a,b);\ \frac{\mathrm{d}f}{\mathrm{d}x} \in L^2(a,b);\ f(a) = 0\}$$

$$R(A) \subset L^2(a,b)$$

where $\mathrm{d}f/\mathrm{d}x = f'$ is a generalized derivative and v is a given positive constant (v is the *convection speed*). Note that each $f \in D(A)$ is absolutely continuous (see section 1.3) because it is differentiable in the generalized sense. Hence, it makes sense to impose the boundary condition $f(a) = 0$, which is included in the definition of $D(A)$.

As in Example 2.17, we first consider the equation

$$(zI-A)f = 0, \qquad z = \alpha + i\beta \tag{2.74}$$

and we look for those values of the complex parameter z, if any, such that (2.74) has a solution $f \in D(A)$ with $\|f\|_2 \neq 0$. We have from (2.74)

$$f' + \frac{z}{v} f = 0, \qquad f = c \exp\left(-\frac{z}{v} x\right)$$

where c is an arbitrary constant. Since the boundary condition $f(a) = 0$ implies $c = 0$, we conclude that (2.74) has only the trivial solution $f = 0$. Hence, (2.74) implies $\|f\|_2 = 0$ and so $(zI-A)^{-1}$ exists $\forall z \in \mathbb{C}$ (see Lemma 2.1) and $P_\sigma(A)$ is empty.

Now, if g is any given element of $L^2(a,b)$, then the equation

$$(zI-g)f = g \tag{2.75}$$

can be written

$$f' + \frac{z}{v} f = \frac{1}{v} g \tag{2.76}$$

and its general solution has the form

$$f = f(x) = c \exp\left(-\frac{z}{v} x\right) + \frac{1}{v} \int_a^x \exp\left\{-\frac{z}{v}(x-y)\right\} g(y)\, dy$$

where the integral is in the Lebesque sense because f' is a generalized derivative and $g \in L^2(a,b) \subset L^1(a,b)$ ($b-a < \infty$, see Exercise 1.9). Since $f(a) = 0$, the integration constant c must be chosen equal to zero and so

$$f = f(x) = \frac{1}{v} \int_a^x \exp\left\{-\frac{z}{v}(x-y)\right\} g(y)\, dy \tag{2.77}$$

is the solution of (2.76) satisfying the condition $f(a) = 0$. We have from (2.77) $\forall\, \alpha = \mathrm{Re}\, z \neq 0$

$$|f(x)| \leq \frac{1}{v} \int_a^x \exp\left\{-\frac{\alpha}{2v}(x-y)\right\} \exp\left\{-\frac{\alpha}{2v}(x-y)\right\} |g(y)|\, dy \tag{2.78}$$

$$|f(x)|^2 \leq \frac{1}{v^2} \int_a^x \exp\left\{-\frac{\alpha}{v}(x-y)\right\} dy \int_a^x \exp\left\{-\frac{\alpha}{v}(x-y)\right\} |g(y)|^2 dy$$

$$= \frac{1}{\alpha v}\left[1-\exp\left\{-\frac{\alpha}{v}(x-a)\right\}\right] \int_a^x \exp\left\{-\frac{\alpha}{v}(x-y)\right\} |g(y)|^2\, dy$$

$$\leq \frac{1}{\alpha v}\left[1-\exp\left\{-\frac{\alpha}{v}(b-a)\right\}\right] \int_a^x \exp\left\{-\frac{\alpha}{v}(x-y)\right\} |g(y)|^2 dy$$

where we used Schwarz inequality (1.8) and where we took into account that

$$\left[1-\exp\left\{-\frac{\alpha}{v}(x-a)\right\}\right]\alpha^{-1} \leq \left[1-\exp\left\{-\frac{\alpha}{v}(b-a)\right\}\right]\alpha^{-1}$$

$$\forall x \in [a,b], \quad \forall \alpha \neq 0.$$

Thus

$$[\|f\|_2]^2 = \int_a^b |f(x)|^2 dx$$

$$\leq \frac{1}{\alpha v}\left[1-\exp\left\{-\frac{\alpha}{v}(b-a)\right\}\right] \int_a^b dx \int_a^x \exp\left\{-\frac{\alpha}{v}(x-y)\right\} |g(y)|^2 dy$$

$$= \frac{1}{\alpha v}\left[1-\exp\left\{-\frac{\alpha}{v}(b-a)\right\}\right] \int_a^b dy\, |g(y)|^2 \int_y^b \exp\left\{-\frac{\alpha}{v}(x-y)\right\} dx$$

$$= \frac{1}{\alpha v}\left[1-\exp\left\{-\frac{\alpha}{v}(b-a)\right\}\right] \int_a^b |g(y)|^2 \left[1-\exp\left\{-\frac{\alpha}{v}(b-y)\right\}\right] \frac{v}{\alpha}\, dy$$

$$\leq \left[\frac{1-\exp\{-(\alpha/v)(b-a)\}}{\alpha}\right]^2 [\|g\|_2]^2 .$$

In an analogous way, if $\alpha = 0$, then we obtain from (2.78)

$$|f(x)| \leq \frac{1}{v} \int_a^x |g(y)|\, dy$$

$$|f(x)|^2 \leq \frac{1}{v^2} \int_a^x 1\, dy \int_a^x |g(y)|^2 dy$$

$$\leq \frac{b-a}{v^2} \int_a^x |g(y)|^2 dy$$

$$[\|f\|_2]^2 \leq \frac{b-a}{v^2} \int_a^b dx \int_a^x |g(y)|^2 dy = \frac{b-a}{v^2} \int_a^b dy |g(y)|^2 \int_y^b dx$$

$$\leq \left(\frac{b-a}{v}\right)^2 [\|g\|_2]^2 .$$

Hence,

$$\|f\|_2 \leq \eta(\alpha)\|g\|_2, \quad \forall z = \alpha+i\beta, \quad \forall g \in L^2(a,b) \tag{2.79}$$

where

$$\left.\begin{aligned} \eta(\alpha) &= \frac{1}{\alpha}\,[1-\exp\{-\frac{\alpha}{v}\,(b-a)\}] \quad &\text{if } \alpha \neq 0 \\ \eta(\alpha) &= \frac{b-a}{v} \quad &\text{if } \alpha = 0 . \end{aligned}\right\} \tag{2.80}$$

We conclude that the element f given by (2.77) belongs to $L^2(a,b)$ $\forall z = \alpha+i\beta$ and $\forall g \in L^2(a,b)$. On the other hand, since it is easy to verify that $f(x)$ is differentiable a.e. on $[a,b]$ (see Royden 1963, p.89) and that it satisfies (2.76), then $Af = -vf' = zf-g$ and so

$$\|Af\|_2 \leq |z|\ \|f\|_2 + \|g\|_2 \leq [|z|\eta(\alpha)+1]\|g\|_2 < \infty$$

and f given by (2.77) is the unique solution of (2.75) belonging to $D(A)$. Thus, f is uniquely determined by g and so $f = (zI-A)^{-1}g$, where the inverse operator is defined for any $g \in L^2(a,b)$ and for any $z = \alpha+i\beta \in \mathbb{C}$. Furthermore, we have from inequality (2.79)

$$\|(zI-A)^{-1}g\|_2 \leq \eta(\alpha)\ \|g\|_2, \quad \forall g \in L^2(a,b), \quad \forall z = \alpha+i\beta \tag{2.81}$$

and so $(zI-A)^{-1} = R(z,A) \in \mathscr{B}(L^2(a,b))$ and the resolvent set $\rho(A)$ coincides with the whole complex plane $\mathbb{C}$. Note that $A \in \mathscr{C}(L^2(a,b))$ because $-(zI-A)^{-1}$ belongs to $\mathscr{B}(L^2(a,b))$ as well (see (2.35) and the discussion that follows).

If in particular $z = \alpha+i\beta$ with $\alpha > 0$, then we obtain from (2.81)

$$\|R(z,A)g\|_2 \leq \alpha^{-1}\|g\|_2, \quad \forall g \in L^2(a,b), \quad \forall \alpha = \text{Re } z > 0 \tag{2.82}$$

because $\eta(\alpha) \le \alpha^{-1}$ $\forall\, \alpha > 0$. As will be made clear in Chapter 4, inequality (2.82), *with the simple factor* α^{-1}, is of fundamental importance. This justifies the trick (2.78) of writing

$$\exp\left\{-\frac{\alpha}{v}(x-y)\right\} = \exp\left\{-\frac{\alpha}{2v}(x-y)\right\}\exp\left\{-\frac{\alpha}{2v}(x-y)\right\}$$

which allowed us to get rid of the factor $1/v^2$ that multiplies the integral on the right-hand side of the inequality for $|f(x)|^2$. □

Remark 2.11.

Similar results can be derived if $X = L^1(a,b)$; for instance, we have from (2.77)

$$\begin{aligned}\|f\|_1 &= \int_a^b |f(x)|\,dx \le \frac{1}{v}\int_a^b dx \int_a^x \exp\{-\frac{\alpha}{v}(x-y)\}|g(y)|\,dy \\ &= \frac{1}{v}\int_a^b dy\,|g(y)| \int_y^b \exp\{-\frac{\alpha}{v}(x-y)\}\,dx \\ &= \frac{1}{\alpha}\int_a^b |g(y)|[1-\exp\{-\frac{\alpha}{v}(b-y)\}]dy \le \eta(\alpha)\,\|g\|_1 .\end{aligned}$$

We also mention that the norm $\|\cdot\|_1$ may have a simple physical meaning. To see this, assume that $f(x) \ge 0$ is the density of particles of a certain type in a medium bounded by the planes $x = a$ and $x = b$. In other words, suppose that $f(x)\,dx$ is the number of particles whose position is between x and $x+dx$. Then $\|f\|_1$ is the total number of particles in the medium under consideration. □

Example 2.19. The heat-diffusion operator

Let $X = L^2(a,b)$ with $-\infty < a < b < +\infty$ and let A be the densely defined self-adjoint heat-diffusion operator of Example 2.14:

$$Af = kf'', \quad D(A) = \{f : f \in L^2(a,b);\ f'' \in L^2(a,b);\ f(a) = f(b) = 0\}$$

$$R(A) \subset L^2(a,b)$$

where we recall that the heat-diffusion coefficient k is a given positive constant. As usual, we first consider the equation

$$(zI-A)f = 0 \tag{2.83}$$

with $z = \alpha+i\beta$, where the unknown f must be sought in $D(A)$. The general solution of the second-order linear differential equation (2.83) reads as follows

$$f = f(x) = c_1 \exp(\mu x)+c_2 \exp(-\mu x), \quad \text{a.e. on } [a,b]$$

where μ is the principal square root of the complex number z/k, i.e. where $\mu = \surd(z/k)$ satisfies the characteristic equation

$$z-k\mu^2 = 0.$$

If we impose the boundary conditions $f(a) = f(b) = 0$, then we obtain the following homogeneous system for c_1 and for c_2

$$\left.\begin{aligned} \exp(\mu a)c_1+\exp(-\mu a)c_2 &= 0 \\ \exp(\mu b)c_1+\exp(-\mu b)c_2 &= 0 \,. \end{aligned}\right\} \tag{2.84}$$

System (2.84) has non-trivial solutions provided that

$$\Delta = \Delta(\mu) = \exp\{\mu(b-a)\}-\exp\{-\mu(b-a)\} = 0$$

i.e.

$$\exp\{2\mu(b-a)\} = 1$$

where we recall that μ is a complex number. Hence,

$$2\mu(b-a) = \log 1 = 2\pi n\mathrm{i}, \quad n = 0,1,2,\ldots \quad (\mathrm{i} = \surd{-1})$$

and so

$$z = k\mu^2 = z_n = -\frac{k\pi^2 n^2}{\delta^2}, \quad n = 0,1,2,\ldots, \quad \delta = b-a.$$

If we substitute $\mu = \mu_n = \surd(z_n/k)$ into system (2.84), then we obtain

$$c_2^{(n)} = -\exp(2\mu_n a)c_1^{(n)}$$

where $c_1^{(n)}$ may be chosen arbitrarily. Correspondingly,

$$f = f^{(n)} = c_1^{(n)}\{\exp(\mu_n x)-\exp(2\mu_n a)\exp(-\mu_n x)\}$$

$$= c_1^{(n)}\left[\exp\left(\frac{\pi n i}{\delta}x\right)-\exp\left\{\frac{\pi n i}{\delta}(2a-x)\right\}\right]$$

and it is easy to verify that $f^{(n)}$ belongs to $D(A)$ and satisfies (2.83) with $z = z_n \; \forall n = 0,1,2,\ldots$ and that $\|f^{(0)}\|_2 = 0$, $\|f^{(n)}\|_2 \neq 0 \;\; \forall n = 1,2,\ldots$. Thus,

$$\{z:z = z_n = -\frac{k\pi^2 n^2}{\delta^2},\quad n = 1,2,3,\ldots\} \subset P_\sigma(A) \tag{2.85}$$

and

$$f^{(n)}(x) = c_1^{(n)}\left[\exp\left(\frac{\pi n i}{\delta}x\right)-\exp\left\{\frac{\pi n i}{\delta}(2a-x)\right\}\right],\quad n = 1,2,\ldots \tag{2.86}$$

is an eigenfunction corresponding to eigenvalue z_n. As far as $z_0 = 0$ is concerned, we shall prove later that $0 \in \rho(A)$.

To investigate the structure of the resolvent set $\rho(A)$, we consider the equation

$$(zI-A)f = g \tag{2.87}$$

where $g \in L^2(a,b)$ and $z \neq z_n \; \forall n = 0,1,2,\ldots$. Equation (2.87) and the condition $f \in D(A)$ lead to the linear system

$$\left.\begin{aligned} &f'' - \frac{z}{k} = -\frac{1}{k}g\,, \quad a < x < b \\ &f(a) = f(b) = 0\,. \end{aligned}\right\} \tag{2.88}$$

From the elementary theory of linear differential equations with constant coefficients, it is known that the general solution of the first part of (2.88) has the form

$$f(x) = c_1\exp(\mu x)+c_2\exp(-\mu x) - \frac{1}{2\mu k}\int_a^x \exp\{\mu(x-y)\}g(y)\,dy + \frac{1}{2\mu k}\int_a^x \exp\{-\mu(x-y)\}g(y)\,dy$$

where c_1 and c_2 are integration constants and where $\mu = \surd(z/k) \neq \surd(z_n/k)\ \forall n = 0,1,2,\ldots$. If we take into account the boundary conditions $f(a) = f(b) = 0$, then we obtain

$$\left.\begin{aligned} \exp(\mu a)c_1+\exp(-\mu a)c_2 &= 0 \\ \exp(\mu b)c_1+\exp(-\mu b)c_2 &= J_1+J_2 \end{aligned}\right\}$$

where

$$J_1 = \frac{1}{2\mu k}\int_a^b \exp\{\mu(b-y)\}g(y)\,dy$$

$$J_2 = -\frac{1}{2\mu k}\int_a^b \exp\{-\mu(b-y)\}g(y)\,dy \ .$$

Hence,

$$c_1 = \frac{\exp(-\mu a)}{\Delta(\mu)}\,[J_1+J_2]$$

$$v_2 = -\frac{\exp(\mu a)}{\Delta(\mu)}\,[J_1+J_2]$$

and so

$$f = f(x) = \frac{1}{2k\mu\Delta(\mu)}\,[B_1(\mu)g+B_2(\mu)g] \tag{2.89}$$

where

$$\left.\begin{aligned} B_1(\mu)g &= -\int_a^b [\exp\{\mu(b+a-x-y)\}+\exp\{-\mu(b+a-x-y)\}]g(y)\,dy \\ B_2(\mu)g &= \int_a^b [\exp\{\mu(b-a-|x-y|)\}+\exp\{-\mu(b-a-|x-y|\}]g(y)\,dy \end{aligned}\right\} \tag{2.90}$$

and where $\mu\Delta(\mu) \neq 0$ because $\mu \neq \surd(z_n/k)\ \forall n = 0,1,2,\ldots$ by assumption. Using the results of Example 2.4, we have from (2.89)

$$\|f\|_2 \leq \frac{1}{2k|\mu||\Delta(\mu)|}\,[\|B_1(\mu)\|+\|B_2(\mu)\|]\ \|g\|_2 < \infty \tag{2.91}$$

$\forall g \in L^2(a,b)$, $\forall \mu \neq \surd(z_n/k)$ $n = 0,1,2,\ldots$, because $B_1(\mu)$ and $B_2(\mu)$ both belong to $\mathscr{B}(L^2(a,b))$. Hence, the element f given by (2.89) belongs to $L^2(a,b)$. Moreover, as usual, it is not difficult to verify *a posteriori* that such an f satisfies (2.88) a.e. on $[a,b]$

and that it belongs to $D(A)$ because $\|Af\|_2 \le |z|\|f\|_2+\|g\|_2 < \infty$. Thus, f is uniquely determined in $D(A)$ by g and so $f = (zI-A)^{-1}g$, where the inverse operator $R(z,A) = (zI-A)^{-1}$ is defined over the whole space $L^2(a,b)$ and is such that

$$\|R(z,A)g\|_2 \le \frac{1}{2k|\mu||\Delta(\mu)|}\,[\|B_1(\mu)\|+\|B_2(\mu)\|]\|g\|_2 \tag{2.92}$$

$\forall g\in L^2(a,b)$ and $\forall z \ne z_n$ $n = 0,1,2,\ldots$, because of (2.91). We conclude that $R(z,A)\in\mathscr{B}(L^2(a,b))$ and so

$$\left\{z : z \ne z_n = -\frac{k\pi^2 n^2}{\delta^2},\quad n = 0,1,2,\ldots\right\} \subset \rho(A).$$

As far as $z_0 = 0$ is concerned, the results of Example 2.14 show that $0\in\rho(A)$. Hence,

$$\left\{z : z \ne z_n = -\frac{k\pi^2 n^2}{\delta^2},\quad n = 1,2,\ldots\right\} = \rho(A) \tag{2.93}$$

and $R_\sigma(A)$ and $C_\sigma(A)$ are empty because $\rho(A) \cup P_\sigma(A) = \mathbb{C}$.

As we have already remarked at the end of Example 2.18, 'simple' upper bounds for $\|R(z,A)g\|_2$ will play a fundamental role in Chapter 4. In the particular case of the heat-diffusion operator, we have proved inequality (2.92), which might be made more explicit by evaluating upper bounds for $\|B_1(\mu)\|$ and for $\|B_2(\mu)\|$. However, since we already know that $(zI-A)^{-1}\in\mathscr{B}(L^2(a,b))$ provided that $z \ne z_n$ $n = 1,2,\ldots$, and since $L^2(a,b)$ is a Hilbert space, the following *direct* procedure can be used to find a 'simple' upper bound for $\|R(z,A)g\|_2$. Given any $f\in D(A)$ and any $z = \alpha+i\beta$, we have

$$\begin{aligned}((zI-A)f,f) &= z(f,f)-(Af,f) = z\|f\|_2^2-k\int_a^b f''(x)\overline{f(x)}\,dx \\ &= z\|f\|_2^2+k\int_a^b |f'(x)|^2 dx = z\|f\|_2^2+k\|f'\|_2^2\end{aligned}$$

because $f(a) = f(b) = 0$. On the other hand, each $f\in D(A)$ can be written as

$$f(x) = f(a)+\int_a^x f'(y)\,dy = \int_a^x f'(y)\,dy$$

and so

$$|f(x)|^2 \le \left[\int_a^x |f'(y)|\,dy\right]^2$$

$$\le \int_a^x 1\,dy \int_a^x |f'(y)|^2 dy \le (x-a)\|f'\|_2^2$$

$$\|f\|_2^2 = \int_a^b |f(x)|^2 dx$$

$$\le \int_a^b (x-a)\;dx\;\|f'\|_2^2 = \frac{(b-a)^2}{2}\|f'\|_2^2$$

$$\|f\|_2^2 \le \frac{\delta^2}{2}\|f'\|_2^2\;,\quad \forall f\in D(A)\;.$$

Hence,

$$k\|f'\|_2^2 \ge 2k\delta^{-2}\|f\|_2^2,\ \forall f\in D(A)$$

and so

$$\mathrm{Re}((zI-A)f,f) = \alpha\|f\|_2^2 + k\|f'\|_2^2 \ge [\alpha+2k\delta^{-2}]\|f\|_2^2$$

$\forall f\in D(A)$, where as usual $\mathrm{Re}(\cdot,\cdot)$ denotes the real part of the complex number $(\cdot,\cdot)$. Now let g be any non-zero element of $L^2(a,b)$ and assume that $z\in\rho(A)$. Then, a unique $f\in D(A)$ exists such that $(zI-A)f = g$ and $f = R(z,A)g$, and so

$$\|g\|_2\|f\|_2 \ge |(g,f)| \ge \mathrm{Re}(g,f)$$

$$= \mathrm{Re}((zI-A)f,f) \ge [\alpha+2k\delta^{-2}]\|f\|_2^2$$

$$\|f\|_2 \le [\alpha+2k\delta^{-2}]^{-1}\|g\|_2$$

provided that $\alpha+2k\delta^{-2} > 0$. Hence,

$$\|R(z,A)g\|_2 \le [\alpha+2k\delta^{-2}]^{-1}\|g\|_2 \tag{2.94}$$

$\forall g\in L^2(a,b)$ and $\forall z = \alpha+i\beta$ such that $\alpha > -2k\delta^{-2}$ (then, z certainly belongs to $\rho(A)$ because $-k\pi^2n^2\delta^{-2} < -2k\delta^{-2}\ \forall n = 1,2,\ldots$). Inequality (2.94) is the required simple upper bound for $\|R(z,A)g\|_2$. □

Remark 2.12.

The heat-diffusion operator can also be studied in the B-space $L^1(a,b)$, $-\infty < a < b < +\infty$; of course, the domain $D(A)$ is then defined as follows:

$$D(A) = \{f: f\in L^1(a,b);\ f''\in L^1(a,b);\ f(a) = f(b) = 0\}.$$

Since $L^1(a,b)$ is not a Hilbert space, the simple procedure leading to (2.94) cannot be used and inequalities for $\|R(z,A)g\|_1$ can be derived directly from (2.89), which is still valid and can be rewritten as

$$f(x) = \frac{1}{2k\mu\Delta(\mu)}\left\{\int_a^x \chi(x,y)g(y)\,dy + \int_x^b \chi(y,x)g(y)\,dy\right\} \tag{2.95}$$

where

$$\chi(x,y) = [\exp\{\mu(b-x)\}-\exp\{-\mu(b-x)\}][\exp\{\mu(y-a)\}-\exp\{-\mu(y-a)\}].$$

Further, if $z = \alpha > 0$, then $\mu = \sqrt{(z/k)} > 0$ and $\exp(\mu\hat{x})-\exp(-\mu\hat{x}) \geq 0$ $\forall\,\hat{x} \geq 0$. Thus,

$$|f(x)| \leq \frac{1}{2k\mu\Delta(\mu)}\left\{\int_a^x \chi(x,y)\,|g(y)|\,dy + \int_x^b \chi(y,x)\,|g(y)|\,dy\right\}$$

because $\chi(x,y) \geq 0$. If we integrate both sides of the preceding inequality, then we obtain after interchanging the order of integrations

$$\begin{aligned}\|f\|_1 &= \int_a^b |f(x)|\,dx \\ &\leq \frac{1}{k\mu^2\Delta(\mu)}\int_a^b (\Delta(\mu)-[\exp\{\mu(y-a)\} - \exp\{-\mu(y-a)\}]- \\ &\qquad [\exp\{\mu(b-y)\}-\exp\{-\mu(b-y)\}])\,|g(y)|\,dy\end{aligned}$$

and so

$$\|f\|_1 = \|R(z,A)g\|_1 \leq \frac{1}{z}\|g\|_1, \quad \forall\, g\in L^1(a,b),\ \forall z > 0 \tag{2.96}$$

and

$$\{z : z = \alpha > 0\} \subset \rho(A).$$

We finally mention that, if $f(x)$ is the absolute temperature in a rigid body bounded by the planes $x = a$ and $x = b$, then $\|f\|_1$ is, under suitable assumptions, proportional to the total heat energy contained in the body. □

Remark 2.13.

The heat-diffusion operator has already been studied in the Sobolev space $L^{s,2}(R^1)$ in Example 2.15. □

Example 2.20. Resolvent of A+B with $A \in \mathscr{C}(X)$ *and* $B \in \mathscr{B}(X)$

Let $A \in \mathscr{C}(X)$ and $B \in \mathscr{B}(X)$ with $\|B\| = b$ and assume that $\{z : z = \alpha > \alpha_0\} \subset \rho(A)$ with

$$\|R(z,A)g;X\| \le (z-\alpha_0)^{-1}\|g;X\|, \quad \forall g \in X, \ \forall z = \alpha > \alpha_0 \tag{2.97}$$

where α_0 is a suitable real number. To investigate the structure of the resolvent set $\rho(A+B)$ of the *perturbed* operator $A+B$, we first observe that $BR(z,A) = B(zI-A)^{-1} \in \mathscr{B}(X)$ with

$$\|B(zI-A)^{-1}f;X\| \le \|B\| \, \|(zI-A)^{-1}f;X\| \le \frac{b}{z-\alpha_0}\|f;X\|$$

$\forall f \in X$ and $\forall z > \alpha_0$, because the *perturbation* B belongs to $\mathscr{B}(X)$ and the resolvent $R(z,A) = (zI-A)^{-1}$ of the *unperturbed* operator A belongs to $\mathscr{B}(X)$ $\forall z > \alpha_0$. Hence, if we choose $z > \alpha_0 + b$, then $b(z-\alpha_0)^{-1} < 1$ and so $\|B(zI-A)^{-1}\| \le b(z-\alpha_0)^{-1} < 1$. As a consequence, the inverse operator $[I-B(zI-A)^{-1}]^{-1}$ exists and belongs to $\mathscr{B}(X)$ with

$$\|[I-B(zI-A)^{-1}]^{-1}f;X\| \le \frac{1}{1-\|B(zI-A)^{-1}\|}\|f;X\|$$

$$\le \frac{1}{1-b(z-\alpha_0)^{-1}}\|f;X\| = \frac{z-\alpha_0}{z-(\alpha_0+b)}\|f;X\|$$

$$\forall f \in X, \quad \forall z > \alpha_0 + b$$

where we took into account the first of (2.62), with $|z| = 1$ and with $B(zI-A)^{-1}$ instead of A. On the other hand, we have $\forall z > \alpha_0 + b > \alpha_0$

$$(zI-A-B)^{-1}f = [\{I-B(zI-A)^{-1}\}(zI-A)]^{-1}f$$

$$= (zI-A)^{-1}\{I-B(zI-A)^{-1}\}^{-1}f \quad \forall f \in X$$

(see Exercise 2.12), and so

$$\|(zI-A-B)^{-1}f;X\| \leq \|(zI-A)^{-1}\| \, \|\{I-B(zI-A)^{-1}\}^{-1}f;X\|$$

$$\leq \frac{1}{z-\alpha_0}\,\frac{z-\alpha_0}{z-(\alpha_0+b)}\,\|f;X\| = \frac{1}{z-(\alpha_0+b)}\,\|f;X\|$$

$\forall f \in X$, $\forall z > \alpha_0 + b = \alpha_0 + \|B\|$. We conclude that the resolvent $R(z,A+B) = (zI-A-B)^{-1}$ of the perturbed operator $A+B$ exists and belongs to $\mathscr{B}(X)$ with

$$\|R(z,A+B)\| \leq \{z-(\alpha_0+\|B\|\}^{-1}, \quad \forall z > \alpha_0 + \|B\| \,. \tag{2.98}$$

Thus,

$$\{z : z = \alpha > \alpha_0 + \|B\|\} \subset \rho(A+B) \,.$$

Example 2.2 1.

Let A be a self-adjoint operator with domain and range contained in the Hilbert space X (see section 2.6). Given any $f \in D(A)$ and $z = \alpha + i\beta$, we have $(\|\cdot\| = \|\cdot;X\|)$

$$\|(zI-A)f\|^2 = ((zI-A)f,(zI-A)f)$$

$$= ((\alpha I-A)f+i\beta f,(\alpha I-A)f+i\beta f)$$

$$= ((\alpha I-A)f,(\alpha I-A)f) - i\beta((\alpha I-A)f,f)$$

$$+ i\beta(f,(\alpha I-A)f) + (i\beta f, i\beta f)$$

$$= \| (\alpha I - A) f \|^2 + \beta^2 \| f \|^2$$

because $((\alpha I-A)f,f) = (f,(\alpha I-A)f)$, $\forall f \in D(A) = D(A^*)$ owing to (2.41). Hence,

$$\| (zI-A)f \| \geq |\beta| \, \| f \|, \quad \forall f \in D(A), \quad \forall z = \alpha + i\beta \ . \tag{2.99}$$

Inequality (2.99) shows that $(zI-A)f = 0$ implies $\| f \| = 0$ if $\beta \neq 0$ and so $(zI-A)^{-1}$ exists with domain

$$D_0 = D((zI-A)^{-1}) = R(zI-A), \quad z = \alpha + i\beta, \ \beta \neq 0$$

because of Lemma 2.1. It also follows that any $g \in D_0$ is the image under the mapping $(zI-A)$ of a unique $f \in D(A)$ with $f = (zI-A)^{-1} g$. Thus, (2.99) can be put into the form

$$\| (zI-A)^{-1} g \| \leq \frac{1}{|\beta|} \| g \|, \quad \forall \ g \in D_0, \quad \forall z = \alpha + i\beta, \quad \beta \neq 0 \tag{2.100}$$

and the inverse operator $(zI-A)^{-1}$ is bounded over its domain D_0, provided that $\beta = \operatorname{Im} z \neq 0$. However, $A = A^* \in \mathscr{C}(X)$ according to (2.39) and consequently $(zI-A)^{-1}$ is bounded and closed because of (*b*) and (*c*) of Theorem 2.3. It follows that D_0 is a closed (linear) subset of X because of (*a*) of Theorem 2.3. To prove that in fact $D_0 = X$, let $\gamma \in X$ be such that

$$((zI-A)f, \gamma) = 0, \quad \forall f \in D(A) \ . \tag{2.101}$$

Then

$$(Af, \gamma) = z(f, \gamma) \ , \quad (Af, \gamma) = (f, \gamma^*), \quad \forall f \in D(A)$$

with $\gamma^* = \bar{z}\gamma$. If we recall the definition of the adjoint operator A^* (see (2.36) and (2.38)), then we conclude that $\gamma \in D(A^*)$ and $\gamma^* = A^*\gamma$. However, $A = A^*$ by assumption and so $\gamma \in D(A) = D(A^*)$ and $\gamma^* = A\gamma$. Thus,

$$z(f, \gamma) = (f, A\gamma) \ , \quad \forall f \in D(A)$$

$$(f, (\bar{z}I - A)\gamma) = 0, \quad \forall f \in D(A)$$

where $\bar{z} = \alpha - i\beta$. The preceding relation implies that $(\bar{z}I-A)\gamma = \theta_X$ because $D(A)$ is a dense subset of X (see the proof that follows (2.37)). However, $-\beta \neq 0$ and, as we have shown above, $(\bar{z}I-A)^{-1}$ exists and is defined over the linear set $D((\bar{z}I-A)^{-1} = R(\bar{z}I-A)$, which certainly contains θ_X, the zero element of X. Thus, $\gamma = (\bar{z}I-A)^{-1}\theta_X = \theta_X$ and (2.101) leads to $\gamma = \theta_X$. On the other hand, each $g\in D_0 = R(zI-A)$ can be written as $g = (zI-A)f$ with $f\in D(A)$, as we have already observed in order to derive (2.100). Hence, we have proved that if

$$(g,\gamma) = 0, \quad \forall\, g\in D_0$$

then $\gamma = \theta_X$.

Finally, since X is a Hilbert space and D_0 is a closed linear subset of X, each $f_0\in X$ can be *uniquely* written as

$$f_0 = \phi+\gamma$$

with $\phi\in D_0$ and with γ such that $(g,\gamma) = 0, \forall\, g\in D_0$ (i.e. with γ *orthogonal* to D_0 (Kato 1966, p.252)). Thus, $\gamma = \theta_X$ and so $f_0 = \phi\in D_0$. In other words, each $f_0\in X$ also belongs to D_0, i.e. $X \subset D_0$; but, since obviously $D_0 \subset X$, we conclude that $X_0 = D_0$ as announced.

Inequality (2.100) then becomes

$$\|R(z,A)g\| \leq \frac{1}{|\beta|}\|g\|, \quad \forall g\in X, \quad \forall z = \alpha+i\beta, \quad \beta \neq 0 \tag{2.102}$$

and

$$\{z : z = \alpha+i\beta,\ \beta \neq 0\} \subset \rho(A)$$

because $R(z,A) = (zI-A)^{-1}\in\mathscr{B}(X)$, $\forall\, z = \alpha+i\beta$ with $\beta \neq 0$.

Remark 2.14.

If $A_1 = iA$ with A self-adjoint and with $i = \sqrt{-1}$, then we have $\forall\, z = z_1+iz_2$

$$(zI-A_1) = (z_1+iz_2)I - iA = i\{(z_2-iz_1)I-A\}$$

and so

$$(zI-A_1)^{-1} = -i\{(z_2-iz_1)I-A\}^{-1}$$

exists provided that $z_1 \neq 0$. Moreover, we have from (2.102) with $\alpha = z_2$ and with $\beta = -z_1$

$$\|(zI-A_1)^{-1}g\| \leq \frac{1}{|z_1|}\|g\|, \quad \forall g \in X, \quad \forall z = z_1+iz_2, \quad z_1 \neq 0 \tag{2.103}$$

and $R(z,A_1) = (zI-A_1)^{-1} \in \mathscr{B}(X)$ $\forall z = z_1+iz_2$ with $z_1 \neq 0$. Operators of the form iA with A self-adjoint are fundamental in quantum mechanics. □

EXERCISES

2.1. Given any linear operator A with $D(A) \subset X$ and with $R(A) \subset Y$, prove that (*a*) $R(A)$ is a linear subset of Y, (*b*) A maps the zero element of X into the zero element of Y (hint : write $\theta_X = 0f_0$ with $f_0 \in D(A)$), (*c*) if A^{-1} exists, then it is a linear operator.

2.2. Show that $\|A\|_{f_0} = \nu(A)+\|Af_0;Y\|$, where f_0 is a given element of D, is a norm in $\text{Lip}(D,Y)$.

2.3. Given the operator

$$A(f) = \frac{1}{\{1-f(x)\}^2}, \quad D(A) = D = \{f: f \in C([a,b]);\ |f(x)| \leq m\}$$

$$R(A) \subset C([a,b])$$

where the constant m is such that $0 < m < 1$ and where $C([a,b])$ is here the real B-space of all real-valued continuous functions on $[a,b]$, prove that $A \in \text{Lip}(D,C([a,b]))$ and that D is a closed subset of $C([a,b])$.

2.4. Prove that the integral operator A of Example 2.4 belongs to $\mathscr{B}(C([a,b]))$ and to $\mathscr{B}(C([a,b]), L^2(a,b))$ with $-\infty < a < b < +\infty$, provided that the kernel $k(x,y)$ is continuous over the closed square $[a,b]\times[a,b]$.

2.5. Prove that $A = \mathrm{d}/\mathrm{d}x \in \mathscr{B}(C^1([a,b]), C([a,b]))$, where $C^1([a,b])$ was defined in Exercise 1.4.

2.6. Given $k = k(x,y) \in C([a,b]\times[a,b])$, prove that the (Volterra) integral operator

$$Af = \int_a^x k(x,y)f(y)\,\mathrm{d}y, \quad f \in D(A) = C([a,b]), \quad x \in [a,b]$$

belongs to $\mathscr{B}(C([a,b]))$ with $\|A\| \le (b-a)\hat{k}$, where $\hat{k} = \max\{|k(x,y)|, a \le x, y \le b\}$. Show also that $A^j \in \mathscr{B}(C([a,b]))$ with $\|A^j\| \le (b-a)^j \hat{k}^j / j!$, $\forall j = 1,2,\ldots$. Hint: if $g = g(x) = Af$, then

$$|g(x)| \le \hat{k}\|f\| \int_a^x \mathrm{d}y = (x-a)\hat{k}\|f\| \le (b-a)\hat{k}\,\|f\|$$

$$|A^2 f| = |Af| = \Big|\int_a^x k(x,x')g(x')\,\mathrm{d}x'\Big| \le \hat{k}\int_a^x |g(x')|\,\mathrm{d}x' \le \ldots .$$

2.7. Prove that a bounded operator B is closed if and only if $D(B)$ is a closed subset of X. Hint: assume first that $D(B)$ is a closed subset of X. Hence, if $\{f_n\} \subset D(B)$ is convergent to f, then $f = X\text{-}\lim f_n \in D(B)$; moreover, $\|Bf_n - Bf; Y\| \le \|B\|\,\|f_n - f; X\| \to 0$ and so $Bf = Y\text{-}\lim Bf_n$, ...

2.8. Prove that the Fourier-Plancherel operator $\mathscr{F}$ given by (1.47) belongs to $\mathscr{B}(L^2(R^n))$ with $\|\mathscr{F}\| \le 1$; further, show that the operator that can be obtained from (1.47) by substituting i with -i is the inverse Fourier-Plancherel operator $\mathscr{F}^{-1}$ and that $\|\mathscr{F}^{-1}\| \le 1$.

2.9. Prove that $(zI-A)Bf = f$, $\forall f \in X$, where $A \in \mathscr{B}(X)$ and B is defined by (2.60).

2.10. Study the convection operator A of Example 2.18 in the space $C([a,b])$ and with domain $D = D(A) = \{f : f \in C([a,b]);\ f' \in C([a,b]);\ f(a) = 0\}$. Show also that $P_\sigma(A) = \mathbb{C}$ if the domain of A is defined by $D_1 = \{f: f \in C([a,b]); f' \in C([a,b])\}$, i.e. if no boundary conditions are included in the definition of the domain of A.

2.11. Study the heat-diffusion operator A of Example 2.19 in the

B-space $X = \{f: f \in C([a,b]);\ f(a) = f(b) = 0\}$ (see Exercise 1.6) where now $D(A) = \{f: f \in X; f'' \in X\}$.

2.12. Let A_1 be a linear operator with domain and range contained in X and let $A_2 \in \mathscr{B}(X)$. Under the assumption that A_1^{-1} and A_2^{-1} exist and belong to $\mathscr{B}(X)$, show that

$$(A_2 A_1)^{-1} = A_1^{-1} A_2^{-1} .$$

2.13. Prove that any eigenvalue z of a self-adjoint operator A is a real number. Hint: if $z \in P_\sigma(A)$, then $(zI-A)f = 0$ with $\|f\| \neq 0$ and so $0 = ((zI-A)f,f) = \ldots$.

2.14. Prove the formula

$$AR(z,A) = zR(z,A)-I ,\ \forall\, z \in \rho(A)$$

where $R(z,A) = (zI-A)^{-1}$ is the resolvent of $A \in \mathscr{C}(X)$.

2.15. Prove the formula

$$R(z,A)-R(z_1,A) = (z_1-z)R(z,A)R(z_1,A)$$

where $z,\ z_1 \in \rho(A)$ and $A \in \mathscr{C}(X)$. Hint: $(z_1-z)R(z,A)R(z_1,A) = (zI-A)^{-1}(z_1I-zI)(z_1I-A)^{-1} = (zI-A)^{-1}\{(z_1I-A)-(zI-A)\}(z_1I-A)^{-1}\}$.

2.16. Prove the formula

$$AR(z,A)f = R(z,A)Af,\ \forall\, f \in D(A),\quad z \in \rho(A),\quad A \in \mathscr{C}(X) .$$

Hint: $(zI-A)(zI-A)^{-1}f = (zI-A)^{-1}(zI-A)f,\ \forall\, f \in D(A)$.

3

ANALYSIS IN BANACH SPACES

3.1. STRONG CONTINUITY

Let X be a B-space and let $\{f_n, n = 1,2,\ldots\}$ be a sequence contained in X. In section 1.1 we agreed to call $\{f_n\}$ convergent to $f \in X$ if $X\text{-}\lim f_n = f$, i.e. if $\lim \| f - f_n \| = 0$ as $n \to \infty$, where $\| \cdot \| = \| \cdot ; X \|$ is the norm in X. In other words, the concept of a convergent sequence was given in terms of the norm $\| \cdot \|$. In an analogous way the solution

$$y = y(t) = \begin{Bmatrix} y_1(t) \\ y_2(t) \end{Bmatrix}$$

of the initial-value problem (0.4)+(0.5) was considered as a function from $[0,+\infty)$ into the B-space R^2 and the limit (0.5) was interpreted in the sense of the norm of R^2. The preceding remarks lead quite naturally to the following definition of *strong continuity*.

Let $w = w(t)$ a function from the open interval $\Delta = (\hat{t}_1, \hat{t}_2)$ into the B-space X, i.e. given any $t \in \Delta$, let $w(t)$ be a uniquely defined element of X. We say that $w(t)$ is *strongly continuous* (or continuous in the sense of $\| \cdot \|$, or simply continuous) at $t_0 \in \Delta$ if

$$X\text{-}\lim_{t \to t_0} w(t) = w(t_0) \tag{3.1a}$$

i.e. if

$$\lim_{t \to t_0} \| w(t) - w(t_0) \| = 0 . \tag{3.1b}$$

If $w(t)$ is continuous at any $t_0 \in \Delta$, then we say that $w(t)$ is continuous in Δ.

Remark 3.1.

The definition of strong continuity coincides with the 'classical' definition if $X = R^1$ because $\| \cdot ; R^1 \| = \| \cdot \|$. □

Example 3.1. Strong continuity in R^n

If $X = R^n$, then

$$w(t) = \begin{pmatrix} w_1(t) \\ \cdots \\ w_n(t) \end{pmatrix}, \quad t\in\Delta$$

where the $w_j(t)$'s are real functions of the real variable $t\in\Delta$. According to (3.1b), $w(t)$ is strongly continuous at $t_0\in\Delta$ if

$$\lim_{t\to t_0} \|w(t)-w(t_0);R^n\| = \lim_{t\to t_0} \left\{\sum_{j=1}^{n} |w_j(t)-w_j(t_0)|^2\right\}^{1/2} = 0 \quad (3.2)$$

i.e. if

$$\lim_{t\to t_0} |w_j(t)-w_j(t_0)| = 0, \ \forall j = 1,2,\ldots,n \ . \quad (3.3)$$

Conversely, (3.2) follows from (3.3) and so strong continuity in the B-space R^n is equivalent to component-wise continuity. □

Example 3.2. Strong continuity in l^1

If $X = l^1$, then

$$w(t) = \begin{pmatrix} w_1(t) \\ w_2(t) \\ \cdots \end{pmatrix}, \quad \|w(t)\|_1 = \sum_{j=1}^{\infty} |w_j(t)| < \infty \ , \quad \forall t\in\Delta$$

where the components $w_j(t)$ are complex-valued functions of $t\in\Delta$ (see Example 1.2). The function $w(t)$ is continuous at $t_0\in\Delta$ if, given any $\varepsilon>0$, a $\delta = \delta(\varepsilon,t_0) > 0$ can be found such that

$$\|w(t_0+h)-w(t_0)\|_1 = \sum_{j=1}^{\infty} |w_j(t_0+h)-w_j(t_0)| < \varepsilon \ , \ \forall |h| < \delta \ . \quad (3.4)$$

Hence,

$$|w_j(t_0+h)-w_j(t_0)| < \varepsilon \ , \ \forall |h| < \delta \ , \ \forall j = 1,2,\ldots \quad (3.5)$$

where $\delta = \delta(\varepsilon,t_0)$ does not depend on j. Thus, the strong continuity of $w(t)$ in l^1 implies that the components $w_j(t)$ are continuous in t, uniformly with respect to the index j. □

Example 3.3. Strong continuity in $C([a,b])$

If $X = C([a,b])$, then $w(t) = f(x;t)$ where $f(x;t)$ is continuous in $x \in [a,b]$ in the 'classical' sense for each $t \in \Delta$. According to (3.1b), $w(t)$ is strongly continuous at $t_0 \in \Delta$ if

$$\lim_{t\to t_0} \|w(t)-w(t_0);\ C([a,b])\|$$
$$= \lim_{t\to t_0} \max\{|f(x;t)-f(x;t_0)|, x\in[a,b]\} = 0 \qquad (3.6)$$

i.e., if, given $\varepsilon > 0$, a $\delta = \delta(\varepsilon,t_0) > 0$ can be found such that

$$\max\{|f(x;t_0+h)-f(x;t_0)|,\ x\in[a,b]\} < \varepsilon\ ,\quad \forall\, |h| < \delta\ . \qquad (3.7)$$

Hence,

$$|f(x;t_0+h)-f(x;t_0)| < \varepsilon\ ,\quad \forall\, x\in[a,b],\quad |h| < \delta \qquad (3.8)$$

where $\delta = \delta(\varepsilon,t_0)$ does not depend on $x \in [a,b]$, and so $f(x;t)$ is continuous at t_0 uniformly with respect to $x \in [a,b]$. Conversely, if $w(t) = f(x;t)$ is continuous at t_0, uniformly in $x \in [a,b]$, then we have

$$|f(x;t_0+h)-f(x;t_0)| < \varepsilon/2,\quad \forall\ x\in[a,b],\quad |h| < \delta_1 = \delta_1(\varepsilon,t_0)$$

and so

$$\max\{|f(x;t_0+h)-f(x;t_0)|,\ x\in[a,b]\} \le \frac{\varepsilon}{2} < \varepsilon,\quad |h| < \delta_1$$

and (3.6) is satisfied. We conclude that $w(t) = f(x;t)$ is strongly continuous at $t_0 \in \Delta$ in the sense of the norm of $C([a,b])$ if and only if $f(x;t)$ is continuous at t_0 uniformly with respect to $x \in [a,b]$. □

Example 3.4. Strong continuity in $L^1(a,b)$

If $X = L^1(a,b)$, then $w(t) = f(x;t)$ is such that, for each $t\in\Delta$, $f(x;t)$ is Lebesque measurable on (a,b) and

$$\|w(t)\|_1 = \int_a^b |f(x;t)|\,dx < \infty$$

(see Example 1.4). The function is strongly continuous at $t_0\in\Delta$ if

$$\|w(t_0+h)-w(t_0)\|_1 = \int_a^b |f(x;t_0+h)-f(x;t_0)|\,dx < \varepsilon \tag{3.9}$$

provided that $|h|<\delta = \delta(\varepsilon,t_0)$. Note that (3.9) does not imply that $f(x;t_0+h)$ converges to $f(x;t_0)$ pointwise, i.e. for almost all $x\in(a,b)$. However, if $w(t)$ is continuous at $t_0\in\Delta$ in the sense of the norm of $C([a,b])$, then

$$|f(x;t_0+h)-f(x;t_0)| < \frac{\varepsilon}{b-a}\ ,\ \forall\, x\in[a,b],\quad |h|<\delta_1 = \delta_1(\varepsilon,t_0)$$

(see Example 3.3), and so

$$\int_a^b |f(x;t_0+h)-f(x;t_0)|\,dx < \varepsilon\ ,\qquad |h|<\delta_1$$

and $w(t)$ is continuous at t_0 in the sense of $\|\cdot\|_1$. In other words, strong continuity in $C([a,b])$ implies strong continuity in $L^1(a,b)$ $(b-a<\infty)$. □

Example 3.5. Strong continuity in $\mathscr{B}(X,Y)$

Let $X_1 = \mathscr{B}(X,Y)$; $B = B(t)$ is a function from Δ into X_1 if $B(t)$ is a linear bounded operator with domain $D(B(t)) = X$ and with range $R(B(t)) \subset Y$ for each $t\in\Delta$. $B(t)$ is continuous at $t_0\in\Delta$ if

$$\lim_{h\to 0} \|B(t_0+h)-B(t_0)\| = 0 \tag{3.10}$$

where $\|\cdot\| = \|\cdot;\mathscr{B}(X,Y)\|$ is the norm in the B-space $\mathscr{B}(X,Y)$ (see Theorem 2.1 with $D = X$). On the other hand, given $f\in X$, $w(t) = B(t)f$ is a function from Δ into Y and it is such that

$$\|w(t_0+h)-w(t_0);Y\| = \|[B(t_0+h)-B(t_0)]f;Y\|$$

$$\leq \|B(t_0+h)-B(t_0)\| \, \|f;X\| \to 0 \quad \text{as} \quad h \to 0$$

because of (3.10). Hence, $w(t)$ is continuous in the sense of $\|\cdot;Y\|$, and so strong continuity of $B(t)$ in $\mathscr{B}(X,Y)$ implies strong continuity of $w(t) = B(t)f$ in Y for each $f \in X$. The converse statement is not true in general. However, if $B(t)f$ is continuous at t_0 in the sense of $\|\cdot;Y\|$, *uniformly* with respect to any $g \in X$ such that $\|g;X\| = 1$, then $B(t)$ is continuous at t_0 in the sense of the norm of $\mathscr{B}(X,Y)$. To see this, we observe that, given $\varepsilon>0$, a $\delta = \delta(\varepsilon,t_0)$ can be found so that

$$\|B(t_0+h)g-B(t_0+h)g;Y\| < \varepsilon \ , \quad |h| < \delta$$

for any $g \in X$ with $\|g;X\| = 1$. Then, we have $\forall f \in X$

$$\|[B(t_0+h)-B(t_0)] \frac{f}{\|f;X\|} \, ;Y\| < \varepsilon, \quad |h| < \delta$$

because $g = f/\|f;X\|$ has norm $\|g;X\| = 1$, $(\|f;X\| \neq 0)$. It follows that

$$\|B(t_0+h)f-B(t_0)f;Y\| < \varepsilon \, \|f;X\| \ , \quad |h| < \delta = \delta(\varepsilon,t_0)$$

where f is any element of X (in fact, the preceding inequality is obviously valid even if $\|f;X\| = 0$), and so

$$\|B(t_0+h)-B(t_0)\| = \sup\left\{\frac{\|B(t_0+h)f-B(t_0)f;Y\|}{\|f;X\|} \ , \ \|f;X\| \neq 0\right\} \leq \varepsilon$$

$\forall |h| < \delta$, because of (2.6) and of Theorem 2.1 (with $D = X$ and with $A = B(t_0+h)-B(t_0)$). Thus, (3.10) is satisfied and $B(t)$ is continuous at $t_0 \in \Delta$ in the sense of the norm of $\mathscr{B}(X,Y)$. □

We finally mention that the family of all strongly continuous function $w = w(t)$ over the closed interval $[t_1,t_2] \subset \Delta$ is a vector space with the operations $\alpha\cdot$ and $+$ defined as follows

$$\left.\begin{aligned} (\alpha w_1)(t) &= \alpha w_1(t) \quad \forall t\in[t_1,t_2] \\ (w_1+w_2)(t) &= w_1(t)+w_2(t) \quad \forall t\in[t_1,t_2] \end{aligned}\right\} \tag{3.11}$$

where the operations $\alpha\cdot$ and $+$ on the right-hand sides of the two equations (3.11) are those defined in the space X to which $w_1(t)$ and $w_2(t)$ belong for each $t\in[t_1,t_2]$. The vector space of all strongly continuous functions over $[t_1,t_2]$ can be made into a B-space, denoted by $C([t_1,t_2],X)$, with the norm

$$\| w;C([t_1,t_2];X)\| = \max\{\| w(t);X\|,\ t\in[t_1,t_2]\} . \tag{3.12}$$

The proof that $C([t_1,t_2];X)$ is a B-space will not be given here because it is similar to the one for the space $C([a,b])$ (see Example 1.3); in fact, $C([t_1,t_2]);\ \mathbb{C}) = C([t_1,t_2])$ because $\|\cdot;\mathbb{C}\| = |\cdot|$. Instead, we shall briefly discuss the notion of convergence in $C([t_1,t_2];X)$. To this aim, assume that the sequence $\{w_n, n = 1,2,\ldots\}$ is convergent to w in the sense of the norm (3.12). Then, $w\in C([t_1,t_2];X)$ because $C([t_1,t_2];X)$ is a B-space and

$$\| w_n-w;C([t_1,t_2];X)\| = \max\{\| w_n(t)-w(t);X\|,\ t\in[t_1,t_2]\} < \varepsilon$$

provided that $n>n_0 = n_0(\varepsilon)$. Hence,

$$\| w_n(t)-w(t);X\| < \varepsilon,\quad \forall t\in[t_1,t_2] ,\quad n>n_0(\varepsilon)$$

and $\{w_n(t), n = 1,2,\ldots\}$ converges to $w(t)$ in the sense of $\|\cdot;X\|$ *uniformly* with respect to $t\in[t_1,t_2]$.

3.2. STRONG DERIVATIVE

Let $w(t)$ be a function from Δ into the B-space X, strongly continuous at $t_0\in\Delta$. If an element $f_0\in X$ exists such that

$$\lim_{h\to 0} \left\| \frac{1}{h}\{w(t_0+h)-w(t_0)\}-f_0 \right\| = 0 \tag{3.13}$$

then we say that f_0 is the *strong derivative* (or the

derivative in the sense of $\|\cdot\|$, or simply the derivative) of $w(t)$ at t_0. Of course, the element f_0 is uniquely determined by (3.13), whenever it exists. If $w(t)$ is strongly differentiable at any $t\in\Delta$, then we say that $w(t)$ is strongly differentiable over Δ and we denote its strong derivative by dw/dt or by $w'(t)$. Thus if $w(t)$ is strongly differentiable over Δ, then we have

$$\lim_{h\to 0} \left\| \frac{1}{h}\{w(t+h)-w(t)\} - \frac{dw}{dt}(t)\right\| = 0, \quad \forall t\in\Delta . \tag{3.14}$$

Remark 3.2.

If $X = R^1$, then the strong derivative coincides with the 'classical' derivative of a real function of the real variable $t\in\Delta$. □

Example 3.6. Strong derivative in $C([a,b])$

Assume that $w(t)$ is a function from Δ into $X = C([a,b])$, (see Example 3.3). If $w(t) = f(x;t)$ is strongly differentiable at $t\in\Delta$, then we have from (3.14)

$$\max\left\{\left|\frac{f(x;t+h)-f(x;t)}{h} - g(x;t)\right| , \; x\in[a,b]\right\} < \varepsilon \tag{3.15}$$

provided that $|h|<\delta = \delta(\varepsilon,t)$, where $g(x;t) = w'(t)$. Inequality (3.15) shows that $g(x;t) = \partial f(x;t)/\partial t$ and that

$$\left|\frac{f(x;t+h)-f(x;t)}{h} - \frac{\partial}{\partial t} f(x;t)\right| < \varepsilon , \quad \forall\, x\in[a,b], \; |h| < \delta$$

where $\delta = \delta(\varepsilon,t)$ does not depend on $x\in[a,b]$. Hence, if $w(t) = f(x;t)$ is differentiable with respect to t in the sense of the norm of $C([a,b])$, then $f(x;t)$ is partially differentiable with respect to t, uniformly in $x\in[a,b]$. Conversely, if $w(t) = f(x;t)$ is partially differentiable with respect to t, uniformly in $x\in[a,b]$, then $w'(t) = \partial f/\partial t$.

For example, let

$$w(t) = f(x;t) = \exp(x+vt), \quad x\in[a,b], \quad t\in\Delta = (\hat{t}_1,\hat{t}_2)$$

where v is a given positive constant. Since $f(x;t)$ is continuous in $x \in [a,b]$ for each $t \in \Delta$, $w(t) = f(x;t)$ can be considered as a function from Δ into $C([a,b])$. Furthermore, we have

$$\left| \frac{\exp[x+v(t+h)] - \exp(x+vt)}{h} - v \exp(x+vt) \right|$$

$$= v \exp(x+vt) \left| \frac{\exp(vh)-1}{vh} - 1 \right|$$

$$\leq v \exp(b+v\hat{t}_2) \left| \frac{\exp(vh)-1}{vh} - 1 \right| < v \exp(b+v\hat{t}_2)\ \varepsilon, \forall x \in [a,b]$$

provided that $|h| < \delta = \delta(\varepsilon)$, because

$$\lim_{h \to 0} \left| \frac{\exp(vh)-1}{vh} - 1 \right| = 0 .$$

Hence,

$$\max\left\{ \left| \frac{\exp[x+v(t+h)] - \exp(x+vt)}{h} - v \exp(x+vt) \right|, \ x \in [a,b] \right\}$$

$$\leq v \exp(b+v\hat{t}_2)\varepsilon = \varepsilon_1, \quad |h| < \delta(\varepsilon)$$

and $w(t) = f(x;t)$ is differentiable in the sense of the norm of $C([a,b])$ with $w'(t) = \partial f/\partial t = v \exp(x+vt)$. □

We end this section with a 'generalized' version of the mean-value theorem. Of course, the result stated below is weaker than the classical one (in fact, it is an inequality rather than on equality). This is because of the major complexity of the notion of strong derivative.

Theorem 3.1. (Mean-value theorem). If $w \in C^1([t_1,t_2];X)$, i.e. if $w = w(t)$ has strongly continuous strong-derivative $w'(t)$ over $[t_1,t_2]$, then

$$\| w(t_2) - w(t_1) \| \leq (t_2 - t_1) \| w'(\bar{t}) \| \tag{3.16}$$

where $\bar{t}$ is a suitable value of t belonging to $[t_1,t_2]$.

Proof. Assuming for simplicity that X is a Hilbert space, let

$$\phi(t) = \mathrm{Re}(w(t)-w(t_1), w(t_2)-w(t_1)), \quad t\in[t_1,t_2] \qquad (3.17)$$

where as usual Re$(\cdot,\cdot)$ denotes the real part of the inner product $(\cdot,\cdot)$. Note that the real function $\phi = \phi(t)$ is continuous in t (in the classical sense) because the inner product is continuous with respect to each of its two factors (see section 1.1). In fact, we have from (3.17) and from (1.8)

$$|\phi(t+h)-\phi(t)| = |\mathrm{Re}(w(t+h)-w(t), w(t_2)-w(t_1))|$$

$$\le \|w(t+h)-w(t)\| \; \|w(t_2)-w(t_1)\| \to 0 \quad \text{as } h\to 0$$

because $w(t)$ is strongly continuous by assumption. Similarly,

$$\left|\frac{\phi(t+h)-\phi(t)}{h} - \mathrm{Re}(w'(t), w(t_2)-w(t_1))\right|$$

$$= \left|\mathrm{Re}\left\{\frac{1}{h}[w(t+h)-w(t)]-w'(t), w(t_2)-w(t_1)\right\}\right|$$

$$\le \left\|\frac{1}{h}[w(t+h)-w(t)]-w'(t)\right\| \; \|w(t_2)-w(t_1)\| \to 0 \quad \text{as } h\to 0$$

because $w(t)$ is strongly differentiable. Hence, $\phi(t)$ is differentiable in the classical sense and

$$\phi'(t) = \mathrm{Re}(w'(t), w(t_2)-w(t_1)), \quad t\in[t_1,t_2]$$

where $\phi'(t)$ is continuous because $w'(t)$ is strongly continuous by assumption. It follows that the classical mean-value theorem applies to $\phi(t)$ and so a suitable $\bar{t}\in[t_1,t_2]$ exists, such that

$$\phi(t_2)-\phi(t_1) = [t_2-t_1]\phi'(\bar{t}). \qquad (3.18)$$

Thus, we have from (3.17) and from (3.18)

$$\mathrm{Re}(w(t_2)-w(t_1), w(t_2)-w(t_1)) = [t_2-t_1]\mathrm{Re}(w'(\bar{t}), w(t_2)-w(t_1)).$$

$$\|w(t_2)-w(t_1)\|^2 = [t_2-t_1]\mathrm{Re}(w'(\bar{t}), w(t_2)-w(t_1))$$

$$\|w(t_2)-w(t_1)\|^2 \le [t_2-t_1]\,\|w'(\bar{t})\|\,\|w(t_2)-w(t_1)\|$$

and inequality (3.16) is proved. □

3.3. STRONG RIEMANN INTEGRAL

Given a function $w = w(t)$ from $\Delta = (\hat{t}_1,\hat{t}_2)$ into the B-space X, the procedure to define the strong Riemann integral of $w(t)$ over the interval $[t_1,t_2] \subset \Delta$ is quite similar to the one usually employed in the classical case. Thus, we first denote by π the *partition* $\{t_1 = s_0<s_1<\ldots<s_j<\ldots<s_n = t_2\}$ of the interval $[t_1,t_2]$ together with the point $\bar{s}_j$, such that $s_j\le\bar{s}_j\le s_{j+1}$, $j = 0,1,\ldots,n-1$. We then consider the following *Riemann sum* just as in the classical case

$$S(\pi,w) = \sum_{j=0}^{n-1} [s_{j+1}-s_j]w(\bar{s}_j)$$

where $S(\pi,w)$ is a suitable element of X because $w(\bar{s}_j)\in X\ \forall j = 0,1,\ldots,n-1$. If the strong limit

$$X - \lim_{|\pi|\to 0+} S(\pi,w)\ ,\quad |\pi| = \max\{[s_{j+1}-s_j],\ j = 0,1,\ldots,n-1\}$$

exists and is independent of the partition π, then we say that $w(t)$ is *strongly Riemann integrable* (or Riemann integrable in the sense of $\|\cdot\|$, or simply Riemann integrable) over $[t_1,t_2]$ and we write

$$X - \lim_{|\pi|\to 0+} S(\pi,w) = \int_{t_1}^{t_2} w(s)\,ds \tag{3.19}$$

i.e.

$$\lim_{|\pi|\to 0+} \|S(\pi,w) - \int_{t_1}^{t_2} w(s)\,ds\| = 0. \tag{3.20}$$

Remark 3.3.

If $X = R^1$, then the strong Riemann integral coincides with the classical Riemann integral of a real function of the real variable t. If $X = R^n$, then strong Riemann integration is equivalent to componentwise classical Riemann integration (see Example 3.1). Finally, the meaning of strong Riemann integration in $C([a,b])$

and in $L^1(a,b)$ can be discussed by taking into account the definition of the norms in such spaces (see Examples 3.3, 3.4, and 3.6). □

The strong Riemann integral of $w(t)$ retains most of the properties of the classical integral and the proofs are similar to the classical ones, provided that $\|\cdot\|$ is used rather than $|\cdot|$. Thus, we have the following theorem.

Theorem 3.2. If $w = w(t)\in C[(t_1,t_2];X)$, then the strong Riemann integral (3.19) exists and the following relations hold:

$$\left\|\int_{t_1}^{t_2} w(s)\,\mathrm{d}s\right\| \le \int_{t_1}^{t_2} \|w(s)\|\,\mathrm{d}s$$

$$\le (t_2-t_1)\max\{\|w(s)\|, s\in[t_1,t_2]\}$$

$$= (t_2-t_1)\|w;C([t_1,t_2];X)\| \tag{3.21}$$

$$\frac{\mathrm{d}}{\mathrm{d}t}\int_{t_1}^{t} w(s)\,\mathrm{d}s = w(t)\;,\quad t_1<t\le t_2\;. \tag{3.22}$$

moreover, if $w(t)$ is strongly differentiable and $w'\in C([t_1,t_2];X)$, then

$$\int_{t_1}^{t_2} w'(s)\,\mathrm{d}s = w(t_2)-w(t_1).\quad \square \tag{3.23}$$

Remark 3.4.

The second integral in (3.21) is obviously a classical Riemann integral and the integrand $\phi(s) = \|w(s)\|\in C([t_1,t_2])$ because $w = w(s)\in C([t_1,t_2];X)$ (see Exercise 3.1). □

Further properties of strong Riemann integral are listed in the following theorems.

Theorem 3.3. If the sequence $\{w_n, n = 1,2,\ldots\}\subset C([t_1,t_2];X)$ converges to w in the sense of the norm of $C([t_1,t_2];X)$ then

$$X\text{-}\lim_{n\to\infty} \int_{t_1}^{t_2} w_n(s)\,ds = \int_{t_1}^{t_2} w(s)\,ds \,. \tag{3.24}$$

Proof. Note that w certainly belongs to $C([t_1,t_2];X)$ because $C([t_1,t_2];X)$ is a B-space (see also the discussion after definition (3.12)). Using (3.21), we have

$$\| \int_{t_1}^{t_2} w_n(s)\,ds - \int_{t_1}^{t_2} w(s)\,ds \| \le \int_{t_1}^{t_2} \| w_n(s) - w(s) \| ds$$

$$\le \max\{\| w_n(s) - w(s) \|, s\in[t_1,t_2]\} \int_{t_1}^{t_2} ds$$

$$= (t_2 - t_1)\| w_n - w; C([t_1,t_2]-X)\| \to 0 \quad \text{as } n\to\infty$$

because $w_n \to w$ in $C([t_1,t_2];X)$ and (3.24) is proved. □

Theorem 3.4. Let $A\in\mathscr{C}(X,Y)$ *and let* $w = w(t)$ *be a function from* $[t_1,t_2]$ *into* X*, such that (i)* $w(t)\in D(A)$ $\forall t\in[t_1,t_2]$*, (ii)* $w\in C([t_1,t_2];X)$*, (iii)* $u = u(t) = Aw(t)\in C([t_1,t_2];Y)$*. Then*

$$A \int_{t_1}^{t_2} w(s)\,ds = \int_{t_1}^{t_2} Aw(s)\,ds$$

where the integral on the left-hand side is in the sense of $\| \cdot ; X\|$ *and the one on the right-hand side is in the sense of* $\| \cdot ; Y \|$.

Proof. If

$$f_n = \sum_{j=0}^{n-1} [s_{j+1} - s_j] w(\bar{s}_j) = S(\pi,w) \,, \quad n = 1,2,\ldots$$

then $f_n\in D(A)$ because $D(A)$ is a linear subset of X and $w_j(\bar{s}_j)\in D(A)$ $\forall j = 0,1,\ldots,n-1$ owing to (i). It follows that

$$Af_n = \sum_{j=0}^{n-1} [s_{j+1} - s_j] Aw(\bar{s}_j) \,, \quad n = 1,2,\ldots \,.$$

On the other hand, (ii) and (iii) ensure that the integrals

$$\int_{t_1}^{t_2} w(s)\,ds \;, \qquad \int_{t_1}^{t_2} Aw(s)\,ds$$

both exist and so

$$X\text{-}\lim_{n\to\infty} f_n = f = \int_{t_1}^{t_2} w(s)\,ds \;,$$

$$Y\text{-}\lim_{n\to\infty} Af_n = g = \int_{t_1}^{t_2} Aw(s)\,ds \;.$$

According to Theorem 2.2, we then conclude that $f \in D(A)$ and that $g = Af$ and (3.25) is proved. □

Remark 3.5.

Relation (3.25) shows that, under the assumptions (i), (ii), and (iii), a closed operator commutes with the strong integral symbol (see also Remark 2.5). □

We finally point out that, if $A \in \mathscr{B}(X,Y)$, then $A \in \mathscr{C}(X,Y)$ because of (2.34). Moreover, (i) is now obviously satisfied since $D(A) = X$, whereas (iii) follows from (ii) because

$$\| Aw(t+h) - Aw(t);Y \| \le \| A \| \; \| w(t+h) - w(t);X \| \;.$$

Hence, (3.25) is valid for any $A \in \mathscr{B}(X,Y)$ under the only additional assumption (ii).

Example 3.7.

The properties of the linear operator

$$Aw = \frac{dw}{dt}$$

$$D(A) = \{w : w = w(t) \in C([0,t_0];X);\frac{dw}{xt} \in C([0,t_0];X);w(0) = \theta_X\}$$

$$\subset C([0,t_0];X), R(A) \subset C([0,t_0];X)$$

where X is a given B-space and d/dt is a strong derivative, can be studied by methods similar to those of section 2.8. Of course, $\|\cdot;X\|$ and strong Riemann integrals are to be used rather than $|\cdot|$ and classical Riemann integrals. □

Example 3.8. Differentiation under the integral sign

Let $f = f(t,s)$ be a function from the square $Q = [t_1,t_2]\times[t_1,t_2]$ into the B-space X and assume that

(i) $f(t,s)$ is strongly continuous over Q:

$$\|f(t+h,s+k)-f(t,s)\| \to 0 \quad \text{as} \quad h^2+k^2 \to 0;$$

(ii) the strong partial derivative $\partial f/\partial t = g(t,s)$ exists and it is strongly continuous over Q:

$$\|\frac{1}{h}\{f(t+k,s)-f(t,s)\} - g(t,s)\| \to 0 \quad \text{as} \quad h\to 0$$

$$\|g(t+h,s+k)-g(t,s)\| \to 0 \quad \text{as} \quad h^2+k^2 \to 0 \quad .$$

Then,

$$\frac{d}{dt}\int_{t_1}^{t_2} f(t,s)\,ds = \int_{t_1}^{t_2} \frac{\partial f}{\partial t}(t,s)\,ds \quad . \tag{3.26}$$

Formula (3.26) can be proved by using the Mean-value Theorem 3.1; the proof is similar to the classical one and it will not be given here. □

3.4. THE DIFFERENTIAL EQUATION $du/dt = F(u)$

Let $F = F(f)$ be an operator belonging to $\mathrm{Lip}(X,X)$ (see section 2.4); hence, $D(F) = X$, $R(F) \subset X$ and a positive constant k exists such that

$$\|F(f)-F(g)\| \le k\|f-g\| \quad \forall f,g\in X \tag{3.27}$$

where $\|\cdot\| = \|\cdot;X\|$.

Given the *initial-value problem*

$$\frac{d}{dt}u(t) = F(u(t)), \quad t_1<t\leq t_2 \tag{3.28a}$$

$$X\text{-}\lim_{t\to t_1+} u(t) = u_0 \tag{3.28b}$$

where u_0 is an assigned element of the B-space X, we say that $u = u(t)$ is a *solution* of (3.28a)+(3.28b) over the interval $[t_1,t_2]$ if

(i) $u(t)$ is strongly continuous over $[t_1,t_2]$,
(ii) $u(t)$ is strongly differentiable over $(t_1,t_2]$,
(iii) $u(t)$ satisfies the *evolution equation* (3.28a) and the *initial condition* (3.28b).

The following proof of the existence of a unique solution of system (3.28) provides a simple example of how strong derivatives and strong Riemann integrals work. We have the following lemma.

Lemma 3.1. If $u(t)$ is a solution of system (3.28) satisfying (i), (ii), and (iii), then

$$u(t) = u_0 + \int_{t_1}^{t} F(u(s))ds, \quad t\in[t_1,t_2]. \tag{3.29}$$

Conversely, if the integral equation (3.29) has a solution $u = u(t)\in Y = C([t_1,t_2];X)$, then $u(t)$ is also a solution of system (3.28).

Proof. If $u(t)$ is strongly continuous over $[t_1,t_2]$, then $v(t) = F(u(t))$ has the same property because

$$\| F(u(t+k))-F(u(t)) \| \leq k\| u(t+h)-u(t) \| \tag{3.30}$$

owing to (3.27). Hence, we can integrate both sides of (3.28a) and (3.29) follows using formula (3.23) and taking into account condition (3.28b).

Conversely, if $u\in Y$ satisfies (3.29), then $F(u(s))$ is a strongly continuous function of s and (3.28a) can be obtained by differentiating both sides of (3.29) (see (3.22)).

Moreover, we have from (3.29)

$$\begin{aligned}\|u(t)-u_0\| &= \left\| \int_{t_1}^{t} F(u(s))\,ds \right\| \\ &\le \int_{t_1}^{t} \|F(u(s))\|\,ds \\ &\le \int_{t_1}^{t} \|F(u(s))-F(u_0)+F(u_0)\|\,ds \\ &\le \int_{t_1}^{t} [k\|u(s)-u_0\|+\|F(u_0)\|]\,ds \\ &\le \int_{t_1}^{t} [k\|u-u_0;Y\|+\|F(u_0)\|]\,ds\end{aligned}$$

where $\|u-u_0;Y\| = \max\{\|u(s)-u_0\|, s\in[t_1,t_2]\}$ is the norm in $Y = C([t_1,t_2];X)$. Thus

$$\|u(t)-u_0\| \le (t-t_1)[k\|u-u_0;Y\|+\|F(u_0)\|] \to 0 \quad \text{as } t\to t_1+$$

and $u(t)$ also satisfies (3.28b). □

Remark 3.6.

Transforming the differential system into an integral equation is a standard step in existence and uniqueness proofs. □

The integral equation (3.29) will now be studied by the classical method of successive approximations. If we define

$$\left.\begin{aligned} u^{(0)}(t) &= u_0 \quad \forall t\in[t_1,t_2] \\ u^{(n)}(t) &= u_0 + \int_{t_1}^{t} F(u^{(n-1)}(s))\,ds\ , \quad t\in[t_1,t_2],\ n = 1,2,\ldots \end{aligned}\right\} \tag{3.31}$$

then each of the $u^{(n)}(t)$ is a continuous function of $t\in[t_1,t_2]$ as it is not difficult to prove by a procedure similar to that of Lemma 3.1. Further, we have from (3.31) with $n = 1,2$

$$\|u^{(1)}(t)-u^{(0)}(t)\| \le \int_{t_1}^{t} \|F(u_0)\|\,ds = (t-t_1)\|F(u_0)\|$$

$$\begin{aligned}
\|u^{(2)}(t)-u^{(1)}(t)\| &= \left\|\int_{t_1}^{t} [F(u^{(1)}(s))-F(u^{(0)}(s))]\,ds\right\| \\
&\le \int_{t_1}^{t} \|F(u^{(1)}(s))-F(u^{(0)}(s))\|\,ds \\
&\le k\int_{t_1}^{t} \|u^{(1)}(s)-u^{(0)}(s)\|\,ds \\
&\le k\int_{t_1}^{t} (s-t_1)\|F(u_0)\|\,ds \\
&= \frac{k}{2}(t-t_1)^2\,\|F(u_0)\|
\end{aligned}$$

and by induction

$$\|u^{(n)}(t)-u^{(n-1)}(t)\| \le \eta\,\frac{k^n}{n!}(t-t_1)^n, \quad \forall t\in[t_1,t_2] \qquad (3.32)$$

where $\eta = \|F(u_0)\|/k$ and $n = 1,2,\ldots$. Thus, $\forall m \ge n+1$,

$$\begin{aligned}
\|u^{(m)}(t)-u^{(n)}(t)\| &\le \|u^{(m)}(t)-u^{(m-1)}(t)\|+\ldots+\|u^{(n+1)}(t)-u^{(n)}(t)\| \\
&\le \eta\sum_{j=n+1}^{m}\frac{k^j}{j!}(t-t_1)^j \\
&\le \eta\sum_{j=n+1}^{\infty}\frac{k^j}{j!}(t-t_1)^j \\
&\le \eta\,\frac{k^{n+1}}{(n+1)!}(t_2-t_1)^{n+1}\sum_{i=0}^{\infty}\frac{k^i}{i!}(t-t_1)^i \\
&= \eta\,\frac{k^{n+1}}{(n+1)!}\exp[k(t-t_1)]
\end{aligned}$$

$\forall t\in[t_1,t_2]$, because

$$\frac{k^j}{j!}(t-t_1)^j = \frac{k^{n+1}}{(n+1)!}(t-t_1)^{n+1}\left\{k^i(t-t_1)^i\,\frac{(n+1)!}{j!}\right\}$$

$$\leq \frac{k^{n+1}}{(n+1)!}(t-t_1)^{n+1}\left\{\frac{k^i}{i!}(t-t_1)^i\right\}$$

where $j \geq n+1$ and $i = j-n-1$. As a consequence, we obtain

$$\begin{aligned}\|u^{(m)}-u^{(n)};Y\| &= \max\{\|u^{(m)}(t)-u^{(n)}(t)\|\ , \\ t\in[t_1,t_2]\} &\leq \eta\,\frac{k^{n+1}}{(n+1)!}(t_2-t_1)^{n+1}\exp\{k(t_2-t_1)\}\end{aligned} \tag{3.33}$$

and so

$$\|u^{(m)}-u^{(n)};Y\| < \varepsilon\ , \quad \forall\, m>n>n_\varepsilon$$

because $k^{n+1}(t_2-t_1)^{n+1}/(n+1)! \to 0$ as $n\to\infty$. Hence, $\{u^{(n)}, n = 1,2,\ldots\}$ is a Cauchy sequence in the B-space Y and a suitable $u\in Y$ exists such that $\|u-u^{(n)};Y\| \to 0$ as $n\to\infty$, i.e. such that $\{u^{(n)}(t), n = 1,2,\ldots\}$ converges to $u(t)$ in the sense of $\|\cdot\|$, uniformly in $t\in[t_1,t_2]$ (see the discussion of (3.12)). Using (3.24), we then have from the second of (3.31)

$$u(t) = X\text{-}\lim_{n\to\infty} u^{(n)}(t) = u_0 + \int_{t_1}^{t} X\text{-}\lim_{n\to\infty} F(u^{(n-1)}(s))\,ds\ .$$

However

$$\|F(u(s))-F(u^{(n-1)}(s))\| \leq k\|u(s)-u^{(n-1)}(s)\| \to 0 \quad \text{as } n\to\infty$$

and so

$$X\text{-}\lim_{n\to\infty} F(u^{(n-1)}(s)) = F(u(s))$$

and $u(t)$ satisfies (3.25). According to Lemma 3.1, $u = u(t)\in Y$ is also a solution of the initial-value problem (3.28). Note that, if $F(u_0) = 0$ (i.e. $\eta = 0$), then $u(t) \equiv u_0$ is a time-independent solution of system (3.28).

It remains to show that the element $u = u(t)\in Y$, found by the successive approximations method (3.31), is the only solution of (3.28) satisfying (i), (ii), and (iii). To prove such a uniqueness property we need the following lemma.

Lemma 3.2. (Gronwall's inequality) Let ϕ_0 be a non-negative constant and $k(t)$ be a continuous non-negative function defined over $[t_1,t_2]$. Then any continuous non-negative function $\phi = \phi(t)$ that satisfies the integral inequality.

$$\phi(t) \le \phi_0 + \int_{t_1}^{t} k(s)\phi(s)\,ds\ , \quad t\in[t_1,t_2] \tag{3.34}$$

is such that

$$0 \le \phi(t) \le \psi(t)\ , \quad \forall\, t\in[t_1,t_2] \tag{3.35}$$

where $\psi(t)$ is the unique continuous solution of the equation

$$\psi(t) = \phi_0 + \int_{t_1}^{t} k(s)\psi(s)\,ds, \quad t\in[t_1,t_2]$$

i.e.

$$\psi(t) = \phi_0 \exp\left\{\int_{t_1}^{t} k(s)\,ds\right\} .$$

Proof. If we put

$$\phi_1(t) = \phi_0 + \int_{t_1}^{t} k(s)\phi(s)\,ds\ , \quad t\in[t_1,t_2]$$

then $\phi_1(t_1) = \phi_0$ and $\phi(t) \le \phi_1(t)$ because of (3.34). Moreover,

$$\frac{d}{dt}\phi_1(t) = k(t)\phi(t) \le k(t)\phi_1(t)$$

because $k(t) \ge 0$, and so

$$\begin{aligned}\frac{d}{dt}&\left[\phi_1(t)\exp\left\{-\int_{t_1}^{t} k(s)\,ds\right\}\right]\\ &= \left\{\frac{d}{dt}\phi_1(t)-k(t)\phi_1(t)\right\}\exp\left\{-\int_{t_1}^{t} k(s)\,ds\right\}\\ &\le 0\end{aligned}$$

Integrating from t_1 to t, we then obtain

$$\phi_1(t)\exp\left\{-\int_{t_1}^{t} k(s)\mathrm{d}s\right\}-\phi_1(t_1)\leq 0$$

i.e.

$$\phi_1(t)\leq \phi_1(t_1)\exp\left\{\int_{t_1}^{t} k(s)\mathrm{d}s\right\}$$

$$= \phi_0\exp\left\{\int_{t_1}^{t} k(s)\mathrm{d}s\right\} = \psi(t)$$

and (3.35) is proved because $\phi(t) \leq \phi_1(t)$ as remarked above. □

Going back to system (3.28), assume that $w = w(t)$ also satisfies (i), (ii), and (iii), and so

$$\frac{\mathrm{d}}{\mathrm{d}t}\{u(t)-w(t)\} = F(u(t))-F(w(t)), \quad t_1<t\leq t_2$$

$$X\text{-}\lim_{t\to t_1+} [u(t)-w(t)] = 0 .$$

Then,

$$u(t)-w(t) = \int_{t_1}^{t} [F(u(s))-F(w(s))]\mathrm{d}s$$

$$\|u(t)-w(t)\| \leq k \int_{t_1}^{t} \|u(s)-w(s)\|\,\mathrm{d}s$$

and Lemma 3.2 with $\phi_0 = 0$, $k(s) = k =$ constant, $\phi(t) = \|u(t)-w(t)\|$ shows that $\phi(t) = \|u(t)-w(t)\| \equiv 0$ $\forall t\in[t_1,t_2]$. Thus, $u(t) \equiv w(t)$ and uniqueness is proved.

3.5. HOLOMORPHIC FUNCTIONS

Let Ω be an open set of the complex plane and let $w = w(z)$ be a function from Ω into the (complex) B-space X. Strong continuity and strong differentiability can be defined in the usual way, using the norm $\|\cdot\|$ of X (see (3.1b) and (3.13)). The function $w(z)$ is said to be *strongly holomorphic* (or, simply, holomorphic) in Ω if it is strongly differentiable at any $z\in\Omega$. Since most of the properties of classical holomorphic functions carry over to strongly holomorphic functions, we shall only state without proof (and under simple but rather restrictive assumptions) some results that will be used in what follows.

If $w(z)$ is holomorphic in Ω, then it is strongly differentiable any number of times and the following *Taylor expansion* is valid:

$$w(z) = w(z_0) + \sum_{n=1}^{\infty} \frac{(z-z_0)^n}{n!} w^{(n)}(z_0), \quad z_0 \in \Omega \tag{3.36}$$

provided that $|z-z_0|$ is small enough. In (3.36), $w^{(n)}(z_0)$ denotes that nth strong derivative of $w(z)$ at z_0 and the series is convergent in the sense of $\|\cdot\|$:

$$\|w(z)-w(z_0)-\sum_{n=1}^{m} \frac{(z-z_0)^n}{n!} w^{(n)}(z_0)\| \to 0 \quad \text{as} \quad m\to\infty.$$

Further, let γ be a curve contained in Ω and defined by the *parametric* equation $z = z(s)$, $s\in[0,1]$, i.e.

$$\gamma = \{z : z = x+iy\in\Omega;\ x = x(s),\ y = y(s), s\in[0,1]\} \tag{3.37}$$

where $i = \sqrt{-1}$ and where $x(s)$ and $y(s)$ are real-valued continuously differentiable functions. If $w(z)$ is strongly continuous in Ω, then the integral of $w(z)$ *along* γ can be defined as follows:

$$\int_\gamma w(z)\,dz = \int_0^1 W(s)x'(s)\,ds + i\int_0^1 W(s)y'(s)\,ds \tag{3.38}$$

where the integrals on the right-hand side are strong Riemann integrals and $W(s) = w(x(s)+iy(s))$ is a strongly continuous function from $[0,1]$ into X because $w(z)$ is strongly continuous at any $z = x+iy\in\Omega$. If $\hat{\gamma}$ is composed of a finite number $\gamma_1,\gamma_2,\ldots,\gamma_J$ of curves of the form (3.37) with $x(s)$ and $y(s)$ continuously differentiable, then the integral of $w(s)$ along $\hat{\gamma}$ is defined by

$$\int_{\hat{\gamma}} w(z)\,dz = \sum_{j=1}^{J} \int_{\gamma_j} w(z)\,dz . \tag{3.39}$$

Finally, assume that $w(z)$ is holomorphic in Ω and that γ is a curve such that

(i) γ is contained in Ω and it is defined by (3.37) with $x(s)$ and $y(s)$ continuously differentiable in $[0,1]$;

(ii) $x(0) = x(1)$, $y(0) = y(1)$, i.e. γ is a *closed curve*;

(iii) $x(s)+iy(s) \neq x(s_1)+iy(s_1)$ $\forall s\in(0,1)$, $s_1\in(0,1)$ with $s \neq s_1$;

(iv) γ encompasses only points belonging to Ω.

Note that (ii) and (iii) imply that γ has no *multiple point* other than $x(0)+iy(0) = x(1)+iy(1)$, i.e. γ does not cross itself. For instance, if $\Omega = \{z:z = x+iy;\ |z|^2 = x^2+y^2<a^2\}$, then the circle $\gamma = \{z:z = x+iy,\ x^2+y^2 = a^2/4\}$ satisfies assumptions (i) - (iv). Under the assumptions (i) - (iv), *Cauchy's integral formula* holds:

$$\int_\gamma w(z)\,dz = 0\ . \tag{3.40}$$

Example 3.9. The resolvent of a closed operator

The resolvent $R(z,B) = (zI-B)^{-1}$ of an operator $B\in\mathscr{C}(X)$ is a holomorphic function from $\Omega = \rho(B)$ into $\mathscr{B}(X)$. Note that $\rho(B)$ is an open subset of the complex plane; to see this, assume that $z\in\rho(B)$ and $z_1 = z+h$ is such that $|h|\|R(z,B)\| < 1$. Then, the inverse operator $[I+hR(z,B)]^{-1}$ exists and is such that

$$\|[I+hR(z,B)]^{-1}\| \leq \frac{1}{1-|h|\|R(z,B)\|} \tag{3.41}$$

(see Example 2.16 with $z = 1$ and with $A = hR(z,B)$). Hence,

$$(z_1I-B) = (zI-B)+hI = [I+hR(z,B)](zI-B)$$

$$(z_1I-B)^{-1} = R(z,B)\{I+hR(z,B)\}^{-1} \tag{3.42}$$

$$\|(z_1I-B)^{-1}\| \leq \frac{\|R(z,B)\|}{1-|h|\|R(z,B)\|} \tag{3.43}$$

where $|z_1-z| = |h| < 1/\|R(z,B)\|$, and so

$$\{z_1:z_1 = z+h\in\mathbb{C},\ |z_1-z|<1/\|R(z,B)\|\} \subset \rho(B).$$

We conclude that each $z\in\rho(B)$ is an interior point and $\rho(B)$ is an open set (see Example 1.5).

Now we show that $R(z,B)$ is strongly differentiable at any $z\epsilon\rho(B)$ and that

$$\frac{\mathrm{d}}{\mathrm{d}z} R(z,B) = -[R(z,B)]^2 , \quad z\epsilon\rho(B) . \tag{3.44}$$

In fact, as we have proved above, $z_1 = z+h\epsilon\rho(B)$ provided that $|h|\|R(z,B)\| < 1$ and so we have from Exercise 2.15

$$R(z+h,B)-R(z,B) = -hR(z+h,B)R(z,B) . \tag{3.45}$$

Thus,

$$\frac{1}{h}[R(z+h,B)-R(z,B)]+[R(z,B)]^2$$

$$= -R(z+h,B)R(z,B)+[R(z,B)]^2$$

$$= -[R(z+h,B)-R(z,B)]R(z,B)$$

and using again (3.45)

$$\frac{1}{h}[R(z+h,B)-R(z,B)]+[R(z,B)]^2 = hR(z+h,B)[R(z,B)]^2$$

$$\left\| \frac{1}{h}[R(z+h,B)-R(z,B)]+[R(z,B)]^2 \right\|$$

$$\le |h|\|R(z+h,B)\|\|R(z,B)\|^2$$

$$\le |h| \frac{\|R(z,B)\|^3}{1-|h|\|R(z,B)\|} \to 0$$

as $|h| \to 0$ because of (3.43), and (3.45) is proved. Finally, differentiating both sides of (3.44) n-1 times

$$\frac{\mathrm{d}^n}{\mathrm{d}z^n} R(z,B) = (-1)^n n! [R(z,B)]^{n+1} , \quad n = 1,2,\ldots . \quad \square \tag{3.46}$$

3.6. FRÉCHET DERIVATIVE

Let $F = F(f)$ be an operator with domain $D(F) \subset X$ and range $R(F) \subset Y$ (see for instance Examples 2.8-2.11), and let $f\epsilon X_0$ where X_0 is an open set contained in $D(F)$. If an operator

$F_f \in \mathscr{B}(X,Y)$ can be found such that

$$F(f+g) = F(f)+F_f g+G(f,g), \quad \forall\, f+g \in X_0 \tag{3.47}$$

where the *remainder* $G(f,g)$ satisfies the condition

$$\lim \frac{\| G(f,g);Y\|}{\| g;X\|} = 0 \quad \text{as} \quad \| g;X\| \to 0 \tag{3.48}$$

then we say that $F(f)$ is *Fréchet differentiable* (or F differentiable) at $f \in X_0$ and that F_f is the *Fréchet derivative* of F at f. The operator F(f) is said to be F differentiable in X_0 if it is F differentiable at any $f \in X_0$.

Remark 3.7.

Note that $F(f+g)$ and $F(f)$ are elements of Y because F is a function from $D(F) \subset X$ into Y. Hence, $G(f,g)$ is an element of Y and F_f maps the *increment* g into $F_f g \in Y$. Thus, the fact that the derivative F_f is an element of $\mathscr{B}(X,Y)$ is quite natural. □

Remark 3.8.

The Fréchet derivative F_f is in general non-linearly dependent on the differentiation point $f \in X_0$ and so F_f is usually a non-linear mapping from X_0 into $\mathscr{B}(X,Y)$. □

Note that the definition of the F derivative generalizes the notion of a strong derivative. To see this, assume that $X_0 = (t_1,t_2) \subset R^1 = X$; then, $\| \cdot;X \| = |\cdot|$ and $F = F(t)$ is a function from (t_1,t_2) into the B-space Y. According to (3.47) and (3.48), F is F differentiable at $t \in (t_1,t_2)$ if

$$F(t+h) = F(t)+F_t h+G(t,h) \quad , \quad t+h \in (t_1,t_2) \tag{3.49}$$

where $\lim[\| G(t,h);Y\| / |h|] = 0$ as $|h| \to 0$ and where $F_t \in \mathscr{B}(R^1,Y)$. Since F_t is a linear and bounded operator with domain $D(F_t) = R^1$ and with range $R(F_t) \subset Y$, then $F_t h$ is a suitable element of Y, $\forall\, h \in R^1$. It follows that $y = y(t) = 1/h\ F_t h$ also belongs to Y, $\forall\, h \neq 0$, and so we have from (3.49)

$$\frac{1}{h}\,[F(t+h)-F(t)]-y(t) = \frac{1}{h}\,G(t,h)\ , \quad h \neq 0$$

$$\left\|\frac{1}{h}\,[F(t+h)-F(t)]-y(t);Y\right\| = \frac{1}{|h|}\,\|G(t,h);Y\| \to 0 \quad \text{as} \quad |h| \to 0$$

and $y(t)$ is the derivative of $F(t)$ in the sense of $\|\cdot;Y\|$.

Example 3.10. F derivative of $F(f) = f_0+Bf$ with $B\in\mathscr{B}(X)$.

If

$$F(f) = f_0+Bf, \quad D(F) = X_0 = X, \quad R(F) \subset X$$

where f_0 is a given element of X and $B\in\mathscr{B}(X)$, then

$$F(f+g) = f_0+B[f+g] = f_0+Bf+Bg = F(f)+Bg$$

and so

$$F_f = B, \quad G(f,g) = 0, \quad \forall\, f,g\ X.$$

Hence, in this particular case, the F derivative of $F(f)$ is the linear operator B, independent of the differentiation point f ('the slope of a straight line is constant'). □

Example 3.11. F derivative of $F(f) = f^2$ with $D(F) = L^2(a,b)$, $R(F) \subset L^1(a,b)$

If

$$F(f) = f^2$$

$$D(F) = X_0 = X = L^2(a,b)$$

$$R(F) \subset Y = L^1(a,b)$$

then we have

$$F(f+g) = (f+g)^2 = f^2+2fg+g^2 = F(f)+F_f g+G(f,g)$$

where

$$F_f = 2fI, \quad G(f,g) = g^2 .$$

Note that $F_f \in \mathscr{B}(X,Y)$ and that condition (3.48) is satisfied because

$$\| F_f g;Y \|^2 = [\int_a^b |2f(x)g(x)|dx]^2 \le 4 \int_a^b |f(x)|^2 dx \int_a^b |g(x)|^2 dx$$

$$= 4\| f;X \|^2 \| g;X \|^2 ; \| G(f,g);Y \| = \int_a^b |g^2(x)|dx = \int_a^b |g(x)|^2 dx = \| g;X \|^2$$

and so

$$\| F_f g;Y \| \le (2\| f;X \|)\| g;X \|, \quad \forall g \in X; \| F_f \| \le 2\| f;X \|$$

$$\frac{\| G(f,g);Y \|}{\| g;X \|} = \| g;X \| \to 0 \quad \text{as} \quad \| g;X \| \to 0 .$$

Remark 3.9.

F_f is the F derivative of $F(f)$ only if *both* the condition $F_f \in \mathscr{B}(X,Y)$ and (3.48) are satisfied. □

EXERCISES

3.1. Prove that, if $w(t)$ is strongly continuous in Δ, then the real function $\phi(t) = \| w(t) \|$ is continuous in Δ in the classical sense. Hint: use (1.4).

3.2. Discuss strong continuity in the space $C^1([a,b])$.

3.3. Prove that the family of operators

$$B(t)f = \int_a^b h(x,y;t)f(y)\,dy , \quad f \in C([a,b])$$

where $h(x,y;t)$ is continuous over $[a,b]\times[a,b]\times[t_1,t_2]$, is continuous in the sense of the norm of $\mathscr{B}(C([a,b]))$, i.e. $\| B(t+h)-B(t) \| < \varepsilon , \quad |h| < \delta = \delta(\varepsilon,t)$.

3.4. Prove that the vector space $C([t_1,t_2];X)$ is a B-space with norm given by (3.12). Hint: follow the procedure of Example 1.3, with

$\| \cdot ;X\|$ instead of $|\cdot|$.

3.5. Discuss strong differentiability in the spaces R^n, l^1, and $L^1(a,b)$.

3.6. Prove that

$$\frac{d}{dt}\{\alpha w_1(t)+\beta w_2(t)\} = \alpha \frac{d}{dt} w_1(t)+\beta \frac{d}{dt} w_2(t)$$

provided that w_1 and w_2 are strongly differentiable.

3.7. Show that the family of all strongly continuous functions with strongly continuous derivatives over $[t_1,t_2]$ is a B-space, denoted by $C^1([t_1,t_2];X)$, with norm

$$\| w;C^1([t_1,t_2];X)\| = \max\{\| w;C([t_1,t_2];X)\|;\|w';C([t_1,t_2];X)\|\}$$

where w' is the strong derivative of w.

3.8. Show that

$$\frac{d}{dt}[Bw(t)] = B\frac{dw}{dt}$$

where $B\in\mathcal{B}(X)$ and $w(t)$ is strongly differentiable.

3.9. Prove that

$$\int_{t_1}^{t_2}\{\alpha w_1(s)+\beta w_2(s)\}ds = \alpha\int_{t_1}^{t_2} w_1(s)ds+\beta\int_{t_1}^{t_2} w_2(s)ds$$

$$\int_{t_1}^{t_2} w(s)ds = \int_{t_1}^{\tau} w(s)ds+\int_{\tau}^{t_2} w(s)ds, \quad t_1<\tau<t_2$$

provided that w_1, w_2, and w are strongly integrable over $[t_1,t_2]$.

3.10. Establish the *chain-rule* for Fréchet derivatives. Hint: for simplicity that $F(f)$ is an operator with domain $D(F) = X$ and with range $R(F) \subset X_1$ and that $F_1(f_1)$ is an operator with $D(F_1) = X_1$ and with $R(F_1) \subset Y$, where X, X_1, and Y are B-spaces. Then, $F_1(F(f)) = \Phi(f)$ has domain $D(\Phi) = X$ and range $R(\Phi) \subset Y$, ...

4

SEMIGROUPS

4.1. LINEAR INITIAL-VALUE PROBLEMS

Let S be a rigid conductor of heat, bounded by the planes $x = a$ and $x = b$, and denote by $\tau = \tau(x;t)$ the temperature in S. Under suitable assumptions, $\tau(x;t)$ satisfies the following system (Carslaw and Jaeger 1959)

$$\frac{\partial}{\partial t}\tau(x;t) = k\frac{\partial^2}{\partial x^2}\tau(x;t)+\phi(x;t), \quad a<x<b, \ t>0 \tag{4.1a}$$

$$\tau(a;t) = \tau(b;t) = 0 \ , \ t>0 \tag{4.1b}$$

$$\tau(x;0) = \tau_0(x) \ , \quad a<x<b \tag{4.1c}$$

where $k>0$ is the heat-diffusion coefficient and $\phi(x;t)$ and $\tau_0(x)$ are known functions. If $\phi(x;t) \not\equiv 0$, then a temperature-independent source or sink of thermal energy is present in S. Further, the boundary conditions (4.1b) show that the boundary planes $x = a$ and $x = b$ are kept at zero temperature, whereas the initial condition (4.1c) means that the temperature in S at $t = 0$ is known. System (4.1) is a mathematical model for the evolution of the *state vector* $\tau = \tau(x;t)$ of the physical system S (see (0.1), (0.2), and the discussion of them in the Introduction). Note that the integral

$$J = \int_a^b \tau(x;t)\,dx$$

is a relevant physical quantity because it is proportional to the variation of the *total* thermal energy contained in S (under suitable assumptions). This suggests that the abstract counterpart of system (4.1) should be studied in the B-space $X = L^1(a,b)$ (see Example 1.4). Thus, if we define the heat-diffusion operator as

$$Af = kf''(x)$$

$$D(A) = \{f : f = f(x) \in X = L^1(a,b);\ f'' \in X;\ f(a) =$$

$f(b) = 0\} \subset X, \quad R(A) \subset X$

(see Example 2.19 and Remark 2.12), then (4.1a) *and* (4.1b) can be rewritten as follows:

$$\frac{d}{dt} u(t) = Au(t)+g(t) , \quad t > 0 \tag{4.2a}$$

where the unknown $u(t) = \tau(x;t)$ must now be interpreted as a function from $[0,+\infty)$ into the B-space X and $g(t) = \phi(x;t)$ is a known function from $[0,+\infty)$ into X. Further, the abstract version of the initial condition (4.1c) becomes

$$X\text{-}\lim_{t\to 0+} u(t) = u_0 \tag{4.2b}$$

with $u_0 = \tau_0(x) \in X$.

More generally, if A is a given linear operator with domain and range contained in a B-space X, then the *evolution equation* (4.2a) and the *initial condition* (4.2b) give rise to a typical *non-homogeneous initial-value problem* in the B-space X. The function $u = u(t)$ is said to be a *strict solution* (or, simply, a solution) of system (4.2a)+(4.2b) if

(i) $u(t)$ is strongly continuous at any $t \geq 0$
(ii) $u(t) \in D(A)$ and it is strongly differentiable at any $t>0$
(iii) $u(t)$ satisfies (4.2a) and (4.2b).

Note that boundary conditions, such as the (4.1b), are included in the definition of $D(A)$ and that they are taken 'automatically' into account because the solution $u(t)$ is sought in $D(A)$ (see (ii)).

If $g(t) \equiv \theta_X$, $\forall t \geq 0$, then system (4.2) becomes a *homogeneous initial-value problem*:

$$\frac{d}{dt} u(t) = Au(t) , \quad t > 0 \tag{4.3a}$$

$$X\text{-}\lim_{t\to 0+} u(t) = u_0 \tag{4.3b}$$

and if $u_1(t)$ is a solution of (4.3a)+(4.3b) and $u_2(t)$ satisfies (4.2a) and the initial condition $X\text{-}\lim u_2(t) = \theta$ as $t \to 0+$, then $u(t) = u_1(t)+u_2(t)$ is a solution of (4.2a)+(4.2b), because A is a *linear* operator (and $D(A)$ is a *linear* subset of X). This is the abstract version of a well-known result of the classical theory of linear differential equations.

4.2. THE CASE $A \in \mathscr{B}(X)$

The homogeneous initial-value problem (4.3) can be solved for *any* $u_0 \in X$ if $A \in \mathscr{B}(X)$, i.e. if $D(A) = X$, $R(A) \subset X$, and $\|A\| = \|A;\mathscr{B}(X)\| = b < \infty$ (see section 2.2). To see this, consider the sequence of operators $\{T_n(t),\ n = 0,1,2,\ldots\}$:

$$T_n(t) = I + \frac{t}{1!}A + \frac{t^2}{2!}A^2 + \ldots + \frac{t^n}{n!}A^n, \quad n = 0,1,2,\ldots, t \in R^1 \tag{4.4}$$

where $T_n(t) \in \mathscr{B}(X)$ because $A^j \in \mathscr{B}(X)$ with $\|A^j\| \le \|A\|^j = b^j < \infty$, $j = 0,1,2,\ldots$. If $m \ge n+1$, then we have from (4.4)

$$\|T_m(t)-T_n(t)\| = \|\sum_{j=n+1}^{m} \frac{t^j}{j!}A^j\| \le \sum_{j=n+1}^{m} \frac{|t|^j}{j!}\|A^j\|$$

$$\le \sum_{j=n+1}^{m} \frac{|t|^j b^j}{j!} \le \sum_{j=n+1}^{\infty} \frac{|t|^j b^j}{j!} \le \frac{(b|t|)^{n+1}}{(n+1)!}\exp(b|t|)$$

(see the discussion after (3.32)), and so

$$\|T_m(t)-T_n(t)\| \le \frac{(b|t|)^{n+1}}{(n+1)!}\exp(b|t|), \quad m \ge n+1,\ t \in R^1. \tag{4.5}$$

Since $(b|t|)^{n+1}/(n+1)!\ 0$ as $n \to \infty$, inequality (4.5) leads to

$$\|T_m(t)-T_n(t)\| < \varepsilon, \quad \forall m > n > n_0(\varepsilon;t)$$

and $\{T_n(t)\}$ is a Cauchy sequence in the B-space $\mathscr{B}(X)$ for each real t (and even for each complex t). Hence, a unique $T(t) \in \mathscr{B}(X)$ exists such that

$$\lim_{n\to\infty} \|T(t)-T_n(t)\| = 0, \quad t \in R^1; \tag{4.6}$$

moreover, passing to the limit as $m \to \infty$ in (4.5), we have

$$\|T(t)-T_n(t)\| \leq \frac{(b|t|)^{n+1}}{(n+1)!}\exp(b|t|)\ , \quad t\in R^1 \tag{4.7}$$

showing that $T(t)$ can be 'approximated' by $T_n(t)$ with an error of the order of $|t|^{n+1}$. Note that (4.6) is equivalent to

$$T(t) = \mathscr{B}(X)\text{-}\lim_{n\to\infty}\sum_{j=0}^{n}\frac{t^j}{j!}A^j = \sum_{j=0}^{\infty}\frac{t^j}{j!}A^j \tag{4.8}$$

which justifies the symbol $\exp(tA)$ for the operator $T(t)$. That $T(t)$ 'behaves as an exponential' is clearly shown by the following theorem.

Theorem 4.1. If $A\in\mathscr{B}(X)$, then the one-parameter family of operators $\{T(t),\ -\infty<t<+\infty\}\subset\mathscr{B}(X)$ is such that

$$T(0) = I \tag{4.9a}$$

$$T(t)T(s) = T(t+s),\quad \forall\, t,s\in R^1 \tag{4.9b}$$

$$\lim_{h\to 0}\|T(t+h)-T(t)\| = 0,\quad \forall\, t\in R^1 \tag{4.9c}$$

$$\frac{d}{dt}T(t) = AT(t) = T(t)A,\quad \forall\, t\in R^1 \tag{4.9d}$$

$$\|T(t)\| \leq \exp(t\|A\|),\quad \forall\, t\in R^1 \tag{4.9e}$$

Proof.

(a) Since $T_n(0) = I\ \forall n = 0,1,2,\ldots,$ (4.9a) follows from (4.6) with $t = 0$.

(b) We have from (4.4)

$$T_n(t)T_n(s) = \left[\sum_{j=0}^{n}\frac{t^j}{j!}A^j\right]\left[\sum_{i=0}^{n}\frac{t^i}{i!}A^i\right]$$

and after expanding and ordering with respect to increasing powers of A

$$T_n(t)T_n(s) = \sum_{r=0}^{n}\sum_{j=0}^{r}\frac{t^j s^{r-j}}{j!(r-j)!}A^r + \sum_{r=n+1}^{2n}\sum_{j=r-n}^{n}\frac{t^j s^{r-j}}{j!(r-j)!}A^r\ .$$

On the other hand,

$$T_n(t+s) = \sum_{r=0}^{n} \frac{(t+s)^r}{r!} A^r = \sum_{r=0}^{n} \frac{1}{r!} \left[\sum_{j=0}^{r} \frac{r!}{j!(r-j)!} t^j s^{r-j}\right] A^r$$

$$= \sum_{r=0}^{n} \sum_{j=0}^{r} \frac{t^j s^{r-j}}{j!(r-j)!} A^r$$

and so

$$T_n(t)T_n(s) - T_n(t+s) = \sum_{r=n+1}^{2n} \left\{\sum_{j=r-n}^{n} \frac{t^j s^{r-j}}{j!(r-j)!}\right\} A^r .$$

On passing to the norms, we obtain

$$\|T_n(t)T_n(s) - T_n(t+s)\| \le \sum_{r=n+1}^{n} \left[\sum_{j=r-n}^{n} \frac{|t|^j |s|^{r-j}}{j!(r-j)!}\right] b^r$$

$$\le \sum_{r=n+1}^{2n} \frac{1}{r!} \left\{\sum_{j=0}^{r} \frac{r!}{j!(r-j)!} |t|^j |s|^{r-j}\right\} b^r$$

$$= \sum_{r=n+1}^{2n} \frac{(|t|+|s|)^r}{r!} b$$

$$\le \sum_{r=n+1}^{\infty} \frac{(|t|+|s|)^r}{r!} b^r \le \frac{(|t|+|s|)^{n+1}}{(n+1)!} b^{n+1} \exp[(|t|+|s|)b] \to 0$$

as $n \to \infty$ and (4.9b) is proved.

(c) Note first that $T_n = T_n(t)$ is continuous in t in the sense of the norm of $\mathscr{B}(X)$ and so $T_n \in C([-\hat{t},\hat{t}];\mathscr{B}(X))$, where $\hat{t}>0$ is a finite but otherwise arbitrary value of t. Now, if $\varepsilon>0$, then an integer $n_1 = n_1(\varepsilon;\hat{t})$ can be found such that

$$0 \le \frac{(b|t|)^{n+1}}{(n+1)!} \exp(b|t|) \le \frac{(b\hat{t})^{n+1}}{(n+1)!} \exp(b\hat{t}) < \varepsilon, \quad \forall\, t \in [-\hat{t},\hat{t}]$$

provided that $n>n_1$, and (4.7) gives

$$\|T(t) - T_n(t)\| < \varepsilon, \quad n>n_1(\varepsilon;\hat{t}), \quad \forall\, t \in [-\hat{t},\hat{t}].$$

Thus, the sequence of $\{T_n(t)\}$ converges to $T(t)$ in the sense of the norm of $\mathscr{B}(X)$, *uniformly* in $t \in [-\hat{t},\hat{t}]$, and so $T(t)$ is continuous as well. In other words, $\{T_n\}$

converges to T in the sense of the norm of the B-space $C([-\hat{t},\hat{t}];\mathscr{B}(X))$ (see the discussion at the end of section 3.1, with $\mathscr{B}(X)$ instead of X), and consequently $T\in C([-\hat{t},\hat{t}];\mathscr{B}(X))$.

(d) On differentiating both sides of (4.4), we have

$$\frac{\mathrm{d}}{\mathrm{d}t} T_n(t) = AT_{n-1}(t), \quad t\in R^1$$

and, since $T_n(0) = I$,

$$T_n(t) = I + \int_0^t AT_{n-1}(s)\,\mathrm{d}s$$

where both the derivative and the integral are in the sense of $\|\cdot\| = \|\cdot;\mathscr{B}(X)\|$. However, $\{T_n\}$ converges to T in the sense of the norm of $C([-\hat{t},\hat{t}];\mathscr{B}(X))$ and so we can pass to the $\mathscr{B}(X)$-limit in both sides of the preceding equality because of Theorem 3.3 (with $\mathscr{B}(X)$ instead of X):

$$T(t) = I + \int_0^t \mathscr{B}(X)\text{-}\lim_{n\to\infty} AT_{n-1}(s)\,\mathrm{d}s = I + \int_0^t AT(s)\,\mathrm{d}s .$$

The first part of (4.9d) then follows by differentiation. As far as the commutation relation $AT(t) = T(t)A$ is concerned, see Exercise 4.1.

(e) See Exercise 4.2. □

Remark 4.1.

If $t = -s$ in (4.9b), then we have that $T(-s)T(s) = T(-s+s) = T(0) = I$ and so $T(-s) = [T(s)]^{-1}$. Thus, the family of operators $\mathscr{F} = \{T(t), -\infty<t<+\infty\}$ is such that (i) $T(0) = I$, (ii) $T(t)T(s) = T(t+s) = T(s)T(t)$, (iii) $[T(t)]^{-1} = T(-t)$, and $\mathscr{F}$ is an *abelian group generated* by $A\in\mathscr{B}(X)$. □

Using the results of Theorem 4.1, we now show that the group $T(t) = \exp(tA)$ can be employed to find the solution of the following initial-value problem:

$$\frac{d}{dt} u(t) = Au(t), \quad t \in R^1 \tag{4.10a}$$

$$X\text{-}\lim_{t \to 0} u(t) = u_0 \tag{4.10b}$$

with $A \in \mathscr{B}(X)$ and $u_0 \in X$.

Theorem 4.2. If $A \in \mathscr{B}(X)$, $T(t) = \exp(tA)$ is the group generated by A and u_0 is an assigned element of X, then

$$u(t) = T(t)u_0, \quad t \in R^1 \tag{4.11}$$

is the unique solution of (4.10a)+(4.10b) that is differentiable in the sense of $\| \cdot ;X \|$ any number of times.

Proof. We have from (4.11) and from (4.9d)

$$\| \frac{1}{h} \{u(t+h)-u(t)\} - Au(t);X \| = \| \frac{1}{h} \{T(t+h)-T(t)\}u_0 - AT(t)u_0;X\|$$

$$\leq \| \frac{1}{h} \{T(t+h)-T(t)\} - AT(t)\| \| u_0;X\| \to 0 \quad \text{as } h \to 0$$

and $u(t)$ is differentiable (and hence continuous) in the sense of $\| \cdot ;X \|$, with

$$\frac{du}{dt} = Au(t), \quad t \in R^1 .$$

Moreover,

$$\| u(t) - u_0;X \| = \| [T(t) - T(0)]u_0;X \| \leq \| T(t) - T(0)\| \| u_0;X\| \to 0$$

as $h \to 0$, because of (4.9a) and (4.9c), and so the function $u(t)$, defined by (4.11), is strongly differentiable and satisfies (4.10a) and (4.10b). To see that $u(t)$ is the *unique* strongly differentiable solution of system (4.10), assume that $w(t)$ is another strongly differentiable solution of (4.10). Then,

$$\frac{d}{dt}\{u(t)-w(t)\} = A\{u(t)-w(t)\} , \quad t \in R^1$$

$$X\text{-}\lim_{t\to 0}\{u(t)-w(t)\} = \theta_X$$

and so

$$u(t)-w(t) = \int_0^t A\{u(s)-w(s)\}ds, \quad t\in R^1$$

$$\|u(t)-w(t);X\| \le \|A\| \int_0^{|t|} \|u(s)-w(s);X\|\, ds, \quad t\in R^1 .$$

Using Gronwall's inequality (Lemma 3.2), we conclude that $\|u(t)-w(t);X\| \equiv 0$, i.e. $u(t) \equiv w(t)$.

Finally, since $A\in\mathscr{B}(X)$ (see Exercise 3.8),

$$\frac{d^2}{dt^2}u(t) = \frac{d}{dt}\{Au(t)\} = A\frac{du}{dt} = A\{Au(t)\} = A^2u(t) ,$$

and in general

$$\frac{d^n}{dt^n}u(t) = A^nu(t) = A^nT(t)u_0 = T(t)A^nu_0, \quad n = 0,1,2,\ldots,t\in R^1$$

i.e. $u(t)$ is strongly differentiable any number of times. □

Remark 4.2.

The function $u = u(t)$ defined by (4.11) is obviously a strict solution of the initial-value problem (4.3) with $A\in\mathscr{B}(X)$, in the sense stated in section 4.1. □

Example 4.1. $A\in\mathscr{B}(R^n)$

If $X = R^n$ and A is the matrix operator of Example 2.2 with $m = n$, then $A\in\mathscr{B}(R^n)$ with

$$\|A\| = b \le \left[\sum_{i,j=1}^{n} |a_{ij}|^2\right]^{1/2} .$$

Hence, A generates the group $\{T(t) = \exp(tA),\ t\in R^1\}$ and the results of Theorem 4.2 hold. Note that, since $X = R^n$,

$$u(t) = \begin{pmatrix} u_1(t) \\ \cdots \\ u_n(t)\end{pmatrix}, \quad u_0 = \begin{pmatrix} u_{10} \\ \cdots \\ u_{n0}\end{pmatrix}$$

where the $u_j(t)$'s are real-valued functions and the u_{j0}'s are given real numbers, and the initial-value problem (4.10) becomes

$$\left.\begin{aligned} \frac{\mathrm{d}}{\mathrm{d}t}u_1(t) &= a_{11}u_1(t)+\ldots+a_{1n}u_n(t) \\ &\ldots\ldots\ldots\ldots \\ \frac{\mathrm{d}}{\mathrm{d}t}u_n(t) &= a_{n1}\,u_1(t)+\ldots+a_{nn}u_n(t) \end{aligned}\right\}$$

$$\lim_{t\to 0} u_j(t) = u_{j0}, \quad j = 1,2,\ldots,n$$

which is a linear differential system with constant coefficients. Note also that the solution $u = u(t)$ can be 'approximated' by a *finite* sum as follows:

$$\begin{aligned} u(t) &= T(t)u_0 = T_m(t)u_0+\{T(t)-T_m(t)\}u_0 \\ &\sum_{j=1}^{m} \frac{t^j}{j!}A^j u_0+w(t) \end{aligned}$$

where the *remainder* $w(t)$ satisfies the inequality

$$\begin{aligned} \|w(t);X\| &= \|\{T(t)-T_m(t)\}u_0;X\| \le \|T(t)-T_m(t)\|\ \|u_0;X\| \\ &\le \frac{(b|t|)^{m+1}}{(m+1)!}\exp(b|t|), \quad t\in R^1 \end{aligned}$$

because of (4.7). □ .

Example 4.2. An integro-differential system

If

$$Af = \int_0^1 (x-y)f(y)\mathrm{d}y\ , \quad f\in X = C([0,1]) \tag{4.12}$$

then $g = g(x) = Af$ is obviously continuous in $[0,1]$; moreover

$$\begin{aligned} |g(x)| = |Af| &\le \int_0^1 |x-y||f(y)|\mathrm{d}y \le \max\{|f(y)|, y\in[0,1]\}\int_0^1 |x-y|\mathrm{d}y \\ &= \|f;X\|\int_0^1 |x-y|\mathrm{d}y \le \|f;X\| \qquad \forall x\in[0,1] \end{aligned}$$

and so

$$\|g;X\| = \|Af;X\| = \max\{|g(x)|, x\in[0,1]\} \le \|f;X\|, \quad \forall f\ X$$

and $A\in \mathscr{B}(X)$ with $\|A\| = b\le 1$ (see Remark 2.1). Note that, since $X = C([0,1])$, $u(t) = \Psi(x;t)$ satisfies the integro-differential system (see (4.10) with A given by (4.12))

$$\left.\begin{aligned} &\frac{\partial}{\partial t}\Psi(x;t) = \int_0^1 (x-y)\Psi(y;t)\,dy\ , \qquad t\in R^1 \\ &\lim_{t\to 0} |\Psi(x;t)-u_0(x)| = 0 \end{aligned}\right\}$$

where the partial derivative and the limit are *uniform* in $x\in[0,1]$ (see Examples 3.3 and 3.6).

The explicit expression of the group $T(t) = \exp(tA)$ can be found in this particular case because of the *simple* form (4.12) of the generator A. In fact, if we put $g = g(x) = Af$, then we have from (4.12)

$$\begin{aligned} A^2 f = Ag &= \int_0^1 (x-x')g(x')\,dx' \\ &= \int_0^1 (x-x')\left\{\int_0^1 (x'-y)f(y)\,dy\right\}dx' \\ &= \int_0^1 f(y)\left\{\int_0^1 (x-x')(x'-y)\,dx'\right\}dy \\ &= \int_0^1 \left(\frac{x+y}{2} - xy - \frac{1}{3}\right) f(y)\,dy = g_1(x) \end{aligned}$$

$$A^3 f = Ag_1 = \int_0^1 (x-x')g_1(x')\,dx' = \int_0^1 f(y)\left\{\int_0^1 (x-x')\left(\frac{x'+y}{2} - x'y - \frac{1}{3}\right)dx'\right\}dy$$

$$= -\frac{1}{12}\int_0^1 (x-y)f(y)\,dy = -\frac{1}{12}Af.$$

By similar 'manipulations', we obtain

$$A^{2j} = \left(-\frac{1}{12}\right)^{j-1} A^2,\ A^{2j+1} = \left(-\frac{1}{12}\right)^{j} A, \quad j = 1,2,3,\ldots$$

and so

$$T_{2n}(t) = I+\sqrt{12}\left[\sum_{j=1}^{n} \frac{(-1)^{j-1}}{(2j-1)!}\left(\frac{t}{\sqrt{12}}\right)^{2j-1}\right] A+$$

$$12\left\{\sum_{j=1}^{n} \frac{(-1)^{j-1}}{(2j)!}\left(\frac{t}{\sqrt{12}}\right)^{2j}\right\}A^2$$

$$T(t) = I+\sqrt{12}\sin\left(\frac{t}{\sqrt{12}}\right)A + 12\left\{1-\cos\left(\frac{t}{\sqrt{12}}\right)\right\}A^2$$

because of (4.4) and (4.6). Hence, the unique solution of (4.10) with A given by (4.12) has the form

$$u(t) = u_0+\sqrt{12}\sin\left(\frac{t}{\sqrt{12}}\right)Au_0+12\left\{1-\cos\left(\frac{t}{\sqrt{12}}\right)\right\}A^2u_0, \quad t\in R^1. \quad \square \tag{4.13}$$

Example 4.3. Best approximation of a non-linear problem by means of a linear problem.

Assume that $u = u(t)$, $t\in[0,\hat{t}]$, is the solution of the *non-linear* initial-value problem (3.28), with $t_1 = 0$ and $\hat{t}_2 = \hat{t}$, and that $w = w(t;\lambda)$ is the solution of the *linear* problem

$$\frac{\mathrm{d}}{\mathrm{d}t} w(t;\lambda) = B(\lambda)w(t;\lambda), \quad t\in R^1$$
$$X\text{-}\lim_{t\to 0} w(t;\lambda) = u_0 \tag{4.14}$$

where $B(\lambda)$ belongs to $\mathscr{B}(X)$ for any value of the real parameter $\lambda\in[\lambda_1,\lambda_2]$. The solution of system (4.14) has the form

$$w(t;\lambda) = \exp\{tB(\lambda)\}u_0, \quad t\in R^1 \tag{4.15}$$

and obviously depends on λ. In what follows, we give a simple procedure to choose λ so that $w(t;\lambda)$ is the 'best approximation' over $[0,\hat{t}]$ of the solution $u(t)$ of the non-linear problem (3.28). As usual, we first transform both (3.28) and (4.14) into integral equations:

$$u(t) = u_0 + \int_0^t F(u(s))\mathrm{d}s, \quad t\in[0,\hat{t}]$$

$$w(t;\lambda) = u_0 + \int_0^t B(\lambda)w(s;\lambda)\mathrm{d}s, \quad t\in R^1$$

and so

$$u(t)-w(t;\lambda) = \int_0^t \{F(u(s))-B(\lambda)w(s;\lambda)\}ds =$$

$$\int_0^t \{F(w(s;\lambda))-B(\lambda)w(s;\lambda)\}ds+ \int_0^t \{F(u(s))-F(w(s;\lambda))\}ds .$$

Thus, if we put

$$\gamma(t;\lambda) = \| u(t)-w(t;\lambda);X \|$$

$$\gamma_0(\lambda) = \int_0^{\hat{t}} \| F(w(s;\lambda))-B(\lambda)w(s;\lambda);X\| ds$$

then we obtain

$$\gamma(t;\lambda) \le \gamma_0(\lambda)+k \int_0^t \gamma(s;\lambda)ds$$

because of (3.27), and using Gronwall's inequality (Lemma 3.2)

$$\gamma(t;\lambda) = \| u(t)-w(t;\lambda);X\| \le \gamma_0(\lambda)\exp(kt), \quad t\in[0,\hat{t}].$$

It is then natural to choose $\lambda\in[\lambda_1,\lambda_2]$ so that

$$\gamma_0(\lambda) = \int_0^{\hat{t}} \| F(\exp\{sB(\lambda)\}u_0)-B(\lambda)\exp\{sB(\lambda)\}u_0;X\| ds$$

has a *minimum*. If λ_0 is such a value of λ, then we say that $w(t;\lambda_0)$ is the 'best approximation' of $u(t)$ over $[0,\hat{t}]$. Correspondingly, the *relative error* satisfies the inequality

$$\varepsilon(\lambda_0;t) = \frac{\| u(t)-w(t;\lambda_0);X\|}{\| w(t;\lambda_0);X\|} \le \frac{\gamma_0(\lambda_0)\exp(kt)}{\| \exp\{tB(\lambda_0)\}u_0;X\|} .$$

The preceding method of approximating a non-linear initial-value problem over a finite time interval by means of a suitable linear problem often gives surprisingly good numerical results. Of course, the value of λ that minimizes $\gamma_0(\lambda)$ must usually be found by a numerical procedure. □

4.3. THE CASE $A\in\mathscr{C}(X)$

Let A be a closed operator with domain $D(A)$ and range $R(A)$

contained in the B-space X. Since, in general, $D(A)$ does not coincide with the whole X, some $f \in D(A)$ may exist such that $Af \notin D(A)$ and so

$$f \notin D(A^2) = \{g : g \in D(A);\ Ag \in D(A)\}$$

i.e. $D(A^2)$ is usually *strictly* contained in $D(A)$. In other words, if $g \in D(A^n)$ with

$$D(A^n) = \{g : g \in D(A^{n-1});\ A^{n-1}g \in D(A)\}, \quad n = 1,2,\ldots$$

then $A^n g$ is a uniquely defined element of X, but the sequence $D(A)$, $D(A^2)$, $D(A^3)$,... usually 'shrinks' as n increases. As a consequence, the operators $T_n(t)$ given by (4.4) are no long longer defined for *any* $f \in X$ and the procedure of section 4.2 does not apply. On the other hand, we can give two equivalent definitions of the elementary exponential function:

$$\exp(at) = \sum_{j=0}^{\infty} \frac{(at)^j}{j!} \tag{4.16a}$$

$$\exp(at) = \lim_{n\to\infty} \left(1 - \frac{at}{n}\right)^{-n} \tag{4.16b}$$

The definition (4.16a) suggested the procedure of section 4.2 to introduce $T(t) = \exp(tA)$ with $A \in \mathscr{B}(X)$; similarly, (4.16b) will suggest to us how to proceed if $A \in \mathscr{C}(X)$. However, as will be clear in what follows, the family $\mathscr{C}(X)$ is 'too large' for our purposes and so we define some suitable sub-families of $\mathscr{C}(X)$. If

$$A \in \mathscr{C}(X) \tag{4.17a}$$

$D(A)$ is dense in X, i.e. $\mathrm{cl}[D(A)] = X$ (see Example 1.5 and the discussion after Remark 1.3) (4.17b)

any $z > \beta$ belongs to the resolvent set $\rho(A)$ of A (see section 2.8) and

$$\|R(z,A)\| = \|(zI-A)^{-1}\| \leq \frac{1}{z-\beta}, \quad \forall z > \beta \tag{4.17c}$$

where β is a real number, then we say that A belongs to the family $\mathcal{G}(1,\beta) \subset \mathcal{C}(X)$ and we write $A\in\mathcal{G}(1,\beta)$ or $A\in\mathcal{G}(1,\beta;X)$ if it is necessary to point out that the domain and the range of A are contained in the B-space X (Kato 1966, Chapter 9). Condition (4.17a) has been discussed in Remarks 2.5 and 3.5 and (4.17b) means that Af is defined for 'sufficiently many' $f\in X$. As far as (4.17c) is concerned, if we *formally* Laplace transform system (0.1)+(0.2), we obtain

$$(zI-A)\tilde{\Psi} = \Psi_0$$

where $\tilde{\Psi} = \tilde{\Psi}(x;z)$ is the Laplace transform of $\Psi(x;t)$, z is the Laplace variable, and where we assume that the operator on the right-hand side of (0.1) is linear. Hence, (4.17c) basically means that $\tilde{\Psi}$ can be uniquely expressed in terms of Ψ_0 if $z>\beta$.

Example 4.4. The convection operator

The *convection* operator $A = -v\mathrm{d}/\mathrm{d}x$ of Example 2.17 belongs to $\mathcal{G}(1,0;L^2(a,b))$. In fact, it satisfies (4.17c) because of (2.82) and it is closed (see the discussion after (2.81)). Finally, A is densely defined because $D(A)$ contains $C_0^\infty([a,b])$, which is a dense subset of $L^2(a,b)$, (Kato 1966, p.130). The convection operator also belongs to $\mathcal{G}(1,0;\ L^1(a,b))$ (see Remark 2.11). □

Example 4.5. The heat-diffusion operator

The *heat-diffusion* operator $A = k\ \mathrm{d}^2/\mathrm{d}x^2$ of Example 2.19 belongs to the family $\mathcal{G}(1,-2k\delta^{-2};L^2(a,b))$ and to $\mathcal{G}(1,0;L^1(a,b))$ (see Remark 2.12). The 'generalized' heat diffusion operator of Example 2.15 belongs to $\mathcal{G}(1,0;L^{s,2}(R^1))$. □

Example 4.6. An operator in l^1.

The matrix operator A of Example 2.17 belongs to $\mathcal{G}(1,0;l^1)$. In fact, it is densely defined because $D(A) \supset D_0$, where D_0 is the dense subset of l^1 composed of all vectors with a finite number of non-zero components. A satisfies (4.17c) because of (2.71) and belongs to $\mathscr{C}(l^1)$ since $(zI-A)^{-1} \in \mathscr{B}(X)$, $\forall z>0$ (see (2.71) and (2.35)). □

Since most of the operators arising from (linear or linearized) mathematical models of 'physical' systems belong to the family $\mathscr{G}(1,\beta;X)$, the following criterion is quite useful in practice because it often avoids long and tedious calculations.

Theorem 4.3. Let X be a Hilbert space and assume that A is a linear operator with domain D(A) dense in X and with range R(A) ⊂ X. Then, $A \in \mathscr{G}(1,\beta;X)$ *if and only if the range of* $(zI-A)$ *coincides with the whole space X for any* $z>\beta$:

$$R(zI-A) = X, \quad \forall z>\beta \tag{4.18}$$

and

$$\mathrm{Re}(Af,f) \le \beta \| f;X \|^2, \quad \forall f \in D(A) \tag{4.19}$$

where $\mathrm{Re}(\cdot,\cdot)$ *is the real part of the inner product* $(\cdot,\cdot)$.

Proof. If $A \in \mathscr{G}(1,\beta;X)$, then each $z>\beta$ belongs to $\rho(A)$ and so $R(z,A) = (zI-A)^{-1} \in \mathscr{B}(X)$ $\forall z>\beta$ and $R(zI-A) = D((zI-A)^{-1}) = X$. Moreover, $f = R(z,A)(zI-A)f$, $\forall f \in D(A)$, $z>\beta$ and

$$\| f;X \|^2 = \| R(z,A)(zI-A)f;X \|^2 \le \| R(z,A) \|^2 \| (zI-A)f;X \|^2$$

$$\le \frac{1}{(z-\beta)^2}((zI-A)f, (zI-A)f) = \frac{1}{(z-\beta)^2}\{z^2\| f;X \|^2 + \| Af;X \|^2 - (zf,Af) - (Af,zf)\}$$

because of (4.17c). Hence,

$$(z^2+\beta^2-2z\beta)\| f;X \|^2 \le z^2\| f;X \|^2 + \| Af;X \|^2 - 2\,\mathrm{Re}\{z(Af,f)\}$$

because

$$(zf,Af)+(Af,zf) = (\overline{Af,zf})+(Af,zf) = 2\ \mathrm{Re}(Af,zf)$$

and so

$$2z\ \mathrm{Re}(Af,f) \le \|Af;X\|^2+2z\beta\|f;X\|^2-\beta^2\|f;X\|^2$$

$$\mathrm{Re}(Af,f) \le \frac{1}{2z}\|Af;X\|^2+\beta\|f;X\|^2-\frac{\beta^2}{2z}\|f;X\|^2$$

$\forall f\in D(A)$ and $\forall z>|\beta|$. On passing to the limit as $z\to+\infty$, we obtain (4.19). Conversely, assume that (4.18) and (4.19) hold. Then, if g is any given element of X, at least one $f\in D(A)$ exists such that $(zI-A)f = g$ with $z>\beta$ because the range of $(zI-A)$ is the whole X. Such an element f is *uniquely* determined by g; to see this, assume that another $f_1\in D(A)$ can be found such that $(zI-A)f_1 = g$. Then,

$$(zI-A)f_1 = (zI-A)f, \quad z(f_1-f) = A(f_1-f)$$

and so

$$z\|f_1-f;X\|^2 = (z(f_1-f),(f_1-f)) = (A(f_1-f),f_1-f)$$

$$= \mathrm{Re}(A(f_1-f),f_1-f) \le \beta\|f_1-f;X\|^2 \quad \forall z>\beta$$

i.e.

$$0 \le (z-\beta)\|f_1-f;X\|^2 \le 0, \quad \forall z>\beta$$

which leads to $\|f_1-f;X\| = 0$, $f_1 = f$. Because of Lemma 2.1, we conclude that the inverse operator $(zI-A)^{-1}$ exists, with domain $D((zI-A)^{-1}) = R(zI-A) = X$. Further,

$$\|g;X\|\|f;X\| \ge |(g,f)| \ge \mathrm{Re}(g,f) = \mathrm{Re}((zI-A)f,f)$$

$$= \mathrm{Re}[z\|f;X\|^2-(Af,f)] = z\|f;X\|^2-\mathrm{Re}(Af,f) \ge (z-\beta)\|f;X\|^2$$

because of (1.8) and (4.19). Hence, if $\|f;X\| \ne 0$,

$$\|f;X\| \le \frac{1}{z-\beta}\|g;X\|, \quad \forall z>\beta, \quad g\in X$$

$$\|(zI-A)^{-1}g;X\| \le \frac{1}{z-\beta}\|g;X\|, \quad \forall\, z>\beta, \quad g\in X$$

and so $R(z,A) = (zI-A)^{-1}\in\mathscr{B}(X)$ with $\|R(z,A)\| \le (z-\beta)^{-1}$ $\forall z>\beta$ and (4.17c) is proved. Finally $A\in\mathscr{C}(X)$ because $-(zI-A)^{-1}\in\mathscr{B}(X) \subset \mathscr{C}(X)\ \forall\, z>\beta$ (see (2.35)). □

Remark 4.3.

It can be proved that it is sufficient to assume that (4.18) holds for at least one $z_0>\beta$.

Remark 4.4.

Condition (4.19) with $\beta<0$ is usually related to some form of dissipation of energy, whereas energy is either conserved or dissipated if $\beta = 0$; correspondingly, A is said to be a *dissipative operator*. Finally, if $\beta>0$, energy cannot be generated at a rate larger than 2β. To understand this, assume that the state vector $u = u(t)$ of a physical system S satisfies (4.3a)+(4.3b), and that $(u(t),u(t))/2 = \|u(t);X\|^2/2$ is proportional to some form of energy contained in S. Then, we have from (4.3a) and from (4.3b)

$$\left(\frac{du}{dt}, u(t)\right) = (Au(t),u(t)), \quad \lim_{t\to 0+}\|u(t);X\|^2 = \|u_0;X\|^2;$$

but

$$\frac{d}{dt}\|u(t);X\|^2 = \frac{d}{dt}(u(t),u(t)) = \left(\frac{du}{dt},u\right) + \left(u,\frac{du}{dt}\right)$$

$$= 2\,\mathrm{Re}\left(\frac{du}{dt}, u\right)$$

and so

$$\left.\begin{aligned} \frac{1}{2}\frac{d}{dt}\|u(t);X\|^2 &= \mathrm{Re}(Au(t),u(t)) \le \beta\|u(t);X\|^2 \\ \lim_{t\to 0+}\|u(t);X\|^2 &= \|u_0;X\|^2 . \end{aligned}\right\}$$

Using Lemma 3.2, we finally obtain

$$\|u(t);X\|^2 \le \|u_0;X\|^2 \exp(2\beta t), \quad t\ge 0. \tag{4.20}$$

Inequality (4.20) gives an upper bound for the 'energy' $\|u(t);X\|^2/2$ contained in the physical system S at time t and the procedure leading to (4.20) is called the *energy method*. □

If the linear operator A satisfies (4.17a) and (4.17b) and is such that

$$\left.\begin{array}{l}\text{any } z>\beta \text{ belongs to the resolvent set } \rho(A)\\ \text{and, for } \textit{any} \text{ integer } j = 1,2,\ldots\\ \|[R(z,A)]^j\| = \|[(zI-A)^{-1}]^j\| \le \dfrac{M}{(z-\beta)^j}, \quad \forall z>\beta\end{array}\right\} \tag{4.17c'}$$

where $M>1$ and β is a real number, there we say that A belongs to the family $\mathscr{G}(M,\beta) \subset \mathscr{C}(X)$ and we write $A\in\mathscr{G}(M,\beta)$, (or $A\in\mathscr{G}(M,\beta;X)$). Note that (4.17c') is *equivalent* to (4.17c) if $M = 1$ because (4.17c) obviously follows from (4.17c') with $j = 1$; conversely, if (4.17c) holds, then

$$\|[R(z,A)]^j\| \le \|R(z,A)\|^j \le (z-\beta)^{-j}, \quad \forall j = 1,2,\ldots$$

and (4.17c') follows. If $0<M<1$, then (4.17c') with $j = 1$ leads again to (4.17c) that is much simpler than (4.17c') since a *sequence* of inequalities must be checked in order to fulfil condition (4.17c'). This justifies the assumption $M>1$ in (4.17c').

Finally, if the linear operator A satisfies (4.17a) and 4.17b) and if

$$\left.\begin{array}{l}\text{any real } z \text{ such that } |z|>\beta \text{ belongs to}\\ \rho(A) \text{ and, for any integer } j = 1,2,\ldots\\ \|[R(z,A)]^j\| = \|[(zI-A)^{-1}]^j\| \le \dfrac{M}{(|z|-\beta)^j}, \forall |z|>\beta\end{array}\right\} \tag{4.17c''}$$

where $\beta\ge 0$ then we say that A belongs to the family $\mathscr{G}'(M,\beta) \subset \mathscr{C}(X)$ and we write $A\in\mathscr{G}'(M,\beta)$ or $A\in\mathscr{G}'(M,\beta;X)$.

Example 4.7. The case A $\mathscr{B}(X)$

If $A \in \mathscr{B}(X)$, then $A \in \mathscr{C}(X)$ because of (2.34) and (4.17b) is satisfied since $D(A) = X$. Moreover, it follows from (2.62)

$$\| [R(z,A)]^j \| \le \| R(z,A) \|^j \le \frac{1}{(|z| - \|A\|)^j}, \quad j = 1,2,\ldots$$

and so $A \in \mathscr{G}'(1, \|A\|)$. □

Example 4.8. The case A_1 = iA with A self-adjoint

Assume that X is a Hilbert space and that A is a densely defined self-adjoint operator (see section 2.6). If $A_1 = \mathrm{i}A$ with $\mathrm{i} = \sqrt{-1}$, then $D(A_1) = D(A)$ because $A_1 f = \mathrm{i}Af$ and so $A_1 f$ is defined if and only if $f \in D(A)$. Further, $A = A^* \in \mathscr{C}(X)$ because of (2.39) and $A_1 = \mathrm{i}A$ is closed as well (see Theorem 2.2). Hence, A_1 is a densely defined closed operator, i.e. it satisfies (4.17a) and (4.17b). Finally, if $z = z_1$ = a real number, we have from (2.103)

$$\| R(z, A_1) \| \le |z|^{-1} \; \forall |z| > 0$$

and so $A_1 \in \mathscr{G}'(1,0;X)$. □

4.4. THE CASE $A \in \mathscr{G}(1,0;X)$: TWO PRELIMINARY LEMMAS

If $A \in \mathscr{G}(1,0;X)$, then A is a densely defined closed operator and (4.17c) holds with $\beta = 0$. As a consequence,

$$R(z,A) = (zI - A)^{-1} \in \mathscr{B}(X), \quad \|R(z,A)\| \le z^{-1}, \; \forall z > 0 \tag{4.21}$$

and the operator

$$\left(I - \frac{t}{n} A\right)^{-1} = \frac{n}{t} \left(\frac{n}{t} I - A\right)^{-1} = \frac{n}{t} R\left(\frac{n}{t}, A\right), \quad t > 0$$

$$\left(I - \frac{t}{n} A\right)^{-1} = I, \quad t = 0$$

belongs to $\mathscr{B}(X)$ $\forall n = 1,2,\ldots$ and so do the operators

$$Z_n(t) = \left\{\left(I - \frac{t}{n} A\right)^{-1}\right\}^n, \quad t \ge 0, \quad n = 1,2,\ldots . \tag{4.22}$$

Definition (4.22) is suggested by the elementary formula (4.16b). To show that the sequence $\{Z_n(t), n = 1,2,\ldots\}$ can be used to introduce the 'exponential operator' $\exp(tA)$, we first prove the following Lemmas.

Lemma 4.1. If $A\in\mathscr{G}(1,0;X)$, then the one-parameter family of operators

$$V(s) = (I-sA)^{-1}, \quad s \geq 0 \tag{4.23}$$

is contained in $\mathscr{B}(X)$ and is such that

$$V(s) = s^{-1} R(s^{-1},A), \quad s > 0 \quad V(0) = I; \tag{4.24a}$$

$$\|V(s)\| = \|V(s);\mathscr{B}(X)\| \leq 1, \quad \forall\, s \geq 0\ ; \tag{4.24b}$$

$$\left.\begin{aligned} &V(s)f\in D(A), \quad \forall f\in X, \quad s > 0;\\ &V(s)f\in D(A), \quad AV(s)f = V(s)Af, \quad \forall f\in D(A), \quad s \geq 0; \end{aligned}\right\} \tag{4.24c}$$

$$\left.\begin{aligned} &\lim_{h\to 0} \|V(s+h)f-V(s)f;X\| = 0, \quad \forall f\in X, \quad s > 0;\\ &\lim_{h\to 0+} \|V(h)f-f;X\| = 0, \quad \forall f\in X \end{aligned}\right\} \tag{4.24d}$$

$$\left.\begin{aligned} &\frac{d}{ds}\{V(s)f\} = A\{V(s)\}^2 f, \quad \forall f\in X \quad \text{if} \quad s > 0;\\ &\frac{d}{ds}\{V(s)f\} = A\{V(s)\}^2 f = \{V(s)\}^2 Af, \quad \forall f\in D(A)\\ &\qquad\qquad \text{if } s \geq 0, \end{aligned}\right\} \tag{4.24e}$$

where the derivative is in the sense of $\|\cdot,X\|$.

Proof. It follows directly from the definition (4.23) that

$$V(0) = I^{-1} = I$$

$$V(s) = \{s(s^{-1}I-A)\}^{-1} = s^{-1}(s^{-1}I-A)^{-1}$$

$$= s^{-1}R(s^{-1},A), \quad s > 0$$

and so

$$\|V(0)\| = \|I\| = 1$$

$$\|V(s)\| = s^{-1}\|R(s^{-1},A)\| \le s^{-1}\frac{1}{s^{-1}} = 1, \quad s > 0$$

because of (4.21). Thus, $V(s) \in \mathscr{B}(X)$ $\forall\, s \ge 0$ and (4.24a) and (4.24b) are proved.

As far as (4.24c) is concerned, note that $V(s)f = (I-sA)^{-1}f$ belongs to $D(A)$ $\forall f \in X$ and $s > 0$ because the range of $(I-sA)^{-1}$ coincides with $D(A)$:

$$R((I-sA)^{-1}) = D(I-sA) = D(I) \cap D(A) = X \cap D(A) = D(A)$$

(see Remark 4.5). Further, the commutation formula (4.24c) is obvious if $s = 0$ because $V(0) = I$, whereas we have for any $s > 0$ and $f \in D(A)$

$$(I-sA)^{-1}(I-sA)f = (I-sA)(I-sA)^{-1}f$$

$$(I-sA)^{-1}f - s(I-sA)^{-1}Af = (I-sA)^{-1}f - sA(I-sA)^{-1}f$$

and so

$$s(I-sA)^{-1}Af = sA(I-sA)^{-1}f$$

$$(I-sA)^{-1}Af = A(I-sA)^{-1}f.$$

Next we prove (4.24d) for $s > 0$; we have from (4.23)

$$V(s+h)f - V(s)f = V(s+h)[\{I-sA\} - \{I-(s+h)A\}]V(s) = hV(s+h)AV(s)f$$

$$\|V(s+h)f - V(s)f;X\| = |h|\,\|V(s+h)AV(s)f;X\|$$

$$\le |h|\,\|V(s+h)\|\,\|AV(s)f;X\| \le |h|\,\|AV(s)f;X\| \to 0$$

as $h \to 0$, $\forall\, f \in X$, where we used (4.24b) and took into account that $V(s)f \in D(A)$ $\forall f \in X$ and $s > 0$ because of (4.24c). Similarly, if $s = 0$ and $g \in D(A)$,

$$\| V(h)g-V(0)g;X\| \le |h|\| AV(0)g;X\| = |h|\| Ag;X\| \to 0$$

as $h \to 0+$. However, $D(A)$ is dense in X because of (4.17b) and so, if $f \in X$ and $\varepsilon > 0$, a suitable g (depending on f and ε) can be found in $D(A)$, such that $\| f-g;X\| < \varepsilon/3$. Thus,

$$V(h)f-V(0)f = V(h)g-V(0)g+V(h)[f-g]-V(0)[f-g]$$

$$\| V(h)f-V(0)f;X\| \le \| V(h)g-V(0)g;X\| + \| V(h)\| \| f-g;X\| +$$

$$\| V(0)\| \| f-g;X\| \le \| V(h)g-V(0)g;X\| + 2\| f-g;X\| < \frac{\varepsilon}{3} + \frac{2}{3}\varepsilon = \varepsilon$$

provided that $0 < h < \delta_{\varepsilon}$, because $\| V(h)g-V(0)g;X\| \to 0$ as $h \to 0+$ as we have shown above, and (4.24d) is proved.

It remains to prove that $V(s)f$ is differentiable and that (4.24e) holds. If $f \in D(A)$ and $s = 0$, we have from (4.23)

$$\| \frac{1}{h}\{V(h)-I\}f-Af;X\| = \| \frac{1}{h} V(h)\{I-(I-hA)\}f-Af;X\|$$

$$= \| V(h)Af-Af;X\| = \| V(h)\phi-\phi;X\| \to 0 \quad \text{as} \quad h \to 0+$$

because of (4.24d) with $\phi = Af$ instead of f. Hence,

$$\left[\frac{\mathrm{d}}{\mathrm{d}s}\{V(s)f\}\right]_{s=0} = Af = A\{V(0)\}^2 f, \quad f \in D(A).$$

Finally, if $s > 0$, then $V(s)f = s^{-1}R(s^{-1},A)f$ and (3.46) with $n = 1$ and with $z = s^{-1}$ gives

$$\begin{aligned}\frac{\mathrm{d}}{\mathrm{d}s}\{V(s)f\} &= -\frac{1}{s^2}R(s^{-1},A)f+s^{-1}\frac{\mathrm{d}}{\mathrm{d}s}\{R(s^{-1},A)f\} \\ &= -\frac{1}{s^2}R(s^{-1},A)f+s^{-1}\frac{\mathrm{d}}{\mathrm{d}z}\{R(z,A)f\}\frac{\mathrm{d}z}{\mathrm{d}s} \\ &= -\frac{1}{s^2}R(s^{-1},A)f-\frac{1}{s^3}(-1)\{R(s^{-1},A)\}^2 f\end{aligned}$$

and so

$$\frac{\mathrm{d}}{\mathrm{d}s}\{V(s)f\} = -s^{-2}R(s^{-1},A)\{f-s^{-1}R(s^{-1},A)f\}$$

$$= -s^{-1}V(s)\{f-V(s)f\} = -s^{-1}V(s)\{(I-sA)f-f\}V(s)f = V(s)AV(s)f$$

and (4.24e) is completely proved. □

Remark 4.5.

If A and B are operators (not necessarily linear) with domains contained in X, then the operator $(A+B)$ is defined by

$$(A+B)(f) = A(f)+B(f), \ \forall f \in D(A+B)$$

where $D(A+B) = D(A) \cap D(B)$ because both $A(f)$ and $B(f)$ must make sense and so f must belong to $D(A)$ *and* to $D(B)$. □

Lemma 4.2. The operators $Z_n(t)$ defined by (4.22) have the following properties:

$$Z_n(t) \in \mathscr{B}(X), \quad \| Z_n(t) \| \le 1, \quad \forall t \ge 0, \quad n = 1,2,\ldots, \tag{4.25a}$$

$$Z_n(0) = I, \quad \forall n = 1,2,\ldots, \tag{4.25b}$$

$$\left.\begin{aligned} &Z_n(t)f \in D(A), \ \forall f \in X, \ t > 0 \\ &Z_n(t)f \in D(A), \quad A Z_n(t)f = Z_n(t)Af, \quad \forall f \in D(A), \quad t \ge 0; \end{aligned}\right\} \tag{4.25c}$$

$$\left.\begin{aligned} &\lim_{h\to 0} \| Z_n(t+h)f - Z_n(t)f; X \| = 0, \quad \forall f \in X, \quad t > 0 \\ &\lim_{h\to 0+} \| Z_n(h)f - f; X \| = 0, \quad \forall f \in X; \end{aligned}\right\} \tag{4.25d}$$

$$\left.\begin{aligned} \frac{d}{dt}\{Z_n(t)f\} &= AZ_n(t)V\left(\frac{t}{n}\right)f, \quad \forall f \in X, \quad t > 0 \\ \frac{d}{dt}\{Z_n(t)f\} &= AZ_n(t)V\left(\frac{t}{n}\right)f \\ &= Z_n(t)V\left(\frac{t}{n}\right)Af, \quad \forall f \in D(A), \quad t \ge 0 \end{aligned}\right\} \tag{4.25e}$$

where the derivative is in the sense of $\| \cdot, X \|$.

Proof. Since

$$Z_n(t) = \left\{\left(I - \frac{t}{n}A\right)^{-1}\right\}^n = \left\{V\left(\frac{t}{n}\right)\right\}^n, \quad n = 1,2,\ldots$$

we have from (4.24a) and (4.24b)

$$Z_n(0) = \{V(0)\}^n = I$$

$$\| Z_n(t) \| \le \| V\left(\frac{t}{n}\right) \|^n \le 1, \quad \forall t \ge 0, \quad n = 1,2,\ldots .$$

Moreover, if $f \in X$ and $t > 0$,

$$Z_n(t)f = V\left(\frac{t}{n}\right)\phi, \quad \phi = \left\{V\left(\frac{t}{n}\right)\right\}^{n-1} f$$

and so $Z_n(t)f \in D(A)$ because of the first of (4.24c) with $s = t/n$ and with ϕ instead of f. On the other hand, if $f \in D(A)$ and $t \ge 0$, we have from the second part of (4.24c)

$$\begin{aligned} Z_n(t)Af &= \left\{V\left(\frac{t}{n}\right)\right\}^n Af = \left\{V\left(\frac{t}{n}\right)\right\}^{n-1} AV\left(\frac{t}{n}\right) f = \ldots \\ &= A\left\{V\left(\frac{t}{n}\right)\right\}^n f = AZ_n(t)f . \end{aligned}$$

Thus, (4.25a), (4.25b), and (4.25c) are proved. As far as (4.25d) is concerned, we have

$$\begin{aligned} Z_n(t+h)f - Z_n(t)f &= \left\{V\left(\frac{t+h}{n}\right)\right\}^n f - \left\{V\left(\frac{t}{n}\right)\right\}^n f \\ &= \left[\sum_{j=0}^{n-1} \left\{V\left(\frac{t+h}{n}\right)\right\}^{n-1-j} \left\{V\left(\frac{t}{n}\right)\right\}^j\right]\left\{V\left(\frac{t+h}{n}\right) - V\left(\frac{t}{n}\right)\right\} f \end{aligned}$$

and so

$$\begin{aligned} &\| Z_n(t+h)f - Z_n(t)f; X \| \\ &\le \left\{\sum_{j=0}^{n-1} \| V\left(\frac{t+h}{n}\right) \|^{n-1-j} \| V\left(\frac{t}{n}\right) \|^j\right. \| V\left(\frac{t+h}{n}\right) f - V\left(\frac{t}{n}\right) f; X \| \\ &\le n \| V\left(\frac{t+h}{n}\right) f - V\left(\frac{t}{n}\right) f; X \| \end{aligned}$$

because of (4.24b), and (4.25d) follows from (4.24d). Finally, if $t > 0$, we obtain from (4.24e) with $s = t/n$

$$\begin{aligned} \frac{d}{dt}\{Z_n(t)f\} &= \frac{d}{dt}\left\{V\left(\frac{t}{n}\right)\right\}^n f = n\left\{V\left(\frac{t}{n}\right)\right\}^{n-1} \frac{d}{dt}\left\{V\left(\frac{t}{n}\right) f\right\} \\ &= \left\{V\left(\frac{t}{n}\right)\right\}^{n-1} A\left\{V\left(\frac{t}{n}\right)\right\}^2 f = AZ_n(t)V\left(\frac{t}{n}\right) f, \quad \forall f \in X. \end{aligned}$$

Since the case $t = 0$ can be dealt with in a similar way, Lemma 4.2 is completely proved. □

4.5. THE SEMIGROUP GENERATED BY $A \in \mathcal{G}(1,0;X)$

Using the results of Lemmas 4.1 and 4.2, we can prove that $\{Z_n(t)g, n = 1,2,\ldots\}$ is a Cauchy sequence in the B-space X for any given $g \in X$ and $t \geq 0$. To see this, assume first that

$$\{f \in D(A^2) = f : f \in D(A); Af \in D(A)\} \subset D(A);$$

then, using (4.24c), (4.25c), and (4.25e) we have

$$\frac{\mathrm{d}}{\mathrm{d}s}\{Z_m(t-s)Z_n(s)f\}$$

$$= Z_m(t-s)\,\frac{\mathrm{d}}{\mathrm{d}s}\{Z_n(s)f\} - \frac{\mathrm{d}}{\mathrm{d}t}\,Z_m(t-s)\{Z_n(s)f\}$$

$$= Z_m(t-s)\,\left\{AZ_n(s)V\left(\frac{s}{n}\right)f\right\} - AZ_m(t-s)V\left(\frac{t-s}{m}\right)\left\{Z_n(s)f\right\}$$

$$= Z_m(t-s)\left\{V\left(\frac{s}{n}\right) - V\left(\frac{t-s}{m}\right)\right\}Z_n(s)f\;.$$

However,

$$V\left(\frac{s}{n}\right) - V\left(\frac{t-s}{m}\right) = V\left(\frac{s}{n}\right)\left\{\left(I - \frac{t-s}{m}\,A\right) - \left(I - \frac{s}{n}\,A\right)\right\}V\left(\frac{t-s}{m}\right)$$

$$= \left(\frac{s}{n} - \frac{t-s}{m}\right)V\left(\frac{s}{n}\right)AV\left(\frac{t-s}{m}\right)$$

and so

$$\frac{\mathrm{d}}{\mathrm{d}s}\left\{Z_m(t-s)Z_n(s)f\right\}$$

$$= \left(\frac{s}{n} - \frac{t-s}{m}\right)Z_m(t-s)V\left(\frac{s}{n}\right)AV\left(\frac{t-s}{m}\right)Z_n(s)Af$$

$$= \left(\frac{s}{n} - \frac{t-s}{m}\right)Z_m(t-s)V\left(\frac{s}{m}\right)V\left(\frac{t-s}{m}\right)Z_n(s)A^2f, \quad f \in D(A^2) \tag{4.26}$$

with $0 \leq s \leq t$. Since (4.26) shows that $\mathrm{d}\{Z_m(t-s)Z_n(s)f\}/\mathrm{d}s$ is a strongly continuous function of $s \in [0,t]$ because of (4.24d) and (4.25d) (see also Exercise 4.7), integrating from 0 to t we obtain

$$Z_n(t)f-Z_m(t)f$$
$$= \int_0^t \left(\frac{s}{n} - \frac{t-s}{m}\right) Z_m(t-s)V\left(\frac{s}{n}\right)V\left(\frac{t-s}{m}\right)Z_n(s)A^2f ds$$

because $Z_m(0) = Z_n(0) = I$. Using (4.24b), (4.25a), and (3.21), we finally obtain

$$\| Z_n(t)f-Z_m(t)f;X\| \le \int_0^t \left(\frac{s}{n} + \frac{t-s}{m}\right)\| A^2f;X\| \, ds$$
$$= \frac{t^2}{2}\left(\frac{1}{n} + \frac{1}{m}\right)\| A^2f;X\|, \quad f\in D(A^2), \quad t \ge 0 \qquad (4.27)$$

and so

$$\| Z_n(t)f-Z_m(t)f;X\| < \frac{\varepsilon}{3}, \ \forall\, n,m > n_0 \qquad (4.28)$$

where n_0 depends on ε, t, and $f\in D(A^2)$. However, $D(A^2)$ is dense in X because $A\in\mathscr{G}(1,0;X)$ (Kato 1966, p.480; see also Exercise 4.8) and inequality (4.28) can be extended to the whole X in the usual way. In fact, if $g\in X$ and $\varepsilon > 0$, then a suitable f (depending on g and on ε) can be found in $D(A)$, such that $\| f-g;X\| < \varepsilon/3$ and so

$$Z_n(t)g-Z_m(t)g = Z_n(t)f-Z_m(t)f+Z_n(t)[g-f]-Z_m(t)[g-f]$$

$$\| Z_n(t)g-Z_m(t)g;X\|$$

$$\le \| Z_n(t)f-Z_m(t)f;X\| + \| Z_n(t)\| \, \| g-f;X\| + \| Z_m(t)\| \, \| g-f;X\|$$

$$< \frac{\varepsilon}{3} + \frac{\varepsilon}{3} + \frac{\varepsilon}{3} = \varepsilon, \quad \forall\, n,m > n_0 \qquad (4.29)$$

where we used (4.25a) and (4.28) and where n_0 depends on ε, t, and f, i.e. on ε, t, and g. Inequality (4.29) shows that $\{Z_n(t)g, n = 1,2,...\}$ is a Cauchy sequence in X for any $g\in X$ and $t \ge 0$ and therefore it converges to a $\phi = \phi(t)\in X$. The t-dependent element of X $\phi = \phi(t)$ is uniquely determined by g and so $\phi(t) = Z(t)g$ with

$$Z(t)g = X\text{-}\lim_{n\to\infty} Z_n(t)g, \quad g\in X, \ t \geq 0 \tag{4.30}$$

and the operator $Z(t)$ is linear for each $t \geq 0$ because the operation X-limit is linear:

$$\begin{aligned} Z(t)(\alpha g+\beta g_1) &= X\text{-}\lim_{n\to\infty} Z_n(t)[\alpha g+\beta g_1] \\ &= \alpha\, X\text{-}\lim_{n\to\infty} Z_n(t)g+\beta X\text{-}\lim_{n\to\infty} Z_n(t)g_1 \\ &= \alpha Z(t)g+\beta Z(t)g_1. \end{aligned}$$

The symbol $\exp(tA)$ is also used for the operator $Z(t)$ because (4.30) with $Z_n(t)$ given by (4.22) is formally similar to (4.16b) and because (4.31b), (4.31e), and (4.31f) of Theorem 4.4 clearly show that $Z(t)$ has an 'exponential-like' behaviour.

Remark 4.6.

$\{Z_n(t)g\}$ converges to $Z(t)g$ in the sense of $\|\cdot;X\|$ for each $g\in X$ and $t \geq 0$, whereas the sequence $\{T_n(t)\}$ defined by (4.4) converges to $T(t)$ in the sense of $\|\cdot\| = \|\cdot;\mathscr{B}(X)\|$ (see 4.6)). □

The properties of the one-parameter family of linear operators $\{Z(t), t \geq 0\}$ can be derived from the corresponding properties of the elements of the sequence $\{Z_n(t)\}$, as is clear from the proof of Theorem 4.4.

Theorem 4.4. If $A\in\mathscr{G}(1,0;X)$, the one-parameter family of linear operators $\{Z(t), t \geq 0\}$ defined by (4.30) is such that

$$\{Z(t), t \geq 0\} \subset \mathscr{B}(X), \quad \|Z(t)\| \leq 1 \quad \forall\, t \geq 0 \tag{4.31a}$$

$$Z(0) = I \tag{4.31b}$$

$$Z(t)f\in D(A), \quad AZ(t)f = Z(t)Af \quad \forall f\in D(A), \quad t \geq 0 \tag{4.31c}$$

$$\left.\begin{aligned}&\lim_{h\to 0}\|Z(t+h)g-Z(t)g;X\| = 0 \quad \forall g\in X, \quad t>0\\ &\lim_{h\to 0+}\|Z(h)g-g;X\| = 0 \quad \forall g\in X\end{aligned}\right\} \tag{4.31d}$$

$$\frac{\mathrm{d}}{\mathrm{d}t}\{Z(t)f\} = AZ(t)f = Z(t)Af \quad \forall f\in D(A), \quad t\geq 0 \tag{4.31e}$$

$$Z(t)Z(s) = Z(s)Z(t) = Z(t+s), \quad \forall t,\ s\geq 0 \tag{4.31f}$$

where the derivative on the left-hand side of (4.31e) is in the sense of $\|\cdot;X\|$.

Proof.

(*a*) We have from (4.25a)

$$0\leq \|Z_n(t)g;X\| \leq \|g;X\|, \ \forall g\in X, \ t\geq 0;$$

passing to the limit as $n\to\infty$ we obtain

$$0\leq \|Z(t)g;X\| \leq \|g;X\|, \quad \forall g\in X, \ t\geq 0$$

and (4.31a) is proved because we already know that $Z(t)$ is linear and defined for any $g\in X$.

(*b*) (4.31b) is obvious since $Z_n(t) = I \ \forall n = 1,2,\ldots$.

(*c*) If $f\in D(A)$ and $t\geq 0$ are given, then $\gamma_n = Z_n(t)f\in D(A)$ and $A\gamma_n = Z_n(t)Af$ because of (4.25c). Using (4.30) we have

$$X\text{-}\lim_{n\to\infty}\gamma_n = Z(t)f$$

$$X\text{-}\lim_{n\to\infty}A\gamma_n = X\text{-}\lim_{n\to\infty}Z_n(t)[Af] = Z(t)Af$$

and so the sequence $\{\gamma_n\}$ is contained in $D(A)$ and is such that $\gamma_n\to Z(t)f$ and $A\gamma_n\to Z(t)Af$. Since $A\in\mathscr{C}(X)$, Theorem 2.2 shows that $Z(t)f\in D(A)$ and that $A[Z(t)f] = Z(t)Af$ and (4.31c) is proved.

(*d*) If $\hat{t}>0$ is any finite value of t, then (4.27) gives

$$\|Z_n(t)f-Z_m(t)f;X\| \leq \frac{\hat{t}^2}{2}\left(\frac{1}{n}+\frac{1}{m}\right)\|A^2f;X\|$$

$\forall f\in D(A^2), \forall t\in[0,\hat{t}]$, and by a procedure similar to that leading to (4.29) we obtain

$$\| Z_n(t)g-Z_m(t)g;X\| < \varepsilon, \quad \forall n,m > n_1, \quad \forall t\in[0,\hat{t}], \; g\in X$$

where $n_1 = n_1(\varepsilon,\hat{t},g)$. On passing to the limit as $m\to\infty$ we obtain

$$\| Z_n(t)g-Z(t)g;X\| \le \varepsilon, \quad \forall n > n_1, \quad \forall t\in[0,\hat{t}], \; g\in X \tag{4.32}$$

Inequality (4.32) proves that $\{Z_n(t)g\}$ converges to $Z(t)g$ *uniformly* in $t\in[0,\hat{t}]$ and so $Z(t)g$ is continuous in t in the sense of $\|\cdot;X\|$ because each $Z_n(t)g$ has such a property (see (4.25d)). In other words, $\{Z_n(t)g\}$ is a Cauchy sequence in the B-space $C([0,\hat{t}]-X)$ and consequently $Z(t)g\in C([0,\hat{t}],X)$ (see the discussion after (3.12)).

(*e*) If $f\in D(A)$ and $t\in[0,\hat{t}]$, we have from (4.25e)

$$Z_n(t)f-f = \int_0^t Z_n(s)V\left(\frac{s}{n}\right)[Af]\mathrm{d}s \, .$$

On the other hand, since $\| Z_n(s)\| \le 1, \; \forall s \ge 0$,

$$Z_n(s)V\left(\frac{s}{n}\right)\phi-Z(s)\phi = Z_n(s)V\left(\frac{s}{n}\right)\phi-Z_n(s)\phi+Z_n(s)\phi-Z(s)\phi$$

$$\| Z_n(s)V\left(\frac{s}{n}\right)\phi-Z(s)\phi;X\|$$

$$\le \| V\left(\frac{s}{n}\right)\phi-\phi;X\| + \| Z_n(s)\phi-Z(s)\phi;X\|$$

$$< \varepsilon+\varepsilon = 2\varepsilon, \quad \forall s\in[0,\hat{t}]$$

provided that $n > n_2 = n_2(\varepsilon,\hat{t},\phi)$ (see (4.32) and the proof of the second of (4.24d) with $h = s/n$). Thus,

$$Z_n(s)V\left(\frac{s}{n}\right)\phi \to Z(s)\phi$$

in the sense of $\|\cdot;X\|$ *uniformly* in $s\in[0,\hat{t}]$ and we can pas to the X-limit under the integral obtaining

$$Z(t)f-f = \int_0^t Z(s)Af\,ds, \quad t \geq 0, \quad f \in D(A) \quad . \tag{4.33}$$

Relation (4.31e) follows from (4.33) on differentiating with respect to t. Note that $d\{Z(t)f\}/dt = Z(t)[Af]$, $f \in D(A)$, is strongly continuous at any $t \geq 0$ because of (4.31d) with $g = Af$.

(*f*) If $f \in D(A)$ and $0 \leq s \leq \tau$, then we have from (4.31e)

$$\frac{d}{ds}\{Z(\tau-s)Z(s)f\}$$

$$= Z(\tau-s)\frac{d}{ds}\{Z(s)f\} + \left[\frac{d}{d(\tau-s)} Z(\tau-s)\{Z(s)f\}\right]\frac{d(\tau-s)}{ds}$$

$$= Z(\tau-s)Z(s)Af - Z(\tau-s)AZ(s)f = \theta_X, \quad \forall\, s \in [0,\tau].$$

Hence, $Z(\tau-s)Z(s)f$ is independent of $s \in [0,\tau]$ and so

$$Z(\tau-s)Z(s)f = \{Z(\tau-s)Z(s)f\}_{s=0}$$

$$Z(\tau-s)Z(s)f = Z(\tau)f$$

i.e.

$$Z(t)Z(s)f = Z(t+s)f, \quad f \in D(A), \quad t,s \geq 0 \tag{4.34}$$

with $t = \tau - s$. Since relation (4.34) can be extended in the usual way from the dense subset $D(A)$ to the whole space X, (4.31f) and Theorem 4.4 are proved. □

Remark 4.7.

The one-parameter family of operators $\{Z(t), t \geq 0\}$ is such that (i) $Z(0) = I$, (ii) $Z(t)Z(s) = Z(s)Z(t) = Z(t+s)$, $\forall\, t,s \geq 0$, and, in general, it is *not* a group because $Z(t)$ is not defined if $t < 0$ (see Remark 4.1). However, $\{Z(t), t \geq 0\}$ is a *commutative semigroup generated* by $A \in \mathcal{G}(1,0;X)$ and A is said to be the *generator* of $Z(t)$. □

4.6. THE CASES $A \in \mathcal{G}(M,0;X)$, $\mathcal{G}_n(M,\beta;X)$, $\mathcal{G}'(M,\beta;X)$

If $A \in \mathcal{G}(M,0,X)$, $Z_n(t)$ can still be defined by (4.22) since $z = t/n \in \rho(A)$, $\forall t > 0$, $n = 1,2,\ldots$, because of (4.17c') with $\beta = 0$. The results of sections 4.4 and 4.5 hold and can be proved by similar techniques; the only modifications needed are an M rather than 1 on the right-hand side of both (4.24b) and (4.25a) and a factor M^2 on the right-hand side of (4.27). As a consequence, the semigroup $\{Z(t) = \exp(tA), t \geq 0\}$ generated by $A \in \mathcal{G}(M,0)$ satisfies (4.31a) with M instead of 1 and (4.31b)-(4.31f).

If $A \in \mathcal{G}(M,\beta;X)$, then the operator

$$A_1 = A - \beta I$$

belongs to the class $\mathcal{G}(M,0;X)$. To see this, note first that $A_1 \in \mathcal{C}(X)$ because of (*b*) of Theorem 2.3 with $B = -\beta I$, and that

$$D(A_1) = D(A) \cap D(-\beta I) = D(A) \cap X = D(A)$$

is dense in X (see Remark 4.5). Further,

$$z_1 I - A_1 = (z_1 + \beta) I - A$$

and the inverse operator

$$(z_1 I - A_1)^{-1} = \{(z_1 + \beta) I - A\}^{-1}$$

exists and is such that

$$\| \{(z_1 I - A_1)^{-1}\}^j \| = \| [\{(z_1 + \beta) I - A\}^{-1}]^j \| \leq M\{(z_1 + \beta) - \beta\}^{-j}$$

$\forall j = 1,2,\ldots$, provided that $z = z_1 + \beta > \beta$ because $A \in \mathcal{G}(M,\beta)$ and so (4.17c') holds. Hence,

$$\| \{R(z_1, A_1)\}^j \| \leq M[z_1]^{-j}, \quad j = 1,2,\ldots, \; z_1 > 0 ;$$

as a consequence, A_1 belongs to $\mathcal{G}(M,0)$ and it generates the semigroup $\{S(t) = \exp(tA_1), t \geq 0\}$ satisfying (4.31b)-

(4.31f) and (4.31a) with M instead of 1.

Theorem 4.5. If $A \in \mathcal{G}(M,\beta;X)$, *then*

$$A_1 = A-\beta I \in \mathcal{G}(M,0;X)$$

and the one-parameter family of linear operators

$$Z(t) = \exp(\beta t)S(t), \quad t \geq 0 \tag{4.35}$$

where $\{S(t) = \exp(tA_1),\ t \geq 0\}$ *is the semigroup generated by* A_1, *has the following properties:*

$$\left.\begin{aligned} &\{Z(t), t \geq 0\} \subset \mathcal{B}(X) \\ &\|Z(t)\| \leq M\exp(\beta t), \quad t \geq 0 \end{aligned}\right\} \tag{4.36a}$$

$$Z(0) = I \tag{4.36b}$$

$$\left.\begin{aligned} &Z(t)f \in D(A) \\ &AZ(t)f = Z(t)Af, \quad \forall f \in D(A), \quad t \geq 0 \end{aligned}\right\} \tag{4.36c}$$

$$\left.\begin{aligned} &\lim_{h\to 0} \|Z(t+h)g-Z(t)g;X\| = 0, \quad \forall g \in X, t > 0 \\ &\lim_{h\to 0+} \|Z(h)g-g;X\| = 0, \quad \forall g \in X; \end{aligned}\right\} \tag{4.36d}$$

$$\frac{\mathrm{d}}{\mathrm{d}t}\{Z(t)f\} = AZ(t)f = Z(t)Af, \quad \forall f \in D(A), \ t \geq 0 \tag{4.36e}$$

$$Z(t)Z(s) = Z(s)Z(t) = Z(t+s), \ \forall t, \ s \geq 0 \tag{4.36f}$$

Proof. Since $S(t) = \exp(tA_1)$ satisfies (4.31b)-(4.31f) and (4.31a) with M instead of 1, relations (4.36a)-(4.36f) follow directly from definition (4.35) (for instance, see Exercise 4.11). □

The family $\{Z(t) = \exp(tA), t \geq 0\}$, defined by (4.35), is said to be the *semigroup generated* by A.

Finally, assume that the linear operator A belongs to the

family $\mathscr{G}'(M,\beta;X)$ defined by (4.17a), (4.17b), and (4.17c"). Then, $A\in\mathscr{G}(M,\beta;X)$ because (4.17c') obviously follows from (4.17c") with $z>\beta\geq 0$. Further, $B = -A\in\mathscr{G}(M,\beta;X)$ as well, since $-A\in\mathscr{C}(X)$ because of Theorem 2.2 and $D(-A) = D(A)$ is dense in X. On the other hand,

$$(z_1 I-B) = -\{(-z_1 I)-A\}\ , \quad (z_1 I-B)^{-1} = -\{(-z_1 I)-A\}^{-1}$$

$$\|\{R(z_1,B)\}^j\| = \|[\{(-z_1)I-A\}^{-1}]^j\| \leq M(|z_1|-\beta)^{-j}$$

$\forall\ |z_1|>\beta$, $j = 1,2,\ldots$, because of (4.17c"). Hence, both A and $B = -A$ belong to $\mathscr{G}(M,\beta)$ and so the semigroups

$$\left.\begin{aligned} Z^{(+)}(t) &= \exp(tA), \quad t\geq 0 \\ Z^{(-)}(t_1) &= \exp(t_1 B) = \exp[t_1(-A)], \quad t_1\geq 0 \end{aligned}\right\} \tag{4.37}$$

are defined and the results of Theorem 4.5 hold for both $Z^{(+)}(t)$, $t\geq 0$, and $Z^{(-)}(t_1)$, $t_1\geq 0$. Now, if we define the one-parameter family of linear operators $\{Z(t), t\in(-\infty,+\infty)\}$ as follows

$$\left.\begin{aligned} Z(t) &= Z^{(+)}(t) \quad \text{if} \quad t\geq 0 \\ Z(t) &= Z^{(-)}(t_1) \quad \text{if} \quad t\leq 0, \quad \text{with } t_1 = -t\geq 0 \end{aligned}\right\} \tag{4.38}$$

then we have the following theorem.

Theorem 4.6. If $A\in\mathscr{G}'(M,\beta;X)$, then the one-parameter family of linear operators $\{Z(t),\ t\in(-\infty,+\infty)\}$ defined by (4.38) is such that

$$\left.\begin{aligned} &\{Z(t), t\in(-\infty,+\infty)\}\subset\mathscr{B}(X) \\ &\|Z(t)\| \leq M\exp(\beta|t|),\ \forall t\in(-\infty,+\infty) \end{aligned}\right\} \tag{4.39a}$$

$$Z(0) = I \tag{4.39b}$$

$$Z(t)f\in D(A), \quad AZ(t)f = Z(t)Af, \quad \forall f\in D(A), \quad t\in(-\infty,+\infty) \tag{4.39c}$$

$$\lim_{h\to 0} \| Z(t+h)g - Z(t)g; X\| = 0, \quad \forall\, g\in X, \quad t\in(-\infty,+\infty) \tag{4.39d}$$

$$\frac{d}{dt}\{Z(t)f\} = AZ(t)f = Z(t)Af, \quad \forall f\in D(A), \quad t\in(-\infty,+\infty) \tag{4.39e}$$

$$Z(t)Z(s) = Z(s)Z(t) = Z(t+s), \quad \forall t,s\in(-\infty,+\infty). \tag{4.39f}$$

Proof. Since $Z(t)$ is defined by means of the semigroups generated by $A\in\mathcal{G}(M,\beta)$ and by $B = -A\in\mathcal{G}(M,\beta)$, (4.39a)-(4.39e) follow from (4.36a)-(4.36e); of course, the cases $t \geq 0$ and $t \leq 0$ should be examined separately. Moreover, (4.39f) follows directly from (4.36f) if $s,t \geq 0$ or if $s,t \leq 0$; however, if $s \leq 0$ and $t \geq 0$, the procedure leading to (4.31f) should be used to prove (4.39f) with $Z^{(+)}(\tau-s)Z^{(-)}(-s)$ instead of $Z(\tau-s)Z(s)$. □

The family $\{Z(t) = \exp(tA), t\in(-\infty,+\infty)\}$ is a *group, generated* by A, and $\{Z(t)\}^{-1} = Z(-t)$ (see Remark 4.1).

Example 4.9. Conservation of the norm of $u(t) = \exp(tA_1)\, u_0$ *with* $A_1\in\mathcal{G}'(1,0,X)$.

If $A_1\in\mathcal{G}'(1,0;X)$ and $u(t) = \exp(tA_1)u_0$ with $u_0\in X$ and $t\in(-\infty,+\infty)$, then we have from (4.39a) with $M = 1$ and $\beta = 0$

$$\|u(t);X\| \leq \|\exp(tA_1)\|\|u_0;X\| \leq \|u_0;X\| \quad \forall\, t\in(-\infty,+\infty).$$

However,

$$\exp(-tA_1)u(t) = \exp(-tA_1)\exp(tA_1)u_0 = u_0$$

and so

$$\|u_0;X\| \leq \|\exp(-tA_1)\|\|u(t);X\| \leq \|u(t);X\|, \quad \forall\, t\in(-\infty,+\infty).$$

We conclude that

$$\|u(t);X\| = \|u_0;X\|, \quad \forall\, t\in(-\infty,+\infty)$$

i.e. the norm of $u(t) = \exp(tA_1)u_0, A_1\in\mathcal{G}'(1,0)$, does *not* depend on $t\in(-\infty,+\infty)$. Such a result holds if in particular A_1 is the operator of Example 4.8, and it is of some importance in quantum mechanics. □

4.7. THE HOMOGENEOUS AND THE NON-HOMOGENEOUS INITIAL-VALUE PROBLEMS

Assume that $A\in\mathcal{G}(M,\beta;X)$ and $u_0\in D(A)$, and consider the following function from $[0,+\infty)$ into X:

$$u(t) = Z(t)u_0, \quad t\geq 0, \quad u_0\in D(A) \tag{4.40}$$

where $Z(t) = \exp(tA)$ is the semigroup generated by A. Then, $u(t)\in D(A)$ $\forall\, t\geq 0$ because of (4.36c) with $f = u_0$, whereas (4.36d) with $g = u_0$ shows that $u(t)$ is continuous in the sense of $\|\cdot;X\|$ at any $t\geq 0$ with

$$X\text{-}\lim_{h\to 0+} u(h) = u_0.$$

Moreover,

$$\frac{d}{dt}u(t) = \frac{d}{dt}\{Z(t)u_0\} = AZ(t)u_0 = Au(t), \quad \forall t\geq 0$$

because of (4.36e); note that $du(t)/dt = Z(t)[Au_0]$ and so $du(t)/dt$ is strongly continuous at any $t\geq 0$. We conclude that the function $u(t)$ defined by (4.40) is a *strict solution* of the homogeneous initial-value problem (4.3) (see (i), (ii), and (iii) of section 4.1); $u(t)$ also satisfies (4.3a) at $t=0$ and its derivative is strongly continuous $\forall t\geq 0$. Furthermore, $u(t)$ is the *unique* strict solution of (4.3); to see this, assume that $w(t)$ is another strict solution of (4.3) and let

$$v(s) = Z(t-s)w(s), \quad 0\leq s\leq t$$

where $t\geq 0$ is assigned. Then, if $t_1 = t-s$, we have

$$\frac{d}{ds}v(s) = Z(t_1)\frac{d}{ds}w(s) + \frac{d}{dt_1}\left\{Z(t_1)w(s)\right\}\frac{dt_1}{ds}$$

$$= Z(t_1)Aw(s) - AZ(t_1)w(s) = \theta_X, \quad 0<s\leq t$$

because $w(s)$ satisfies (4.3a) with s instead of t, at any $s > 0$. Thus, using Theorem 3.1,

$$v(s) = v(s_1) \;\forall\, s, s_1 \in (0,t];$$

in particular, $v(s) = v(t)$ $\forall\, s\in(0,t]$ and on passing to the X-limit as $s\to 0+$ we obtain

$$v(t) = X\text{-}\lim_{s\to 0+} v(s) = v(0) = Z(t)w(0) = Z(t)u_0 = u(t)$$

because $w(s)$ satisfies (4.3b) with s instead of t. However,

$$v(t) = Z(t-t)w(t) = Z(0)w(t) = w(t)$$

and so $w(t) = u(t)$ $\forall\, t \geq 0$. The above results can be summarized as follows.

Theorem 4.7. If $A\in\mathcal{G}(M,\beta;X)$ *and* $u_0\in D(A)$, *then* $u(t) = Z(t)u_0 = \exp(tA)u_0$, $t \geq 0$, *is the unique strict solution of the homogeneous initial-value problem (4.3). The function* $u(t)$ *is also continuously differentiable at any* $t \geq 0$ *and satisfies (4.3a) at* $t = 0$. □

Remark 4.8.

If $A\in\mathcal{G}'(M,\beta;X)$ and $u_0\in D(A)$, then $u(t) = Z(t)u_0$ satisfies (4.3a) at any $t\in(-\infty,+\infty)$ and the limit (4.3b) holds as $t\to 0$. □

Note that (4.40) and (4.36f) give

$$u(t) = Z(t-t_0)Z(t_0)u_0 = Z(t-t_0)u(t_0), \;\forall\, t \geq t_0 \geq 0 \tag{4.41}$$

and so $u(t)$ satisfies (4.3a) at any $t > t_0$:

$$\frac{d}{dt}\, u(t) = Au(t), \quad t > t_0 \tag{4.3a'}$$

and the 'initial' condition

$$X\text{-}\lim_{t\to t_0+} u(t) = u(t_0). \tag{4.3b'}$$

Moreover, if $w(t)$ is continuous at any $t \geq t_0$ and differentiable at any $t > t_0$ and if $w(t)$ satisfies (4.3a')+(4.3b'), then $w(t) = u(t)$ $\forall t \geq t_0$. In other words, $u(t)$ is equivalently defined at any $t \geq t_0$ either by (4.40) or by assuming that it is the strict solution of system (4.3a')+(4.3b').

Example 4.10. Discretization of the time-like variable

If $A \in \mathcal{G}(1,0)$ and $u_0 \in D(A)$, then $u(t) = Z(t)u_0$, $t \geq 0$, is the unique strict solution of problem (4.3). However, since it is usually difficult to derive an *explicit* expression of the semigroup $Z(t) = \exp(tA)$, we consider a method to *discretize* the time-like variable t. If $[0,\hat{t}]$ is given and if $\delta > 0$ is a *time step* and m an integer such that $m\delta = \hat{t}$, then we discretize system (4.3) as follows

$$\frac{1}{\delta}\,[w_j - w_{j-1}] = Aw_j, \quad j = 1,2,\ldots,m; \quad w_0 = u_0 \tag{4.42}$$

where w_j is the 'approximated value' of $u_j = u(t_j)$, with $t_j = j\delta$, $j = 1,2,\ldots,m$. Relation (4.42) can be solved for w_j:

$$(I - \delta A)w_j = w_{j-1}, \quad j = 1,2,\ldots,m; \quad w_0 = u_0$$

$$w_j = (I - \delta A)^{-1}w_{j-1} = V(\delta)w_{j-1}, \quad j = 1,2,\ldots,m; \quad w_0 = u_0 \tag{4.43}$$

with $V(\delta)$ given by (4.23), and so

$$w_j = \{V(\delta)\}^2 w_{j-2}, \quad w_j = \{V(\delta)\}^3 w_{j-3}, \ldots$$

$$w_j = \{V(\delta)\}^j u_0, \quad j = 1,2,\ldots,m\,. \tag{4.44}$$

Since an explicit expression of the operator $V(\delta) = (I - \delta A)^{-1}$ is often much simpler to derive than that of the semigroup $Z(t)$ (see for instance Examples 2.17-2.19), (4.44) may be quite useful both for computational purposes and to get 'some feeling' on how $u(t)$ approximately behaves at $t = t_j$.

The 'exact' relation corresponding to (4.43) follows directly from (4.40):

$$u_j = u(t_j) = Z(t_j)u_0 = Z(\delta + t_{j-1})u_0 = Z(\delta)Z(t_{j-1})u_0$$

$$u_j = Z(\delta)u_{j-1}, \quad j = 1,2,\ldots,m; \quad u_0 = u_0$$

and using (4.36c) it is easy to verify that $u(t_j)\in D(A^2)$ if $u_0\in D(A^2)$. To find an upper bound for the *error* $\Delta_j = u_j - w_j$, we write Δ_j as follows

$$\Delta_j = Z(\delta)u_{j-1} - V(\delta)w_{j-1} = [Z(\delta)-V(\delta)]u_{j-1} + V(\delta)[u_{j-1} - w_{j-1}]$$

and so

$$\begin{aligned}\|\Delta_j;X\| &\le \|[Z(\delta)-V(\delta)]u_{j-1};X\| + \|V(\delta)\|\|\Delta_{j-1};X\| \\ &\le \|[Z(\delta)-V(\delta)]u_{j-1};X\| + \|\Delta_{j-1};X\|\end{aligned}$$

since $\|V(\delta)\| \le 1$ because $A\in\mathcal{G}(1,0)$. Using inequality (4.27) with $n\to\infty$, $m = 1$, and $t = \delta$, we obtain

$$\|\Delta_j;X\| \le \frac{\delta^2}{2}\|A^2u_{j-1};X\| + \|\Delta_{j-1};X\|$$

and since $A^2u_{j-1} = A^2Z(t_{j-1})u_0 = Z(t_{j-1})A^2u_0$,

$$\|\Delta_j;X\| \le \frac{\delta^2}{2}\|A^2u_0;X\| + \|\Delta_{j-1};X\|, \quad j = 1,2,\ldots$$

because $\|Z(t_{j-1})\| \le 1$. Hence,

$$\begin{aligned}\|\Delta_j;X\| &\le \frac{\delta^2}{2}\|A^2u_0;X\| + \left[\frac{\delta^2}{2}\|A^2u_0;X\| + \|\Delta_{j-2};X\|\right] \\ &\ldots\ldots \le j\,\frac{\delta^2}{2}\|A^2u_0;X\|\end{aligned}$$

because $\Delta_0 = u_0 - u_0 = \theta_X$, and finally

$$\|\Delta_j;X\| \le \frac{\hat{t}\delta}{2}\|A^2u_0;X\| \ \forall\ j = 0,1,2,\ldots,m \tag{4.45}$$

since $j\delta \le m\delta = \hat{t}$. Inequality (4.45) gives the required upper bound for the error $\Delta_j = u_j - w_j$ and shows that it is proportional to the time step δ.

Note that if we had started from the following discretized version of (4.3):

$$\frac{1}{\delta}[w_j - w_{j-1}] = Aw_{j-1}, \quad j = 1,2,\ldots,m; \quad w_0 = u_0$$

then we would have obtained

$$w_j = (I+\delta A)w_{j-1}, \quad w_j = (I+\delta A)^j u_0$$

which may be meaningless since u_0 does not necessarily belong to $D(A^j)$, $j = 3,4,\ldots,m$, even if $u_0 \in D(A^2)$.

If $A \in \mathcal{G}(1,\beta)$, the above procedure applies equally well because $A_1 = A - \beta I \in \mathcal{G}(1,0)$ (see the discussion before Theorem 4.5), and the substitution $U(t) = u(t)\exp(-\beta t)$ leads to an initial-value problem for $U(t)$ with A_1 instead of A. □

Finally, consider the non-homogeneous problem (4.2a) +(4.2b): the existence of a strict solution $u = u(t)$, $t \geq 0$, obviously depends on the properties of the source term $g(t)$, as is shown in Theorem 4.8. In the following Lemma, we prove a formula that will be used in the proof of Theorem 4.8.

Lemma 4.3. If $Z(t) = \exp(tA)$ *with* $A \in \mathcal{G}(M,\beta;X)$ *then*

$$\int_{t'}^{t} Z(s)f\,ds \in D(A) \tag{4.46}$$

$$Z(t)f - Z(t')f = A\int_{t'}^{t} Z(s)f\,ds \tag{4.47}$$

$$\forall\, f \in X, \quad 0 \leq t' \leq t.$$

Proof. If $\hat{f} \in D(A)$, we can integrate (4.36e) obtaining

$$Z(t)\hat{f} - Z(t')\hat{f} = \int_{t'}^{t} AZ(s)\hat{f}\,ds, \quad 0 \leq t' \leq t$$

but $A \in \mathcal{C}(X)$ and so

$$\left.\begin{aligned} &\int_{t'}^{t} Z(s)\hat{f}\,ds \in D(A) \\ &Z(t)\hat{f} - Z(t')\hat{f} = A\int_{t'}^{t} Z(s)\hat{f}\,ds \end{aligned}\right\} \tag{4.48}$$

because of (3.25) with $w(s) = Z(s)\hat{f}\in D(A)$, $t_2 = t$ and $t_1 = t'$. If $f\in X$ and $\varepsilon_n = 1/n$, $n = 1,2,\ldots$, a sequence $\{f_n, n = 1,2,\ldots\} \subset D(A)$ can be found such that $\| f-f_n;X\| < 1/n$, because $D(A)$ is dense in X. As a consequence,

$$\| Z(s)f-Z(s)f_n;X\| \le \| Z(s)\| \| f-f_n;X\|$$

$$\le \frac{M}{n}\exp(\beta s) \le \frac{M}{n}\exp(|\beta| t) \quad \forall s\in[t',t]$$

and $\{Z(s)f_n\}$ converges to $Z(s)f$ *uniformly* in $s\in[t',t]$. Now relations (4.48) hold with $f = f_n$ because $\{f_n\} \subset D(A)$:

$$\gamma_n = \int_{t'}^{t} Z(s)f_n ds\in D(A), \quad Z(t)f_n - Z(t')f_n = A\gamma_n ;$$

hence, the sequence $\{\gamma_n\}$ is contained in $D(A)$ and it is such that

$$X\text{-}\lim_{n\to\infty} \gamma_n = \int_{t'}^{t} Z(s)f\, ds$$

because of Theorem 3.3, and

$$X\text{-}\lim_{n\to\infty} A\gamma_n = X\text{-}\lim_{n\to\infty} [Z(t)-Z(t')]f_n = [Z(t)-Z(t')]f.$$

However, $A\in\mathscr{C}(X)$ and so

$$\int_{t'}^{t} Z(s)fds\in D(A), \quad A\int_{t'}^{t} Z(s)fds = [Z(t)-Z(t')]f$$

because of Theorem 2.2. □

Theorem 4.8. If (a) $A\in\mathscr{G}(M,\beta;X)$, *(b)* $u_0\in D(A)$, *(c)* $g(t)$ *is strongly continuous at any* $t \ge 0$, *then each strict solution* $u(t)$, $t\ge 0$, *of system (4.2) has the form*

$$u(t) = Z(t)u_0 + \int_0^t Z(t-s)g(s)ds, \quad t\ge 0. \tag{4.49}$$

Conversely, $u(t)$ *given by (4.49) is the unique strict solution of system (4.2) if, in addition to (a), (b) and (c), we assume that* (d1) $g(t)$ *is strongly differentiable and* $g'(t)$ *is strongly continuous at any* $t \ge 0$, *or that* (d2)

$g(t) \in D(A)$ and $Ag(t)$ is a strongly continuous function of t at any $t \geq 0$.

Proof. If $u(t)$ is a strict solution of (4.2), then we have

$$\frac{d}{ds}\{Z(t-s)u(s)\} = Z(t-s)\frac{d}{ds}u(s) + \frac{d}{dt_1}\{Z(t_1)u(s)\}\frac{dt_1}{ds}$$

with $t_1 = t-s$, and also

$$\frac{d}{ds}\{Z(t-s)u(s)\} = Z(t-s)[Au(s)+g(s)]-AZ(t-s)u(s)$$

$$\frac{d}{ds}\{Z(t-s)u(s)\} = Z(t-s)g(s), \quad 0 < s \leq t \tag{4.50}$$

because $u(s)$ satisfies (4.2a) with s instead of t at any $s > 0$. However, $g(s)$ is continuous at any $s \geq 0$ and so $Z(t-s)g(s)$ is continuous $\forall s \in [0,t]$; hence, if we integrate (4.50) from $\varepsilon > 0$ to $t > \varepsilon$, we obtain

$$u(t)-Z(t-\varepsilon)u(\varepsilon) = \int_{\varepsilon}^{t} Z(t-s)g(s)\,ds$$

and (4.49) follows on passing to the X-limit as $\varepsilon \to 0+$. Conversely, if (*a*), (*b*), (*c*), and (*d*1) are satisfied, then we have from (4.49) with $\hat{s} = t-s$

$$u(t) = Z(t)u_0 + \int_0^t Z(\hat{s})g(t-\hat{s})\,d\hat{s} \tag{4.51}$$

$$\frac{d}{dt}u(t) = AZ(t)u_0+Z(t)g(0) + \int_0^t Z(\hat{s})g'(t-\hat{s})\,d\hat{s} \tag{4.52}$$

because $g(t)$ and $g'(t)$ are strongly continuous and so (3.22) and (3.26) can be used. On the other hand, since

$$g(t-\hat{s}) = g(0) + \int_0^{t-\hat{s}} g'(r)\,dr$$

(4.51) becomes

$$u(t) = Z(t)u_0 + \int_0^t Z(\hat{s})g(0)\,d\hat{s} + \int_0^t d\hat{s} \int_0^{t-\hat{s}} Z(\hat{s})g'(r)\,dr$$

and after interchanging the order of integrations

$$u(t) = Z(t)u_0 + \int_0^t Z(\hat{s})g(0)\,d\hat{s} + \int_0^t dr \int_0^{t-r} Z(\hat{s})g'(r)\,d\hat{s} \,. \quad (4.51')$$

However, $Z(t)u_0 \in D(A)$ because of assumption (b) and (4.36c), and

$$\int_0^t Z(\hat{s})g(0)\,d\hat{s} \in D(A)$$

$$A\int_0^t Z(\hat{s})g(0)\,d\hat{s} = [Z(t)-I]g(0)$$

$$\int_0^{t-r} Z(\hat{s})g'(r)\,d\hat{s} \in D(A)$$

$$A\int_0^{t-r} Z(\hat{s})g'(r)\,d\hat{s} = [Z(t-r)-I]g'(r)$$

because of (4.46) and (4.47). Moreover, using (3.25) we have

$$\int_0^t dr\{\int_0^{t-r} Z(\hat{s})g'(r)\,d\hat{s}\} \in D(A)$$

$$A\int_0^t dr\{\int_0^{t-r} Z(\hat{s})g'(r)\,d\hat{s}\} = \int_0^t A\{\int_0^{t-r} Z(\hat{s})g'(r)\,d\hat{s}\}dr$$

$$= \int_0^t \{Z(t-r)-I\}g'(r)\,dr \,.$$

Thus, (4.51') shows that $u(t) \in D(A)$ and

$$Au(t) = AZ(t)u_0 + \{Z(t)-I\}g(0) + \int_0^t \{Z(t-r)-I\}g'(r)\,dr$$

$$Au(t) = \{AZ(t)u_0 + Z(t)g(0) + \int_0^t Z(t-r)g'(r)\,dr\} - \{g(0) + \int_0^t g'(r)\,dr\}. \quad (4.53)$$

By comparing (4.52) with (4.53) we obtain

$$Au(t) = \frac{d}{dt}u(t) - \{g(0) + \int_0^t g'(r)\,dr\} = \frac{du}{dt} - g(t)$$

and $u(t)$ satisfies (4.2a).

If (a), (b), (c), and $(d2)$ hold, then we have directly from (4.49)

$$\frac{\mathrm{d}}{\mathrm{d}t}\, u(t) = AZ(t)u_0 + Z(0)g(t) + \int_0^t \left\{\frac{\partial}{\partial t}\, Z(t-s)g(s)\right\}\mathrm{d}s$$

$$= AZ(t)u_0 + g(t) + \int_0^t AZ(t-s)g(s)\,\mathrm{d}s$$

$$= A\{Z(t)u_0 + \int_0^t Z(t-s)g(s)\,\mathrm{d}s\} + g(t) = Au(t) + g(t)$$

where we used (3.26) and (3.25), and $u(t)$ satisfies (4.2a). Note that assumptions (a), (b), and (c) ensure that the integral on the right-hand side of (4.49) exists and that $u(t)$ also satisfies (4.2b), as is easy to verify.

Finally, $u(t)$ is the *unique* strict solution of the initial-value problem (4.2) if (a), (b), (c), and $(d1)$ or if (a), (b), (c), and $(d2)$ hold. In fact, if $w(t)$ is another strict solution of (4.2), then $U(t) = u(t) - w(t)$ is a strict solution of the homogeneous problem:

$$\frac{\mathrm{d}}{\mathrm{d}t}\, U(t) = AU(t), \quad t > 0; \qquad X\text{-}\lim_{t\to 0+} U(t) = \theta_X.$$

However, $U(t) \equiv \theta_X \; \forall\, t \geq 0$ satisfies the above system and it is its unique strict solution because of Theorem 4.7. Hence, $u(t) \equiv w(t) \; \forall\, t \geq 0$ and the proof of Theorem 4.8 is complete.

□

Assumptions (a), (b), and (c) do *not* ensure that (4.49) is a strict solution of system (4.2); however, (a), (b), (c), *and* $(d1)$, or (a), (b), (c), *and* $(d2)$, guarantee that the unique strict solution of (4.2) can be written into the form (4.49). Note that the first term on the right-hand side of (4.49) is the solution of the homogeneous problem (4.3) whereas the second term satisfies the non-homogeneous evolution equation (4.2a) and the initial condition (4.2b) with $u_0 = \theta_X$. This generalizes a well-known result of the elementary theory of linear differential equations (see Exercise 4.10).

Example 4.11. Oscillating sources

If $A \in \mathscr{G}(M, \beta; X)$, $u_0 \in D(A)$ and

$$g(t) = g_0 \exp(i\omega t), \quad g_0 \in X, \quad t \geq 0 \tag{4.54}$$

where $i = \sqrt{-1}$ and ω is a real number, then assumptions (*a*), (*b*), (*c*), and (*d*1) of Theorem 4.8 are satisfied and formula (4.49) can be used. However, since $g(t)$ has the simple form (4.54) ($g(t)$ is an *oscillating source*), an explicit expression can be derived for $u(t)$ if $z = i\omega \in \rho(A)$. To see this, consider the function

$$v(t) = v_0 \exp(i\omega t), \quad t \geq 0$$

where $v_0 \in X$ is *a priori* unknown. Substituting into (4.2a) we have

$$i\omega v_0 \exp(i\omega t) = Av_0 \exp(i\omega t) + g_0 \exp(i\omega t)$$

$$(i\omega I - A)v_0 = g_0, \quad v_0 = (i\omega I - A)^{-1} g_0$$

with $(i\omega I - A)^{-1} \in \mathscr{B}(X)$ because $i\omega \in \rho(A)$ by assumption; hence,

$$v(t) = \exp(i\omega t)\,(i\omega I - A)^{-1} g_0, \quad t \geq 0$$

satisfies (4.2a). However, $v(0) = (i\omega I - A)^{-1} g_0$ and so the strict solution of system (4.2) with $g(t)$ defined by (4.54) has the form

$$u(t) = Z(t)\{u_0 - (i\omega I - A)^{-1} g_0\} + \exp(i\omega t)(i\omega I - A)^{-1} g_0, \quad t \geq 0 \tag{4.55}$$

as it is easy to check. Note that the first term on the right-hand side of (4.55) is the solution of the homogeneous problem (4.3) with $\{u_0 - (i\omega I - A)^{-1} g_0\}$ instead of u_0. It also follows from (4.55) that

$$\| u(t) - \exp(i\omega t)(i\omega I - A)^{-1} g_0; X\| \leq \| Z(t)\| \, \| u_0 - (i\omega I - A)^{-1} g_0; X\|$$

$$\leq M \exp(\beta t)\{\| u_0;X\|+\| (i\omega I-A)\|^{-1} \| g_0;X\|\}$$

and so, if $\beta < 0$,

$$\lim_{t\to+\infty} \| u(t)-\exp(i\omega t)(i\omega I-A)^{-1}g_0;X\| = 0$$

i.e. $\exp(i\omega t)(i\omega I-A)^{-1}g_0$ is the *asymptotic form* of $u(t)$ as $t\to+\infty$. If X is a Hilbert space, the condition $\beta < 0$ means that the operator A is related to some form of dissipation of energy (see Remark 4.4 and Example 4.13). Note also that $(i\omega I-A)^{-1}$ may not exist or it may happen that $(i\omega I-A)^{-1}\notin\mathscr{B}(X)$ if $i\omega\notin\rho(A)$.

The above results are similar to those found in the theory of forced oscillations in mechanical systems. □

Example 4.12. Periodic solutions

Consider the non-homogeneous system

$$\frac{d}{dt} u(t) = Au(t)+g(t), \ t > 0 \tag{4.56a}$$

$$X\text{-}\lim_{t\to0+} u(t) = \hat{u} \tag{4.56b}$$

where $A\in\mathscr{G}(1,-a)$ with $a > 0$, $g(t)$ satisfies (c) and $(d1)$ or (c) and $(d2)$ of Theorem 4.8 and where $\hat{u}$ will be chosen in a suitable way. If system (4.56) has the strict solution $u(t)$, then it can be written into the form (4.49):

$$u(t) = Z(t)\hat{u} + \int_0^t Z(t-s)g(s)\,ds, \quad t \geq 0 \tag{4.57}$$

and so at $t = \hat{t}$

$$u(\hat{t}) = \hat{\gamma}+Z(\hat{t})\hat{u} \tag{4.58}$$

with

$$\hat{\gamma} = \int_0^{\hat{t}} Z(\hat{t}-s)g(s)\,ds$$

and where $\hat{t} > 0$ is *assigned*. We now prove that the element $\hat{u}$ can be chosen so that $u(\hat{t}) = \hat{u}$, i.e. so that

$$\hat{u} = \hat{\gamma}+Z(\hat{t})\hat{u} \tag{4.59}$$

(see for instance Barbu 1976, p.138). Since $A\in\mathscr{G}(1,-a)$ with $a>0$, then $\|Z(\hat{t})\| \le \exp(-a\hat{t}) = q<1$ and so (4.59) has one and only one solution $\hat{u}\in X$ because of the contraction-mapping theorem (see Example 2.12) with $D_0 = X = Y$ and with $A(f) = \hat{\gamma}+Z(\hat{t})f)$. Furthermore,

$$\hat{u} = X\text{-}\lim_{n\to\infty} \hat{u}_n \tag{4.60}$$

with

$$\hat{u}_{n+1} = \hat{\gamma}+Z(\hat{t})\hat{u}_n, \quad n = 0,1,2,\ldots; \quad \hat{u}_0 = \hat{\gamma}. \tag{4.61}$$

Now $\hat{u}_0 = \hat{\gamma}\in D(A)$ (see the proof of Theorem 4.8), and so $\hat{u}_1 = \hat{\gamma}+Z(\hat{t})u_0$ $\in D(A)$ because of (4.36c); by induction we conclude that $\hat{u}_n\in D(A)$ $\forall\, n = 0,1,2,\ldots$. Then, we have from (4.61)

$$A\hat{u}_{n+1} = A\hat{\gamma}+Z(\hat{t})A\hat{u}_n$$

$$A\hat{u}_{m+1}-A\hat{u}_m = Z(\hat{t})\,[A\hat{u}_m-A\hat{u}_{m-1}]$$

$$\|A\hat{u}_{m+1}-A\hat{u}_m;X\| \le \|Z(\hat{t})\|\|A\hat{u}_m-A\hat{u}_{m-1};X\|$$

and so

$$\|A\hat{u}_{m+1}-A\hat{u}_m;X\| \le q\|A\hat{u}_m-A\hat{u}_{m-1};X\|$$

$$\|A\hat{u}_{m+1}-A\hat{u}_m;X\| \le q^2\|A\hat{u}_{m-1}-A\hat{u}_{m-2};X\|$$

$$\cdots\cdots$$

$$\|A\hat{u}_{m+1}-A\hat{u}_m;X\| \le q^m\|A\hat{u}_1-A\hat{u}_0;X\|$$

with $q = \exp(-a\hat{t})<1$. Hence, if $m\ge n+1$,

$$\|A\hat{u}_m-A\hat{u}_n;X\|$$

$$\le \|A\hat{u}_m-A\hat{u}_{m-1};X\|+\|A\hat{u}_{m-1}-A\hat{u}_{m-2};X\|+\ldots+\|A\hat{u}_{n+1}-A\hat{u}_n;X\|$$

$$\leq \left[\sum_{j=n}^{m-1} q^j\right] \| A\hat{u}_1 - A\hat{u}_0; X \|$$

$$\leq \frac{q^n}{1-q} \| A\hat{u}_1 - A\hat{u}_0; X \| \to 0 \quad \text{as} \quad n\to\infty$$

and the sequence $\{\hat{u}_n\}$, which is contained in $D(A)$ and is convergent to $\hat{u}$, is such that $\{A\hat{u}_n\}$ is a Cauchy sequence in the B-space X. Then $\{A\hat{u}_n\}$ is also convergent in X and so $\hat{u}\in D(A)$ because of Theorem 2.2.

Going back to system (4.56), assume now that the element $\hat{u}$ on the right-hand side of (4.56b) is defined by (4.60), i.e. that $\hat{u}$ is the unique solution of (4.59). Then, $\hat{u}\in D(A)$, as we have proved above, and conditions (*a*), (*b*), (*c*), and (*d*1) (or (*a*), (*b*), (*c*), and (*d*2)) of Theorem 4.8 are satisfied. As a consequence, the unique strict solution $u(t)$ of system (4.56) is given by (4.57), $u(\hat{t})$ satisfies (4.58), and so $u(\hat{t}) = \hat{u}$ because of (4.59).

If we further assume that $g(t)$ is *periodic with period* $\hat{t}$

$$g(\hat{s}+\hat{t}) = g(\hat{s}) \quad \forall\, \hat{s} \geq 0 \tag{4.62}$$

then we have from (4.57) with $t = \tau+\hat{t}$, $\tau \geq 0$,

$$u(\tau+\hat{t}) = Z(\tau+\hat{t})\hat{u} + \int_0^{\tau+\hat{t}} Z(\tau+\hat{t}-s)g(s)\,ds$$

$$u(\tau+\hat{t}) = Z(\tau)Z(\hat{t})\hat{u} + \int_0^{\hat{t}} Z(\tau)Z(\hat{t}-s)g(s)\,ds + \int_{\hat{t}}^{\tau+\hat{t}} Z(\tau+\hat{t}-s)g(s)\,ds$$

$$= Z(\tau)\{Z(\hat{t})\hat{u} + \int_0^{\hat{t}} Z(\hat{t}-s)g(s)\,ds\} + \int_0^{\tau} Z(\tau-\hat{s})g(\hat{s}+\hat{t})\,d\hat{s}$$

with $\hat{s} = s-\hat{t}$. Using (4.57) with $t = \hat{t}$ and (4.62), we obtain

$$u(\tau+\hat{t}) = Z(\tau)u(\hat{t}) + \int_0^{\tau} Z(\tau-\hat{s})g(\hat{s})\,d\hat{s};$$

but $u(\hat{t}) = \hat{u}$ and so

$$u(\tau+\hat{t}) = Z(\tau)\hat{u} + \int_0^{\tau} Z(\tau-\hat{s})g(\hat{s})\,d\hat{s}$$

i.e.

$$u(\tau+\hat{t}) = u(\tau) \quad \forall \tau \geq 0 \tag{4.63}$$

and the strict solution of system (4.56) is also periodic with periodic $\hat{t}$. □

Example 4.13. Oscillating heat sources

If A is the heat-diffusion operator of Example 2.19

$$Af = k\frac{d^2f}{dx^2}$$

$$D(A) = \{f: f\in L^2(a,b); f''\in L^2(a,b); f(a) = f(b) = 0\}$$

then the system

$$\frac{du}{dt} = Au(t)+g_0 \exp(i\omega t), \quad t>0 \tag{4.64a}$$

$$X\text{-}\lim_{t\to 0+} u(t) = u_0 \tag{4.64b}$$

gives the evolution of the temperature $u(t) = u(x;t)$ in a rigid slab of thickness $b-a$ with an oscillating heat source $g = g_0 \exp(i\omega t)$. Note that g_0 is a given element of $L^2(a,b)$ and so $g_0 = g_0(x)$ with $\|g_0\|_2 = \{\int_a^b |g_0(x)|^2 dx\}^{1/2} < \infty$. Since $A\in\mathscr{G}(1,-2k\delta^{-2};L^2(a,b))$ with $\delta = b-a$ and $i\omega\in\rho(A)$ for any real ω (see (2.93) of Example 2.19), the results of Example 4.11 apply with $M = 1$, $\beta = -2k\delta^{-2} < 0$, and $X = L^2(a,b)$, provided that $u_0\in D(A)$. Thus, the unique strict solution of system (4.64) is given by (4.55) and $\exp(i\omega t)(i\omega I-A)^{-1}g_0$ is the asymptotic form of $u(t)$ as $t\to+\infty$. Further, if

$$u_0 = \hat{u} = (i\omega I-A)^{-1}g_0$$

then (4.55) gives

$$u(t) = \exp(i\omega t)(i\omega I-A)^{-1}g_0, \ \forall\ t\geq 0 \tag{4.65}$$

showing that $u(t)$ is periodic with period $\hat{t} = 2\pi/\omega$. This agrees with the results of Example 4.12. Note that $(i\omega I-A)^{-1}g_0$ is defined by (2.95) with $g_0(y)$ instead of $g(y)$ and with $\mu = (i\omega/k)^{1/2} =$

$i^{1/2}(\omega/k)^{1/2} = (\omega/2k)^{1/2}(1+i)$ and so (4.65) gives the temperature $u(t) = u(x;t)$ *explicitly* as a function of x and of t. □

EXERCISES

4.1. Prove the commutation formula $AT(t) = T(t)A$. Hint: $AT_n(t) = T_n(t)A \;\forall\, n = 1,2,\ldots$.

4.2. Prove inequality (4.9e). Hint: take the norm of both sides of (4.4).

4.3. Verify that $u(t)$ defined by (4.13) satisfies (4.10a) and (4.10b) with A given by (4.12).

4.4. Show that the integral operator

$$Af = \int_0^1 \exp(x-y)f(y)\,dy\;, \quad f\in X = C([0,1])$$

belongs to $\mathscr{B}(X)$ with $\|A\| \le \exp(1)-1$ and that it generates the group

$$T(t) = I+\{\exp(t)-1\}A\;.$$

Hint: prove that $A^j f = Af \;\forall\, j = 1,2,3,\ldots$.

4.5. Prove that $\mathscr{G}(M,\beta) \subset \mathscr{G}(M_1,\beta_1)$ if $M_1 \ge M$ and $\beta_1 \ge \beta$.

4.6. If $A\in\mathscr{G}(1,\beta;X)$ where X is a Hilbert space, then the non-homogeneous initial-value problem (4.2a)+(4.2b) has at most one strict solution $u = u(t)$. Hint: assume that another strict solution $w = w(t)$ exists and then use the energy method of Remark 4.4 for $\delta(t) = u(t)-w(t)$.

4.7. Let $B(t)$ and $B_1(t)$ belong to $\mathscr{B}(X)$ $\forall\, t\in[t_1,t_2]$ and assume that $u(t) = B(t)f$ and $u_1(t) = B_1(t)g$ are continuous at $t_0\in(t_1,t_2)$ in the sense of $\|\cdot;X\|$, $\forall\, f,g\in X$. Prove that $w(t) = B(t)B_1(t)g$, $g\in X$, is continuous at t_0.

4.8. Prove that the domain $D(A^2)$ of the operator A^2 is dense in X if

$A\in\mathcal{G}(1,0;X)$. Hint: if $z>1$, then $(zI-A)^{-1}\in\mathcal{B}(X)$ and its range

$$R((zI-A)^{-1}) = D(zI-A) = D(A)$$

is dense in X. Thus, if $f_0\in X$ and $\varepsilon>0$, a suitable $f\in R((zI-A)^{-1})$ can be found such that $\|f-f_0;X\| < \varepsilon/2$. However, f belongs to the range of $(zI-A)^{-1}$ and so $f = (zI-A)^{-1}g$ with $g\in X$. Further, since $D(A)$ is dense in X, a $g_1\in D(A)$ exists such that $\|g_1-g;X\| < \varepsilon/2$. Hence, if we put $f_1 = (zI-A)^{-1}g_1$, then we have

$$\|f_1-f_0;X\| \le \|f_1-f;X\|+\|f-f_0;X\|$$

$$= \|(zI-A)^{-1}[g_1-g];X\|+\|f-f_0;X\|$$

$$\le \frac{1}{z}\|g_1-g;X\|+\|f-f_0;X\| < \frac{\varepsilon}{2}+\frac{\varepsilon}{2} = \varepsilon$$

because of (4.17c) with $\beta = 0$ and $z>1$. On the other hand, $f_1 = (zI-A)^{-1}g_1$ with $g_1\in D(A)$ belongs to $D(A^2)$ because ... .

4.9. Prove that (4.25a) holds with M instead of 1 if $A\in\mathcal{G}(M,0)$.

4.10. Discuss formula (4.49) under the assumptions $X = R^1$, $A = \alpha I$ with α a given real number.

4.11. Prove (4.36e). Hint: if $f\in D(A) = D(A_1)$ and $t\ge 0$, then

$$\frac{d}{dt}\{Z(t)f\} = \beta\exp(\beta t)S(t)f+\exp(\beta t)\frac{d}{dt}\{S(t)f\}.$$

5

PERTURBATION THEOREMS

5.1. INTRODUCTION

Consider the heat-conducting slab S of section 4.1 and assume that the temperature $\tau(x;t)$ in S satisfies the system

$$\frac{\partial}{\partial t}\tau(x;t) = k\frac{\partial^2}{\partial x^2}\tau(x;t) + \int_a^b h(x,y)\tau(y;t)\,dy, \quad a<x<b, \quad t>0 \tag{5.1a}$$

$$\tau(a;t) = \tau(b;t) = 0, \quad t>0 \tag{5.1b}$$

$$\tau(x;\ 0) = \tau_0(x), \quad a<x<b \tag{5.1c}$$

where $h(x,y)$ is a given function, defined on the square $[a,b]\times[a,b]$. Note that the second term on the right-hand side of (5.1a) is a temperature-dependent source implying that the temperature at $x\in S$ is affected by the temperature at any other $y\in S$ through the 'coupling' function $h(x,y)$. If the heat-diffusion operator A is defined as in section 4.1 and if we put

$$Bf = \int_a^b h(x,y)f(y)\,dy \tag{5.2}$$

then the abstract version of system (5.1) reads as follows

$$\frac{d}{dt}u(t) = Au(t)+Bu(t), \quad t>0, \tag{5.3a}$$

$$X\text{-}\lim_{t\to 0+} u(t) = u_0. \tag{5.3b}$$

In (5.3a) and in (5.3b), $u(t) = \tau(x;t)$ must be interpreted as a function from $[0,+\infty)$ into the B-space $X = L^1(a,b)$ (see Example 3.4) and $A\in\mathcal{G}(1,0;L^1(a,b))$ (see Example 4.5). As far as the operator B is concerned, it is easy to verify that $B\in\mathcal{B}(L^1(a,b))$ if the kernel $h(x,y)$ is continuous over the square $[a,b]\times[a,b]$ (see Exercise 2.4).

System (5.3) provides a simple example of a *bounded perturbation*: the operator $(A+B)$ on the right-hand side of

(5.3a) is obtained from the *unperturbed* A by adding the *perturbing* operator B, which is said to be a *bounded perturbation* because $B\in\mathscr{B}(L^1(a,b))$. More generally, consider the initial-value problem (5.3a)+(5.3b) and assume that $A\in\mathscr{G}(M,\beta;X)$ and $B\in\mathscr{B}(X)$. Then the question arises whether or not the operator $(A+B)$ is still the generator of a semigroup and, if it is, how the *perturbed* semigroup $\exp\{t(A+B)\}$ is related to the *unperturbed* $\exp(tA)$. Section 5.2 will be devoted to answer such a question, which is quite crucial for the applications because it is often difficult to prove directly that $(A+B)$ generates a semigroup whereas it may be relatively simple to show that $A\in\mathscr{G}(M,\beta;X)$ and $B\in\mathscr{B}(X)$.

5.2. BOUNDED PERTURBATIONS

Theorem 5.1. If $A\in\mathscr{G}(M,\beta;X)$ *and* $B\in\mathscr{B}(X)$, *then* (*a*) $(A+B)\in\mathscr{G}(M,\beta+M\|B\|;X)$ *and* (*b*) *the perturbed semigroup* $\{\hat{Z}(t) = \exp\{t(A+B)\},\ t\geq 0$ *is given by the following successive approximation method*

$$\hat{Z}(t)f = X\text{-}\lim_{n\to\infty} Z^{(n)}(t)f,\quad t\geq 0,\quad f\in X, \tag{5.4}$$

with

$$\left.\begin{aligned} Z^{(n+1)}(t)f &= Z(t)f + \int_0^t Z(t-s)BZ^{(n)}(s)f\,ds,\quad n=0,1,2,\ldots \\ Z^{(0)}(t)f &= Z(t)f \end{aligned}\right\} \tag{5.5}$$

where $\{Z(t) = \exp(tA), t\geq 0\}$ *is the semigroup generated by* A.

Proof.
(*a*) For simplicity, we shall only consider the case $A\in\mathscr{G}(1,0;X)$, $B\in\mathscr{B}(X)$ (if $A\in\mathscr{G}(M,\beta;X)$, see Kato 1966, p.495). Now, since A satisfies (4.17a), (4.17b), and (4.17c) with $\beta = 0$ and since $B\in\mathscr{B}(X)$, $A+B\in\mathscr{C}(X)$ because of (*b*) of Theorem 2.3 and $D(A+B)$ is dense in X because $D(A+B) = D(A)\cap D(B) = D(A)\cap X = D(A)$ (see Remark 4.5). Furthermore,

$$\|\{zI-(A+B)\}^{-1}\| \le (z-\|B\|)^{-1}, \quad \forall z > \|B\|$$

because of (2.98) with $\alpha_0 = 0$. Hence, $A+B\in\mathcal{G}(1,\|B\|;X)$ and the semigroup $\{\hat{Z}(t) = \exp\{t(A+B)\}, t\ge 0\}$ is uniquely defined and satisfies (4.36a)-(4.36f) with $M = 1$ and $\beta = \|B\|$.

(*b*) If $u_0\in D(A+B) = D(A)$, then the unique strict solution of the initial-value problem (5.3) has the form

$$u(t) = \hat{Z}(t)u_0 = \exp\{t(A+B)\}u_0, t\ge 0 \tag{5.6}$$

and it is continuously differentiable at any $t\ge 0$ because of Theorem 4.7. Then, the function

$$g(t) = Bu(t), \quad t\ge 0$$

is strongly continuous and its derivative $g'(t) = Bu'(t)$ is also strongly continuous at any $t\ge 0$ (see Exercise 3.8), and the second term on the right-hand side of (5.3a) can be interpreted as a source $g = g(t)$ satisfying conditions (*c*) and (*d*1) of Theorem 4.8. Hence, $u(t)$ is also the strongly continuous solution of the integral equation

$$u(t) = Z(t)u_0 + \int_0^t Z(t-s)Bu(s)\,ds, \quad t\ge 0 \tag{5.7}$$

that follows from (4.49) with $g(s) = Bu(s)$. However, for a reason that will be clear from what follows, we shall rather consider here the equation

$$v(t) = Z(t)f + \int_0^t Z(t-s)Bv(s)\,ds, \quad t\ge 0 \tag{5.8}$$

where f is *any* given element of X. Note that (5.8) coincides with (5.7) if $f = u_0\in D(A+B) = D(A)$. The unique continuous solution of the integral equation (5.8) can be obtained by the usual method of successive approximations, (see for instance section 3.4):

$$v(t) = X\text{-}\lim_{n\to\infty} v_n(t), \quad t\ge 0, \tag{5.9}$$

$$\left.\begin{aligned} v_{n+1}(t) &= Z(t)f + \int_0^t Z(t-s)Bv_n(s)\,ds, \quad n = 0,1,2,\ldots \\ v_0(t) &= Z(t)f, \quad t \geq 0. \end{aligned}\right\} \tag{5.10}$$

We have from (5.10) with $n = 0$:

$$v_1(t) = Z(t)f + \int_0^t Z(t-s)BZ(s)f\,ds, \quad t \geq 0$$

and $v_1(t)$ is uniquely determined by f, i.e. we can put

$$v_1(t) = Z^{(1)}(t)f, \quad t \geq 0$$

where the operator $Z^{(1)}(t)$ is such that

$$Z^{(1)}(t)f = Z(t)f + \int_0^t Z(t-s)BZ(s)f\,ds, \quad t \geq 0, \quad \forall f \in X.$$

Hence, $Z^{(1)}(t) \in \mathscr{B}(X)$ $\forall t \geq 0$ because $Z^{(1)}(t)$ is obviously linear and

$$\begin{aligned} \| Z^{(1)}(t)f; X\| &\leq \| Z(t)\| \| f; X\| + \int_0^t \| Z(t-s)\| \| B\| \| Z(s)\| \| f; X\| \,ds \\ &\leq \| f; X\| + \| B\| \int_0^t ds \| f; X\| = \left(1 + \frac{\| B\| t}{1!}\right) \| f; X\|, \quad \forall f \in X \end{aligned}$$

since $\| Z(t)\| = \| \exp(tA)\| \leq 1$. By induction, we conclude that $v_n(t)$ is uniquely determined by f:

$$v_n(t) = Z^{(n)}(t)f, \quad t \geq 0 \tag{5.11}$$

and that $Z^{(n)}(t) \in \mathscr{B}(X)$ $\forall t \geq 0$ with

$$\| Z^{(n)}(t)f; X\| \leq \left\{ \sum_{j=0}^{n} \frac{(\| B\| t)^j}{j!} \right\} \| f; X\|, \quad \forall t \geq 0, \ f \in X. \tag{5.12}$$

Then, (5.9) shows that $v(t)$ is uniquely determined by f:

$$v(t) = \tilde{Z}(t)f, \quad t \geq 0 \tag{5.13}$$

and that $\tilde{Z}(t) \in \mathscr{B}(X)$ $\forall t \geq 0$ with

$$\|\tilde{Z}(t)f;X\| \le \exp(\|B\|t)\|f;X\|, \quad \forall t \ge 0, \quad f \in X \tag{5.14}$$

as can easily be verified on passing to the limit as $n \to \infty$ in (5.12).

Now if in particular $f = u_0 \in D(A+B)$, then $v(t) = \tilde{Z}(t)u_0$ is the unique continuous solution of equation (5.7) and consequently $u(t) = v(t) = \tilde{Z}(t)u_0 \ \forall t \ge 0$. However, $u(t)$ is given by (5.6) and so

$$\hat{Z}(t)u_0 = \tilde{Z}(t)u_0 \quad \forall t \ge 0, \quad \forall u_0 \in D(A+B). \tag{5.15}$$

Relation (5.15) can be extended to the whole space X because $D(A+B)$ is a dense subset (see Example 2.6 and Exercise 5.1):

$$\hat{Z}(t)f = \tilde{Z}(t)f \quad \forall t \ge 0, \quad \forall f \in X \tag{5.16}$$

and (5.5) is proved since (5.9), (5.10), (5.11), and (5.13) give

$$\tilde{Z}(t)f = X\text{-}\lim_{n\to\infty} Z^{(n)}(t)f$$

$$\left.\begin{aligned} Z^{(n+1)}(t)f &= Z(t)f + \int_0^t Z(t-s)BZ^{(n)}(s)f\,ds, \quad n = 0,1,2, \\ Z^{(0)}(t)f &= Z(t)f. \end{aligned}\right\}$$

This concludes the proof of Theorem 5.1. □

Remark 5.1.

If $A \in \mathcal{G}'(M,\beta;X)$ and $B \in \mathcal{B}(X)$, then $(A+B) \in \mathcal{G}'(M,\beta+M\|B\|;X)$. □

Example 5.1. An integral perturbation of the heat-diffusion operator

If $h(x,y)$ is continuous over the square $[a,b]\times[a,b]$, then the integral operator B defined by (5.2) has domain $D(B) = L^1(a,b)$; furthermore,

$$\|Bf\|_1 = \int_a^b dx \left| \int_a^b h(x,y)f(y)\,dy \right| \le \int_a^b dx \int_a^b |h(x,y)|\,|f(y)|\,dy$$

$$\le \hat{h} \int_a^b dx \int_a^b |f(y)|\,dy = (b-a)\hat{h}\, \|f\|_1$$

with $\hat{h} = \max\{|h(x,y)|,\ a \le x,y \le b\}$, and so $B \in \mathscr{B}(L^1(a,b))$ with $\|B\| \le (b-a)\hat{h}$. On the other hand, the heat-diffusion operator A belongs to $\mathscr{G}(1,0;L^1(a,b))$ (see Remark 2.12 and Example 4.5). Hence,

$$A+B \in \mathscr{G}(1,\|B\|;L^1(a,b)) \subset \mathscr{G}(1,(b-a)\hat{h};\ L^1(a,b))$$

(see Exercise 4.5).

Similarly, if $h = h(x,y) \in L^2((a,b)\times(a,b))$, then $B \in \mathscr{B}(L^2(a,b))$ with

$$\|B\| \le \beta_0 = \left[\int_a^b dx \int_a^b |h(x,y)|^2 dy\right]^{1/2}$$

(see Example 2.4). Further, if the heat-diffusion operator A is defined as in Example 2.19, then

$$A \in \mathscr{G}(1,-2k\delta^{-2};L^2(a,b))$$

with $\delta = b-a$ (see Example 4.5). Hence,

$$A+B \in \mathscr{G}(1,-2k\delta^{-2}+\beta_0;L^2(a,b))$$

and the unique strict solution of the abstract version (5.3) of system (5.1) reads as follows:

$$u(t) = \hat{Z}(t)u_0 = \exp\{t(A+B)\}u_0, \quad t \ge 0, \quad u_0 \in D(A) \tag{5.17}$$

and so

$$\|u(t)\|_2 = \left[\int_a^b |\tau(x;t)|^2 dx\right]^{1/2} \le \|\hat{Z}(t)\|\,\|u_0\|_2$$

$$\le \exp\{(\beta_0-2k\delta^{-2})t\}\left\{\int_a^b |\tau_0(x)|^2 dx\right\}^{1/2}, \quad t \ge 0 \tag{5.18}$$

because of (4.36a) with $M = 1$, $\beta = \beta_0-2k\delta^{-2}$. Note that

$$\lim_{t\to+\infty} \|u(t)\|_2 = 0$$

if $\beta_0 < 2k\delta^{-2}$, i.e. basically if the heat energy produced per unit time by the temperature-dependent source (5.2) is less than the heat energy lost by the slab S per unit time through the boundary planes $x=a$ and $x=b$. □

Example 5.2. $A\in\mathscr{G}(1,0;l^1)$, $B\in\mathscr{B}(l^1)$

The operator A of Example 2.17 belongs to $\mathscr{G}(1,0;l^1)$ (see Example 4.6); moreover, if

$$B = \begin{pmatrix} b_{11}, & b_{12}, & \cdots \\ b_{21}, & b_{22}, & \cdots \\ \cdots, & \cdots, & \cdots \end{pmatrix}$$

with

$$\sum_{i=1}^{\infty} |b_{ij}| \le b, \quad \forall j = 1,2,\ldots$$

then $B\in\mathscr{B}(l^1)$ and $\|B\| \le b$ (see Example 2.3). Hence

$$A+B = \begin{pmatrix} b_{11}, & b_{12}, & b_{13}, & \cdots \\ b_{21}, & b_{22}-1, & b_{23}, & \cdots \\ b_{31}, & b_{32}, & b_{33}-2, & \cdots \\ \cdots, & \cdots, & \cdots, & \cdots \end{pmatrix} \in\mathscr{G}(1,b;l^1) \qquad \square$$

Example 5.3. $A\in\mathscr{G}(M,\beta;X)$, $B = zB_0$ *with* $B_0\in\mathscr{B}(X)$ *and* $z\in\mathbb{C}$.

Assume that $A\in\mathscr{G}(M,\beta;X)$ and that $B = zB_0$, where $B_0\in\mathscr{B}(X)$ and z is a complex parameter. Then, $\|B\| = |z|\|B_0\|$ and so $A+B\in\mathscr{G}(M,\beta+|z|M\|B_0\|;X)$, because of Theorem 5.1. Further, we have from (5.5) with $n = 0$ and with $n = 1$

$$Z^{(1)}(t)f = Z(t)f+z\int_0^t Z(t-s)B_0Z(s)f ds = Z(t)f+z\hat{Z}^{(1)}(t)f$$

where

$$\hat{Z}^{(1)}(t)f = \int_0^t Z(t-s)B_0Z(s)f ds \;;$$

$$Z^{(2)}(t)f = Z(t)f+z\int_0^t Z(t-s)B_0Z^{(1)}(s)fds =$$

$$= Z(t)f+z\int_0^t Z(t-s)B_0\{Z(s)f+z\hat{Z}^{(1)}(s)f\}ds$$

$$= Z(t)f+z\hat{Z}^{(1)}(t)f+z^2\hat{Z}^{(2)}(t)f$$

where

$$\hat{Z}^{(2)}(t)f = \int_0^t Z(t-s)B_0\hat{Z}^{(1)}(s)fds \ .$$

Thus, if $B = zB_0$, relations (5.5) give

$$Z^{(n+1)}(t)f = Z(t)f+z\hat{Z}^{(1)}(t)f+\ldots+z^{n+1}\hat{Z}^{(n+1)}(t)f$$

$$\left.\begin{aligned} \hat{Z}^{(n+1)}(t)f &= \int_0^t Z(t-s)B_0\hat{Z}^{(n)}(s)fds, \quad n = 0,1,\ldots \\ \hat{Z}^{(0)}(t)f &= Z(t)f \end{aligned}\right\} \quad (5.19)$$

and so

$$\exp\{t(A+zB_0)\}f = \sum_{j=0}^{\infty} z^j\hat{Z}^{(j)}(t)f, \quad f\in X, \quad t\geq 0 \quad (5.20)$$

because of (5.4). Formula (5.20) shows that the semigroup generated by $A+zB_0$ is a holomorphic function of the complex variable z, defined for any $z\in\mathbb{C}$ (see section 3.5). □

The symbol $\exp\{t(A+B)\}$ is often used for the semigroup $\hat{Z}(t)$ generated by $A+B$, with $A\in\mathscr{G}(M,\beta;X)$ and $B\in\mathscr{B}(X)$. This does *not* usually imply that $\exp\{t(A+B)\}f = \exp(tA)\exp(tB)f$. However, if A and B *commute*, i.e. if

$$Bf\in D(A), \quad \forall f\in D(A); \quad ABf = BAf, \quad \forall f\in D(A) \quad (5.21)$$

then

$$\exp\{t(A+B)\}f = \exp(tA)\exp(tB)f = \exp(tB)\exp(tA)f \quad (5.22)$$

$\forall f \in X$, $t \geq 0$, where $\exp(tB) = \sum_{j=0}^{\infty} (Bt)^j/j!$ because $B \in \mathscr{B}(X)$ (see section 4.2).

To see this, we first note that the operator $A_1 = A - \beta I$ belongs to $\mathscr{G}(M,0;X)$ (see section 4.6) and commutes with B because $D(A) = D(A_1)$ and

$$A_1 Bf = ABf - \beta Bf = BAf - \beta Bf$$

$$= B(A-\beta I)f, \quad \forall f \in D(A_1)$$

i.e.

$$A_1 Bf = BA_1 f, \quad \forall f \in D(A_1). \tag{5.23}$$

Then, if $t \geq 0$ and $n = 1,2,\ldots$, we obtain from (5.23)

$$\left(I - \frac{t}{n} A_1\right) Bf = B\left(I - \frac{t}{n} A_1\right) f, \quad \forall f \in D(A_1)$$

$$Bf = \left(I - \frac{t}{n} A_1\right)^{-1} B\left(I - \frac{t}{n} A_1\right) f, \quad \forall f \in D(A_1)$$

and so, if $g = (I-(t/n)A_1)f$, we obtain

$$B\left(I - \frac{t}{n} A_1\right)^{-1} g = \left(I - \frac{t}{n} A_1\right)^{-1} Bg \tag{5.24}$$

where g is any element belonging to the range $R(I-(t/n)A_1) = D((I-(t/n)A_1)^{-1}) = X$, because $(I-(t/n)A_1)^{-1} \in \mathscr{B}(X)$. In other words, the commutation formula (5.24) holds for *any* $g \in X$. It then follows from (5.24) that

$$BZ_n(t)g = Z_n(t)Bg, \quad \forall g \in X, \quad t \geq 0$$

with $Z_n(t)$ given by (4.22), and on passing to the X-limit as $n \to +\infty$

$$B \exp(tA_1)g = \exp(tA_1)Bg, \ \forall g \in X, \quad t \geq 0.$$

Finally, since $\exp(tA) = \exp(\beta t)\exp(tA_1)$ because of definition (4.35), we obtain

$$B\exp(tA)g = \exp(tA)Bg, \quad \forall g\in X, \quad t\geq 0. \tag{5.25}$$

Hence, if $A\in\mathscr{G}(M,\beta;X)$ and $B\in\mathscr{B}(X)$ commute according to (5.21), then the semigroup $Z(t) = \exp(tA)$ and B commute as well, because of (5.25). Now, we have from (5.5)

$$\begin{aligned} Z^{(1)}(t)f &= Z(t)f + \int_0^t Z(t-s)Z(s)Bf\,\mathrm{d}s \\ &= Z(t)f + Z(t)\int_0^t Bf\,\mathrm{d}s \\ &= Z(t)[I+tB]f = [I+tB]Z(t)f \end{aligned}$$

$$\begin{aligned} Z^{(2)}(t)f &= Z(t)f + \int_0^t Z(t-s)BZ(s)[I+sB]f\,\mathrm{d}s \\ &= Z(t)f + Z(t)\int_0^t [B+sB^2]f\,\mathrm{d}s \\ &= Z(t)[I+tB+\frac{t^2}{2}B^2]f \\ &= [I+tB+\frac{t^2}{2}B^2]Z(t)f. \end{aligned}$$

where we used (5.25), (4.36f), and (3.25), ($Z(t)\in\mathscr{B}(X)\subset\mathscr{C}(X)$). By induction we obtain

$$Z^{(n)}(t)f = Z(t)\sum_{j=0}^{n}\frac{(tB)^j}{j!}f = \left\{\sum_{j=0}^{n}\frac{(tB)^j}{j!}\right\}Z(t)f$$

and (5.22) is proved on passing to the X-limit as $n\to\infty$ and using (5.4).

5.3. THE CASES $B = B(t)\in\mathscr{B}(X)$ AND B RELATIVELY BOUNDED

Consider the initial-value problem

$$\frac{\mathrm{d}}{\mathrm{d}t}u(t) = Au(t)+B(t)u(t), \quad t>0 \tag{5.26a}$$

$$X\text{-}\lim_{t\to 0+} u(t) = u_0 \tag{5.26b}$$

and assume that $A\in\mathscr{G}(M,\beta;X)$, $u_0\in D(A)$, and $\{B(t), t\geq 0\}\subset\mathscr{B}(X)$ with $\phi(t) = B(t)f$ strongly continuous $\forall t\geq 0$ for *any* given

$f \in X$. Note that $\phi(t) = B(t)f$ is assumed to be continuous in t in the sense of $\|\cdot;X\|$ and so $B = B(t)$ does not need to be continuous in the sense of $\|\cdot;\mathscr{B}(X)\|$ (see Example 3.5). If system (5.26) has the strict solution $u = u(t)$, $t \geq 0$, then

$$g(t+h)-g(t) = B(t+h)u(t+h)-B(t)u(t)$$

$$= B(t+h)\{u(t+h)-u(t)\}+\{B(t+h)-B(t)\}u(t)$$

$$\|g(t+h)-g(t);X\| \leq \|B(t+h)\|\,\|u(t+h)-u(t);X\| +$$

$$\|\{B(t+h)-B(t)\}u(t);X\| \to 0$$

as $h \to 0$ because $\|B(t+h)\|$ is bounded (see Exercise 5.4). Hence, $g(t) = B(t)u(t)$ is strongly continuous and consequently $u(t)$ is also the strongly continuous solution of the integral equation

$$u(t) = Z(t)u_0 + \int_0^t Z(t-s)B(s)u(s)\,\mathrm{d}s, \quad t \geq 0 \tag{5.27}$$

because of Theorem 4.8. Furthermore, if $\phi(t) = B(t)f$ is also strongly differentiable with a strongly continuous derivative at any $t \geq 0$, then each strongly continuous solution of (5.27) is a strict solution of (5.26). This can be proved by a procedure similar to that of the converse part of Theorem 4.8.

The above results can be summarized as follows.

Theorem 5.2. If $A \in \mathscr{G}(M,\beta;X)$, $u_0 \in D(A)$, *and the one-parameter family of linear operators* $\{B(t), t \geq 0\}$ *is such that (i)* $B(t) \in \mathscr{B}(X)$ $\forall t \geq 0$ *and (ii)* $\phi(t) = B(t)f$ *is strongly continuous* $\forall t \geq 0$ *for any* $f \in X$, *then any strict solution of system (5.26) is also a strongly continuous solution of (5.27). If in addition (iii)* $\phi(t)$ *is strongly differentiable and its derivative is strongly continuous at any* $t \geq 0$, *then any strongly continuous solution of (5.27) is a strict solution of system (5.26).* □

If in particular $B(t)$ is continuous in the sense of

$\|\cdot\| = \|\cdot;\mathscr{B}(X)\|$, then assumptions (i) and (ii) are satisfied, (see Example 3.5); hence, each strict solution $u(t)$ of (5.26) is a continuous solution of (5.27) and so

$$\|u(t);X\| \le M\exp(\beta t)\|u_0;X\| + M\int_0^t \exp\{\beta(t-s)\}\|B(s)\|\,\|u(s);X\|\,ds$$

i.e.

$$\gamma(t) \le \gamma_0 + M\int_0^t \|B(s)\|\gamma(s)\,ds, \quad s \ge 0$$

with

$$\gamma(t) = \|u(t);X\|\exp(-\beta t), \quad \gamma_0 = M\|u_0;X\|.$$

Using Gronwall's inequality (Lemma 3.2), we obtain

$$\gamma(t) \le \gamma_0 \exp\{M\int_0^t \|B(s)\|\,ds\}, \quad t \ge 0$$

and consequently

$$\|u(t);X\| \le M\|u_0;X\|\exp\{\beta t + M\int_0^t \|B(s)\|\,ds\}, \quad t \ge 0 \tag{5.28}$$

which gives an upper bound for the norm of each strict solution of system (5.26). Note that, if B belongs to $\mathscr{B}(X)$ and is s independent, (5.28) gives

$$\|u(t);X\| \le M\exp\{(\beta + M\|B\|)t\}\|u_0;X\|, \quad t \ge 0$$

which agrees with the results of Theorem 5.1.

Remark 5.2.

The results stated in Theorem 5.2 are important because the integral equation (5.7) is usually easier to study than the system (5.26). □

Example 5.4. The integral operator $B(t)$ of Exercise 3.3

The integral operator $B(t)$ of Exercise 3.3 is continuous in the norm of $\mathscr{B}(C([a,b]))$ and so $\phi(t) = B(t)f$ satisfies (i) and (ii) of Theorem 5.2 with $X = C([a,b])$. □

Let A and B be operators with domain and range contained in the B-space X; B is said to be *A-bounded* or *relatively bounded with respect to A* if

$$\left. \begin{aligned} D(A) &\subset D(B) \\ \|Bf;X\| &\leq a\|f;X\| + b\|Af;X\|, \quad \forall f \in D(A) \end{aligned} \right\} \tag{5.29}$$

where a and b are non-negative constants, independent of f. We have the following theorem (Kato 1966, p.499).

Theorem 5.3. Let $A \in \mathscr{G}(1,\beta;X)$, $B \in \mathscr{G}(1,\beta';X)$; *if B is A-bounded and if the constant b can be chosen smaller than 1/2, then* $A+B \in \mathscr{G}(1,\beta+\beta';X)$. *If X is a Hilbert space, then it is sufficient that* $b < 1$. □

The proof of Theorem 5.3 will not be given here; however, Example 5.5 shows that B is in some sense 'less singular' than A, if B is A-bounded.

Example 5.5. Relative boundedness of the convection operator with respect to the heat-diffusion operator

If

$$Af = k\frac{\mathrm{d}^2 f}{\mathrm{d}x^2}, \quad D(A) = \{f : f \in L^2(a,b); f'' \in L^2(a,b); f(a) = f(b) = 0\}$$

$$Bf = -v\frac{\mathrm{d}f}{\mathrm{d}x}, \quad D(B) = \{f : f \in L^2(a,b); f' \in L^2(a,b); f(a) = 0\}$$

(see Examples 2.18 and 2.19), then $D(A) \subset D(B)$ and we have for any $f \in D(A)$:

$$0 \leq [\|Bf\|_2]^2 = v^2 \int_a^b |f'(x)|^2 \mathrm{d}x = v^2 \int_a^b f'(x)\overline{f'(x)}\, \mathrm{d}x$$

$$= -v^2 \int_a^b f''(x)\overline{f(x)}\, \mathrm{d}x$$

because $f \in D(A)$ and so $f(a) = f(b) = 0$. Using Schwartz inequality (1.8) we obtain

$$[\|Bf\|_2]^2 \leq v^2 \|f''\|_2 \|f\|_2 \quad \forall f \in D(A).$$

However,

$$\|f''\|_2 \|f\|_2 \leq \frac{1}{4n} [n\|f\|_2 + \|f''\|_2]^2, \quad n = 1,2,\ldots$$

as it is easy to verify, and so

$$\|Bf\|_2 \leq \frac{v}{2} [\sqrt{n}\|f\|_2 + \frac{1}{\sqrt{n}} \|f''\|_2]$$

$$\|Bf\|_2 \leq \frac{v}{2} \sqrt{n}\|f\|_2 + \frac{v}{2k\sqrt{n}} \|Af\|_2, \quad n = 1,2,\ldots, \quad f \in D(A). \tag{5.30}$$

Thus, the convection operator B is bounded with respect to the heat-diffusion operator A, with $a = v\sqrt{n}/2$ and $b = v/2k\sqrt{n}$, where n is any positive integer. On the other hand, $A \in \mathscr{G}(1,-2k(b-a)^{-2};L^2(a,b))$, $B \in \mathscr{G}(1,0;L^2(a,b))$ (see Examples 4.4 and 4.5) and $v/2k\sqrt{n} < 1$ if we choose n large enough. Hence A and B satisfy the assumptions of Theorem 5.3 with $X = L^2(a,b)$, $\beta = -2k(a-b)^{-2}$, and $\beta' = 0$, and so $A+B \in \mathscr{G}(1,-2k(b-a)^{-2};L^2(a,b))$. Note that A and $A+B$ belong to the same class $\mathscr{G}(1,-2k(b-a)^{-2}; L^2(a,b))$, i.e. the convection operator does not change the 'energy dissipation power' of the heat-diffusion operator, characterized by the negative constant $\beta = -2k(b-a)^{-2}$ (see Remark 4.4). □

5.4. THE SEMILINEAR CASE

In section 3.4 we studied the simple non-linear initial-value problem (3.28) with $F \in \mathrm{Lip}(X,X)$; since $\mathscr{B}(X) \subset \mathrm{Lip}(X,X)$ because of (2.21), system (3.28) may be considered as the simplest non-linear analogue of (4.3) with $A \in \mathscr{B}(X)$. Similarly, the following *semilinear* initial-value problem

$$\frac{d}{dt} u(t) = Au(t) + F(u(t)), \quad t > 0 \tag{5.31a}$$

$$X\text{-}\lim_{t \to 0+} u(t) = u_0 \tag{5.31b}$$

with $A \in \mathscr{G}(M,\beta;X)$ and $F \in \mathrm{Lip}(D,X)$, may be considered as a generalization of (5.3a)+(5.3b) since system (5.31) coincides with (5.3) if $F\in\mathscr{B}(X)\subset \mathrm{Lip}(X,X)$. Note that (5.31a) is said to be a *semilinear equation of evolution* because the non-linear operator F on the right-hand side of (5.31a) belongs to the class $\mathrm{Lip}(D,X)$ and so it is 'well behaved', i.e. in some sense its properties are similar to those of an operator belonging to the class $\mathscr{B}(X)$.

Example 5.6. The non-linear temperature-dependent source $F(\tau) = \mu\tau^2(x;t)\{1+\tau^2(x;t)\}^{-1}$

Consider the heat-conduction problem of section 5.1 and assume that the integral term on the right-hand side of (5.1a) is substituted by the following non-linear temperature-dependent source:

$$F(\tau) = \mu \frac{[\tau(x;t)]}{1+[\tau(x;t)]^2} \tag{5.32}$$

where μ is a constant (see Example 2.9). Then, $F\in\mathrm{Lip}(L^2(a,b), L^2(a,b))$ and (5.1) becomes a semilinear problem of the form (5.31). □

To be more specific, we list the assumptions that will be used to study the semilinear problem (5.31):

$$A\in\mathscr{G}(M,\beta;X); \tag{5.33a}$$

(5.33b): F has domain $D(F)$ and range $R(F)$ contained in X and an *open* and *convex*† set $D\subset D(F)$ exists such that $F\in\mathrm{Lip}(D,X)$ with $\nu(F)=\gamma$:
$\|F(f_1)-F(f)\| \le \gamma\|f_1-f\| \quad \forall f_1, f\in D$;

(5.33c): $F(f)$ is Fréchet differentiable at any $f\in D$ and its Fréchet derivative F_f is such that $\|F_f g\| \le \gamma_1\|g\|$ $\forall f\in D$, $\forall g\in X$, where γ_1 is a positive constant independent of f and g;

† D is *convex* if $\alpha f+(1-\alpha)g\in D$ $\forall f, g\in D$, $\forall \alpha\in[0,1]$.

$\| F_{f_1} g - F_f g \| \to 0$ as $\| f_1 - f \| \to 0$ $\forall g \in X$, with $f_1, f \in D$; (5.33d)

u_0 belongs to both $D(A)$ and $D: u_0 \in D(A) \cap D$. (5.33e)

The function $u(t)$, $t \in [0, t_0]$ is said to be a *strict solution* of system (5.31) *over* $[0, t_0]$ if

(*a*) $u(t) \in D(A) \cap D(F)$ $\forall t \in [0, t_0]$,
(*b*) $u(t)$ is strongly continuous $\forall t \in [0, t_0]$ and strongly differentiable $\forall t \in (0, t_0]$,
(*c*) $u(t)$ satisfies (5.31a) $\forall t \in (0, t_0]$ and (5.31b).

(Compare this definition with (i)+(ii)+(iii) of Section 4.1.)

Theorem 5.4. Under assumptions (5.33), the semilinear initial-value problem (5.31) has a unique strict solution $u(t)$ *over* $[0, \hat{t}]$ *if* $\hat{t}$ *is suitably chosen; moreover,* $u(t) \in D(A) \cap D$ $\forall t \in [0, \hat{t}]$. □

Since the proof of Theorem 5.4 is rather lengthy, we shall divide it into several Lemmas. Note first that (4.2a) and (4.49) with $g(t) = F(u(t))$ *suggest* to transform system (5.31) into the integral equation

$$u(t) = Z(t) u_0 + \int_0^t Z(t-s) F(u(s)) ds \tag{5.34}$$

where $Z(t) = \exp(tA)$ is the semigroup generated by A.

Lemma 5.1. Under assumptions (5.33a,b,e), the integral equation (5.34) has a unique solution $u(t)$ *that is continuous in the sense of the norm of* X *and belongs to* D *at any* $t \in [0, \hat{t}]$, *provided that* $\hat{t}$ *is suitably small.*

Proof. The *closed* sphere

$$\sigma_0 = \{ f : f \in X ; \| f - u_0 \| \le r \}$$

has centre $u_0 \in D(A) \cap D$ because of (5.33e), radius r, and it is contained in the *open* sphere

$$\sigma_1 = \{f: f \in X; \| f - u_0 \| < r_1\}$$

with $r < r_1$. Hence, the radius r_1 can be chosen so that

$$\sigma_0 \subset \sigma_1 \subset D$$

since D is an open subset of X because of (5.33b) (see Example 1.5). Then the B-space X_1, the closed subset $\Sigma_0 \subset X_1$ and the non-linear operator Q are defined as follows:

$$X_1 = C([0,\hat{t}];X), \ \| v;X_1 \| = \max\{\| v(t) \|, t \in [0,\hat{t}]\}$$

$$\left.\begin{aligned} \Sigma_0 &= \{v: v \in X_1; v(t) \in \sigma_0 \quad \forall\, t \in [0,\hat{t}]\} \\ Q(v)(t) &= Z(t)u_0 + \int_0^t Z(t-s)F(v(s))\,ds \\ D(F) &= \Sigma_0, \ R(F) \subset X_1 \end{aligned}\right\} \qquad (5.35)$$

(see (3.12) and the discussion that follows). Note that Q maps each $v \in \Sigma_0$ into $w = Q(v) \in X_1$; in fact, if $v \in \Sigma_0$, then $v(t) \in \sigma_0 \subset D \subset D(F) \ \ \forall\, t \in [0,\hat{t}]$ and so

$$\| F(v(s+h)) - F(v(s)) \| \leq \gamma \| v(s+h) - v(s) \|, \ \forall\, s, s+h \in [0,\hat{t}]$$

because of (5.33b). Hence, $F(v(t))$ as well as $w(t) = Q(v)(t)$ are continuous in the sense of $\| \cdot \| \forall t \in [0,\hat{t}]$ and consequently $w \in X_1$. Now using (5.33a,b) we have from (5.35) for any v, $v_1 \in \Sigma_0$ and for any $t \in [0,\hat{t}]$

$$\begin{aligned} \| Q(v)(t) - Q(v_1)(t) \| &\leq \gamma M \int_0^t \exp\{\beta(t-s)\} \| v(s) - v_1(s) \| \,ds \\ &\leq \gamma M \| v - v_1; X_1 \| \int_0^t \exp\{\beta(t-s)\}\,ds \\ &= \gamma M \| v - v_1; X_1 \| \ \frac{\exp(\beta t) - 1}{\beta} \end{aligned}$$

because

$$\| v - v_1; X_1 \| = \max\{\| v(s) - v_1(s) \|, s \in [0,\hat{t}]\}$$

and so

$$\|Q(v)-Q(v_1);X_1\| = \max\{\|Q(v)(t)-Q(v_1)(t)\|, t\in[0,\hat{t}]\}$$

$$\leq \gamma M \frac{\exp(\beta\hat{t})-1}{\beta}\|v-v_1;X_1\| \quad \forall v,v_1\in\Sigma_0 = D(Q) \qquad (5.36)$$

since $\{\exp(\beta\hat{t})-1\}/\beta = \max\{\{\exp(\beta t)-1\}/\beta, t\in[0,\hat{t}]\}$. In an analogous way, we obtain from (5.35)

$$\|Q(v)(t)-u_0\| \leq \|Z(t)u_0-u_0\| + M\int_0^t \exp\{\beta(t-s)\}\|F(v(s))\|\,ds$$

$$\leq \|Z(t)u_0-u_0\| + M\{\gamma r+\|F(u_0)\|\}\frac{\exp(\beta t)-1}{\beta}$$

since

$$\|F(v(s))\| \leq \|F(v(s))-F(u_0)\|+\|F(u_0)\|$$

$$\leq \gamma\|v(s)-u_0\|+\|F(u_0)\|$$

$$\leq \gamma r+\|F(u_0)\|, \quad \forall s\in[0,\hat{t}]$$

because of (5.33b) and because $v(s)\in\sigma_0$ $\forall s\in[0,\hat{t}]$. Thus

$$\|Q(v)-u_0;X_1\| \leq \max\{\|Z(t)u_0-u_0\|, \quad t\in[0,\hat{t}]\}+$$

$$M[\gamma r+\|F(u_0)\|]\frac{\exp(\beta\hat{t})-1}{\beta}\forall v\in\Sigma_0. \qquad (5.37)$$

Note that (5.36) and (5.37) hold even if $\beta = 0$ if we substitute $\hat{t}$ for $\{\exp(\beta\hat{t})-1\}/\beta$.

If we put

$$q(\hat{t}) = \frac{1}{r}\max\{\|Z(t)u_0-u_0\|, \quad t\in[0,\hat{t}]\}+$$

$$\frac{M}{r}[\gamma r+\|F(u_0)\|]\frac{\exp(\beta\hat{t})-1}{\beta} \qquad (5.38)$$

then we have from (5.36) and from (5.37)

$$\|Q(v)-Q(v_1);X_1\| \leq q(\hat{t})\|v-v_1;X_1\| \quad \forall v,v_1\in\Sigma_0 \qquad (5.39)$$

$$\|Q(v)-u_0;X_1\| \leq q(\hat{t})r \quad \forall v\in\Sigma_0. \tag{5.40}$$

However, $q(\hat{t}) \to 0+$ as $t \to 0+$ since $\{\exp(\beta\hat{t})-1\}/\beta \to 0+$ and $\|Z(t)u_0-u_0\| \to 0+$ as $t \to 0$ because of (4.36d), and so $0 < q(t) < 1$ if $\hat{t}$ is small enough. Then, (5.39) shows that the non-linear operator Q is strictly contractive over Σ_0, whereas (5.40) proves that Q maps the closed set Σ_0 into itself because

$$\|Q(v)(t)-u_0\| \leq Q(v)-u_0;X_1\| \leq q(\hat{t})r < r \quad \forall t\in[0,\hat{t}]$$

and $Q(v)(t)\in\sigma_0$ $\forall t\in[0,\hat{t}]$, i.e. $Q(v)\in\Sigma_0$. We conclude that the equation

$$u = Q(u) \tag{5.41}$$

has a unique solution $u\in\Sigma_0 \subset X_1 = C([0,\hat{t}];X)$ because of the contraction mapping theorem (see Example 2.12). Thus, $u = u(t)$ is continuous in the sense of $\|\cdot\|$, belongs to $\sigma_0 \subset D \subset D(F)$ for any $t\in[0,\hat{t}]$, and is the unique continuous solution of (5.34) because of the definition (5.35). □

The unique continuous solution $u(t)$, $t\in[0,\hat{t}]$ of the integral equation (5.34) is sometimes called a *mild solution* of system (5.31) and usually has a precise physical meaning since the physical principles from which the evolution equation (5.31a) is derived are often best understood if they are expressed in an integral form.

To proceed with the proof of Theorem 5.4, we transform equation (5.34) by using the substitution $s' = t-s$:

$$u(t) = Z(t)u_0 + \int_0^t Z(s')F(u(t-s'))\mathrm{d}s'. \tag{5.34'}$$

Differentiating *formally* both sides of (5.34') with respect to t and letting $u_1(t) = \mathrm{d}u(t)/\mathrm{d}t$, we obtain

$$u_1(t) = Z(t)\{Au_0+F(u_0)\} + \int_0^t Z(s')F_{u(t-s')}u_1(t-s')\mathrm{d}s' \tag{5.42}$$

because of (4.36d), (3.22), (3.26), and because the chain

rule for Fréchet derivatives gives (see Exercise 3.10),

$$\frac{\partial}{\partial t} F(u(t-s')) = F_{u(t-s')} \frac{\mathrm{d}}{\mathrm{d}t} u(t-s') = F_{u(t-s')} u_1(t-s')$$

where $F_{u(t-s')}$ is the Fréchet derivative of F at $u(t-s')$. Using again the substitution $s' = t-s$, we have from (5.42)

$$u_1(t) = Z(t)\{Au_0+F(u_0)\} + \int_0^t Z(t-s)F_{u(s)} u_1(s)\mathrm{d}s. \tag{5.43}$$

Note that (5.43) is a *linear* integral equation, similar to (5.27) with $\{Au_0+F(u_0)\}$ instead of u_0 and with $B(s) = F_{u(s)} \in \mathscr{B}(X)$ because of (5.33c).

Remark 5.3.

The form (5.34') of equation (5.34) allows us to obtain the linear equation (5.43) for the formal derivative $u_1(t)$ of $u(t)$, whereas formal differentiation of (5.34) gives

$$u_1(t) = Z(t)[Au_0+F(u_0)] + \int_0^t AZ(t-s)F(u(s))\mathrm{d}s$$

which may be meaningless because $F(u(s))$ does not necessarily belong to $D(A)$ (see (4.36e)). □

Lemma 5.2. Under assumptions (5.33a,c,d,e), the linear integral equation (5.34) has a unique strongly continuous solution $u_1 = u_1(t)$ *defined over* $[0,\hat{t}]$.

Proof. The solution of equation (5.43) can be sought by the usual method of successive approximations:

$$\left.\begin{aligned} u_1^{(n+1)}(t) &= Z(t)\{Au_0+F(u_0)\} + \int_0^t Z(t-s)F_{u(s)}u_1^{(n)}(s)\mathrm{d}s, n = 0,1,\ldots, \\ u_1^{(0)}(t) &= Z(t)[Au_0+F(u_0)],\ t\in[0,\hat{t}] \ . \end{aligned}\right\} \tag{5.44}$$

Note that $u_1^{(n+1)}(t)$ is a strongly continuous function of $t\in[0,\hat{t}\,]$ because $u_1^{(0)}(t)$ has such a property and because of (5.33c,d). Moreover, we have from (5.44) for any $t \in [0,\hat{t}]$

$$\|u_1^{(0)}(t)\| \le M\exp(\beta t)\|Au_0+F(u_0)\| \le M_0\|Au_0+F(u_0)\|$$

where $M_0 = M\exp(\beta\hat{t})$ if $\beta > 0$ and $M_0 = M$ if $\beta \le 0$, and so

$$\|u_1^{(1)}(t)-u_1^{(0)}(t)\| \le \int_0^t M\exp\{\beta(t-s)\}\|F_{u(s)}u_1^{(0)}(s)\|\,ds$$

$$\le M_0\gamma_1\int_0^t \|u_1^{(0)}(s)\|\,ds \le (M_0\gamma_1)M_0\|Au_0+F(u_0)\|\int_0^t ds$$

$$= M_0(M_0\gamma_1 t)\|Au_0+F(u_0)\|$$

$$\|u_1^{(2)}(t)-u_1^{(1)}(t)\| \le \int_0^t M\exp\{\beta(t-s)\}\|F_{u(s)}\{u_1^{(1)}(s)-u_1^{(0)}(s)\}\|\,ds$$

$$\le M_0\gamma_1\int_0^t \|u_1^{(1)}(s)-u_1^{(0)}(s)\|\,ds$$

$$\le M_0(M_0\gamma_1)^2\|Au_0+F(u_0)\|\int_0^t s\,ds$$

$$= M_0\frac{(M_0\gamma_1 t)^2}{2}\|Au_0+F(u_0)\|$$

because of (5.33a,c). From this point on, the proof is similar to that after (3.31) of section 3.4. □

The following Lemma is of a rather technical character and it will be used in the proof of Lemma 5.4.

Lemma 5.3. Under assumptions (5.33b,c,d) we have

$$F(\hat{f})-F(f) = H_{\hat{f},f}[\hat{f}-f] \quad \forall\, \hat{f},f \in D \tag{5.45}$$

where $H_{\hat{f},f}$ *is the linear operator defined by*

$$H_{\hat{f},f}g = \int_0^1 F_{f+z(\hat{f}-f)}\,g\,dz \qquad g \in X \tag{5.46}$$

and it is such that

$$\| H_{\hat{f},f} g \| \le \gamma_1 \| g \| \quad \forall \hat{f}, f \in D, \; g \in X \tag{5.47a}$$

$$\| H_{\hat{f},f} g - F_f g \| \to 0 \quad \text{as} \quad \| \hat{f} - f \| \to 0 \tag{5.47b}$$

with $\hat{f}, f \in D$, $g \in X$.

Proof. If $\hat{f}$ and f belong to D, then $f_1 = f + z(\hat{f} - f) = z\hat{f} + (1-z)f \in D$ $\forall z \in [0,1]$ since D is a convex set because of (5.33b) and so

$$\phi(z) = F(f + z(\hat{f} - f)), \qquad z \in [0,1] \tag{5.48}$$

is a function from the closed interval $[0,1]$ into the B-space X: $\phi(z) \in R(F) \subset X$ $\forall z \in [0,1]$. Using the chain rule for Fréchet derivatives we have from (5.48)

$$\frac{d}{dz} \phi(z) = F_{f_1} \frac{d}{dz} f_1 = F_{f+z(\hat{f}-f)} [\hat{f} - f] \tag{5.49}$$

where F_{f_1} is the Fréchet derivative of F at $f_1 = f + z(\hat{f} - f)$ (which exists because $f_1 \in D$ and because of (5.33c)) and where $\hat{f} - f = df_1/dz$. Note that $\phi(z)$ and $d\phi(z)/dz$ are continuous in $z \in [0,1]$ in the sense of $\| \cdot \|$ because of (5.33b,d). Then, integrating (5.49) from 0 to 1 we obtain

$$\phi(1) - \phi(0) = \int_0^1 \{ F_{f+z(\hat{f}-f)} [\hat{f} - f] \} dz$$

and (5.45) is proved because $\phi(1) = F(\hat{f})$ and $\phi(0) = F(f)$. Further, we have from (5.46) and from (5.33c)

$$\| H_{\hat{f},f} g \| \le \int_0^1 \| F_{f+z(\hat{f}-f)} g \| dz$$

$$\le \gamma_1 \| g \| \int_0^1 dz = \gamma_1 \| g \|$$

and (5.47a) is proved. Finally, since $F_f g$ does not depend on the parameter z, definition (5.46) gives

$$H_{\hat{f},f} g - F_f g = \int_0^1 F_{f+z(\hat{f}-f)} g dz - \int_0^1 F_f g dz$$

and so

$$\| H_{\hat{f},f}g - F_f g\| \le \int_0^1 \| F_{f+z(\hat{f}-f)}g - F_f g \| \, dz \ . \tag{5.50}$$

However,

$$\| F_{f_1} g - F_f g\| < \varepsilon, \quad f \in D, \quad g \in X$$

for any $f_1 \in D$ such that $\| f_1 - f\| < \delta = \delta(\varepsilon, f, g)$ because of (5.33d). Thus, if we choose $\hat{f}$ and f in D so that $\| \hat{f} - f \| < \delta$, then for any $z \in [0,1]$

$$\| [f+z(\hat{f}-f)] - f \| = \| z(\hat{f}-f)\| = z\| \hat{f}-f \| \le \| \hat{f}-f \| < \delta$$

and consequently

$$\| F_{f+z(\hat{f}-f)}g - F_f g \| < \varepsilon \quad \forall\, z \in [0,1].$$

Substituting the preceding inequality into (5.50) we have

$$\| H_{\hat{f},f}g - F_f g\| \le \int_0^1 \varepsilon dz = \varepsilon$$

provided that $\| \hat{f} - f\| < \delta$ and (5.74b) is proved. □

Lemma 5.4. Under assumptions (5.33a)-(5.33e), the strongly continuous solution $u(t)$ *of (5.34) is strongly differentiable and* $du(t)/dt = u_1(t) \;\forall\, t \in [0,\hat{t}]$, *where* $u_1(t)$ *is the strongly continuous solution of the linear equation (5.43).*

Proof. If $t \in [0,\hat{t}), t+h \in [0,\hat{t})$ with $h > 0$, then we have from (5.34')

$$u(t+h) - u(t) = Z(t)\{Z(h) - I\}u_0 + \int_t^{t+h} Z(s')F(u(t+h-s'))ds' +$$

$$\int_0^t Z(s')\{F(u(t+h-s')) - F(u(t-s'))\}ds'$$

and also

$$u(t+h)-u(t) = Z(t)[\{Z(h)-I\}u_0+\int_0^h Z(h-\hat{s})F(u(\hat{s}))d\hat{s}] +$$

$$\int_0^t Z(t-s)\{F(u(s+h))-F(u(s))\}ds \tag{5.51}$$

with $\hat{s} = t+h-s'$ and with $s = t-s'$. Hence, if

$$J(t;h) = \frac{1}{h}\{u(t+h)-u(t)\}-u_1(t), \quad h \neq 0 \tag{5.52}$$

$$J_0(t;h) = Z(t)\left[\frac{1}{h}\{Z(h)-I\}u_0+\frac{1}{h}\int_0^h Z(h-\hat{s})F(u(\hat{s}))d\hat{s}-\right.$$

$$\left.Au_0-F(u_0)\right], \quad h \neq 0 \tag{5.53}$$

then we obtain from (5.51) and from (5.43)

$$J(t;h) = J_0(t;h)+\int_0^t Z(t-s)\left[\frac{1}{h}\{F(u(s+h))-F(u(s))\}-F_{u(s)}u_1(s)\right]ds. \tag{5.54}$$

However, (5.45) and (5.52) give

$$\frac{1}{h}\{F(u(s+h))-F(u(s))\} = H_{u(s+h),u(s)}\left\{\frac{u(s+h)-u(s)}{h}\right\}$$

$$= H_{u(s+h),u(s)}J(s;h) + H_{u(s+h),u(s)}u_1(s)$$

and substituting into (5.54) we have

$$J(t;h) = J_0(t;h) + \int_0^t Z(t-s)\{H_{u(s+h),u(s)}u_1(s)-F_{u(s)}u_1(s)\}ds+$$

$$\int_0^t Z(t-s)H_{u(s+h),u(s)}J(s;h)\,ds$$

and also

$$\|J(t;h)\| \le \|J_0(t;h)\| + \hat{M}\int_0^{\hat{t}} \|H_{u(s+h),u(s)}u_1(s)-F_{u(s)}u_1(s)\|ds +$$

$$\hat{M}\gamma_1\int_0^t \|J(s;h)\|ds, \quad \forall t\in[0,\hat{t}) \tag{5.55}$$

because of (5.47a) and because $\|Z(t')\| \leq M \exp(\beta t') \leq \hat{M}$, $\forall\, t' \in [0,\hat{t}]$, with $\hat{M} = M \exp(\beta\hat{t})$ if $\beta > 0$ and with $\hat{M} = M$ if $\beta \leq 0$. On the other hand, we have from (5.53)

$$\|J_0(t;h)\| \leq \hat{M}\Big\{\|\frac{1}{h}\{Z(h)-I\}u_0 - Au_0\| + \|\frac{1}{h}\int_0^h Z(h-\hat{s})F(u(\hat{s}))\,d\hat{s} - F(u_0)\|\Big\}$$

with

$$X\text{-}\lim_{h\to 0+} \frac{1}{h}\{Z(h)-I\}u_0 = \Big\{\frac{d}{dt} Z(t)u_0\Big\}_{t=0} = Z(0)Au_0 = Au_0$$

because $u_0 \in D(A) \cap D$ (see (4.36e) with $t = 0$), and with

$$X\text{-}\lim_{h\to 0+} \frac{1}{h}\int_0^h Z(h-\hat{s})F(u(\hat{s}))\,d\hat{s} = Z(0)F(u(0)) = F(u_0)$$

(see Exericse 5.8). Hence,

$$\|J_0(t;h)\| \leq \hat{M}\varepsilon \quad \forall\, t \in [0,\hat{t}) \tag{5.56}$$

provided that $0 < h < \delta_1 = \delta_1(\varepsilon)$. Finally, consider the second term on the right-hand side of (5.55); for each $s \in [0,\hat{t})$, $\|u(s+h)-u(s)\| \to 0$ as $h \to 0+$ because $u(s)$ is strongly continuous and so

$$b(s;h) = \|H_{u(s+h),u(s)}u_1(s) - F_{u(s)}u_1(s)\| \to 0 \quad \text{as} \quad h \to 0+$$

because of (5.47b) with $\hat{f} = u(s+h)$, $f = u(s)$, and $g = u_1(s)$. Moreover,

$$\begin{aligned} 0 \leq b(s;h) &\leq \|H_{u(s+h),u(s)}u_1(s)\| + \|F_{u(s)}u_1(s)\| \\ &\leq 2\gamma_1\|u_1(s)\| \leq 2\gamma_1 \max\{\|u_1(s)\|, s \in [0,\hat{t}]\} \\ &= 2\gamma_1\|u_1;X_1\| = \alpha_0 \quad \forall\, s \in [0,\hat{t}), \quad h \geq 0 \end{aligned}$$

because of (5.47a) and (5.33c). It follows that

$$\lim_{h\to 0+} \int_0^{\hat{t}} \|H_{u(s+h),u(s)}\, u_1(s) - F_{u(s)}u_1(s)\|\,ds = 0$$

by the *bounded-convervence* theorem† and so

$$\int_0^{\hat{t}} \| H_{u(s+h),u(s)}u_1(s) - F_{u(s)}u_1(s) \| \, ds < \varepsilon \tag{5.57}$$

if $0 < h < \delta_2 = \delta_2(\varepsilon)$. Inequalities (5.55), (5.56), and (5.57) give

$$\| J(t;h) \| \leq 2\hat{M}\varepsilon + \gamma_1 \hat{M} \int_0^t \| J(s;h) \| \, ds \quad \forall\, t \in [0,\hat{t}) \tag{5.58}$$

provided that $0 < h < \delta_0 = \delta_0(\varepsilon) = \min\{\delta_1(\varepsilon), \delta_2(\varepsilon)\}$ and that $t+h \in [0,\hat{t})$. Using Gronwall's inequality (Lemma 3.2), we obtain from (5.58) for any $t \in [0,\hat{t})$

$$\| J(t;h) \| \leq 2\hat{M}\varepsilon \, \exp(\gamma_1 \hat{M} t)$$

and so $\quad \| J(t;h) \| \leq [2\hat{M} \exp(\gamma_1 \hat{M} \hat{t})]\varepsilon \,, \quad 0 < h < \delta_0(\varepsilon)$

i.e. $\quad \lim_{h \to 0+} \| J(t;h) \| = 0 .$

Since a similar result can be derived if $t \in (0,\hat{t}]$, $t+h \in (0,\hat{t}]$ with $h < 0$, Lemma 5.4 is proved because of (5.52). □

Lemma 5.5. Under assumptions (5.33a)-(5.33e), the unique strongly continuous solution $u(t)$ of (5.34) is such that (a) $u(t) \in D(A) \cap D \;\; \forall\, t \in [0,\hat{t}]$, (b) $u(t)$ satisfies (5.31a) and (5.31b).

Proof. The proof of this Lemma is similar to that of the converse part of Theorem 4.8 with $g(t) = F(u(t))$. (*a*) Since $u(t)$ is differentiable with $du(t)/dt = u_1(t)$ because of Lemma 5.4, the chain rule gives

$$\frac{d}{d\hat{s}} F(u(\hat{s})) = F_{u(\hat{s})} \left\{ \frac{d}{d\hat{s}} u(\hat{s}) \right\} = F_{u(\hat{s})} \, u_1(\hat{s}) \tag{5.59}$$

† The following version of the *bounded-convergence* theorem is needed (Royden 1963, p.200); if the real function $b(s;h), s \in [0,\hat{t}],\ h \in [0,h_1]$, is such that (i) $b(s;h)$ is measurable as a function of $s \,\forall h \in [0,h_1]$, (ii) $|b(s;h)| \leq \alpha_0$ for a.e. $s \in [0,\hat{t}]$, $h \in [0,h_1]$, and (iii) $\lim b(s;h) = b_0(s)$ as $h \to 0+$ for a.e. $s \in [0,\hat{t}]$, then $\lim_{h\to 0+} \int_0^{t} b(s;h)\,ds = \int_0^{\hat{t}} b_0(s)\,ds$.

and $F_{u(\hat{s})}\ u_1(\hat{s})$ is strongly continuous in $\hat{s}$ because $u(\hat{s})$ and $u_1(\hat{s})$ have such a property and because of (5.33b,d). Hence, integrating from 0 to s we obtain

$$F(u(s))-F(u_0) = \int_0^s F_{u(\hat{s})}\ u_1(\hat{s})\,d\hat{s}$$

and (5.34) can be written as follows:

$$u(t) = Z(t)u_0 + \int_0^t Z(t-s)F(u_0)\,ds + \int_0^t \left\{Z(t-s) \int_0^s F_{u(\hat{s})}u_1(\hat{s})\,d\hat{s}\right\}ds$$

and interchanging the order of integrations

$$u(t) = Z(t)u_0 + \int_0^t Z(t-s)F(u_0)\,ds + \int_0^t \left\{\int_{\hat{s}}^t Z(t-s)F_{u(\hat{s})}u_1(\hat{s})\,ds\right\}d\hat{s}\ . \tag{5.60}$$

However, Lemma 4.1 with $f = [F_{u(\hat{s})}u_1(\hat{s})]$, $t' = 0$ and with $t-\hat{s}$ instead of t gives

$$\int_{\hat{s}}^t Z(t-s)F_{u(\hat{s})}u_1(\hat{s})\,ds = \int_0^{t-\hat{s}} Z(s')\left\{F_{u(\hat{s})}u_1(\hat{s})\right\}ds' \in D(A)$$

and

$$A\int_{\hat{s}}^t Z(t-s)F_{u(\hat{s})}u_1(\hat{s})\,ds = A\int_0^{t-\hat{s}} Z(s')\left\{F_{u(\hat{s})}u_1(\hat{s})\right\}ds'$$
$$= \{Z(t-\hat{s})-I\}F_{u(\hat{s})}u_1(\hat{s}).$$

Thus, the double integral on the right-hand side of (5.60) belongs to $D(A)$ because of (3.25) and

$$A\int_0^t \{\int_s^t Z(t-s)F_{u(\hat{s})}\ u_1(\hat{s})\,ds\}\ d\hat{s} = \int_0^t A\{\int_s^t Z(t-s)F_{u(\hat{s})}u_1(\hat{s})\,ds\}\,ds$$
$$= \int_0^t \{Z(t-\hat{s})-I\}F_{u(\hat{s})}\ u_1(\hat{s})\ d\hat{s}\ . \tag{5.61}$$

Moreover, using again Lemma 4.1 we have

$$\int_0^t Z(t-s)F(u_0)\,ds = \int_0^t Z(s')F(u_0)\,ds' \in D(A) \qquad \text{and}$$

$$A\int_0^t Z(t-s)F(u_0)\,ds = A\int_0^t Z(s')F(u_0)\,ds' = \{Z(t)-I\}F(u_0) \tag{5.62}$$

and so $u(t)\in D(A)\ \forall t\in[0,t]$ since the term $Z(t)u_0$ of (5.60) also belongs to $D(A)$ because $u_0 \in D(A) \cap D$ and because of (4.36e). However, $u(t) \in D \; \forall t \in [0,\hat{t}]$ because of Lemma 5.1 and consequently $u(t) \in D(A) \cap D$ for any $t \in [0,\hat{t}]$.

(*b*) We have from (5.60), (5.61), and (5.62):

$$Au(t) = Z(t)Au_0 + \{Z(t)-I\}F(u_0) + \int_0^t \{Z(t-\hat{s})-I\}F_{u(\hat{s})}u_1(\hat{s})\,d\hat{s}$$

and using (5.43)

$$Au(t) = u_1(t) - F(u_0) - \int_0^t F_{u(\hat{s})}u_1(\hat{s})\,d\hat{s}..$$

If we take into account that $u_1(t) = du(t)/dt$ and use relation (5.59), we obtain

$$Au(t) = \frac{d}{dt}u(t) - F(u_0) - \int_0^t \frac{d}{d\hat{s}}F(u(\hat{s}))\,d\hat{s}$$

$$Au(t) = \frac{d}{dt}u(t) - F(u(t))$$

and $u(t)$ satisfies (5.31a). Finally, we have from (5.34)

$$\|u(t)-u_0\| \leq \|Z(t)u_0-u_0\| + \hat{M}\int_0^t \|F(u(s))\|\,ds \to 0$$

as $t \to 0+$ and $u(t)$ also satisfies (5.31b). This completes the proof of Lemma 5.5. □

Theorem 5.4 follows from Lemmas 5.4 and 5.5.

Remark 5.4.

The proof of Theorem 5.4 follows the lines of that given by I.Segal (1963) under more general assumptions on F. □

5.5. GLOBAL SOLUTION OF THE SEMILINEAR PROBLEM (5.31)

Under assumptions (5.33), the unique strict solution $u(t)$ of the semilinear problem (5.31) exists over the interval $[0,\hat{t}]$ with $\hat{t}$ *suitably small* and it is said to be a solution *local in time* to point out that $u(t)$ is defined over the 'small' interval $[0,\hat{t}]$. Now if the 'large' interval $[0,t_0]$ is given *a priori*, then system (5.31) may have no strict solution defined over the whole $[0,t_0]$ because for instance $u(t)$ may tend to 'escape' from the set D over which (5.33b, c,d) hold. The following assumptions guarantee that a unique strict solution $u(t)$ of (5.31) exists over the whole $[0,t_0]$; then, $u(t)$ is said to be a *global* solution of (5.31).

$$A\in \mathcal{G}(M,\beta;X); \tag{5.63a}$$

$$\left.\begin{array}{l}\|F(f_1)-F(f)\| \le \alpha(\|f_1\|,\|f\|)\|f_1-f\| \quad \forall f_1,f\in D(F)=X,\\ \text{where } \alpha(\rho_1,\rho_2) \text{ is a non-decreasing function of } \rho_1\\ \text{and of } \rho_2;\end{array}\right\} \tag{5.63b}$$

$$\left.\begin{array}{l}F(f) \text{ is Fréchet differentiable at any } f\in D(F)=X\\ \text{and its Fréchet derivative } F_f \text{ is such that}\\ \|F_f g\| \le \alpha_1(\|f\|)\|g\| \; \forall f,g\in X, \text{ where } \alpha_1(\rho)\\ \text{is a non-decreasing function of } \rho;\end{array}\right\} \tag{5.63c}$$

$$\|F_{f_1}g-F_f g\| \to 0 \quad \text{as} \quad \|f_1-f\|\to 0 \quad \forall g,f_1,f\in X \tag{5.63d}$$

$$u_0\in D(A); \tag{5.63e}$$

$$\left.\begin{array}{l}\textit{if} \text{ a strict solution } w=w(t) \text{ of (5.31) exists over}\\ [0,t_1]\subset[0,t_0], \text{ then } \|w(t)\|\le\eta \;\forall t\in[0,t_1], \text{ where}\\ \eta \text{ is a suitable constant that may depend on } u_0 \text{ and}\\ t_0 \text{ (and, without loss of generality, such that}\\ \|u_0\|\le\eta).\end{array}\right\} \tag{5.63f}$$

Note that $D(F) = X$ and that assumptions (5.63a)-(5.63e) ensure existence and uniqueness of a local strict solution of (5.31). In fact, if we let $D = \sigma = \{f: f\in X; \|f-u_0\| < r\}$, with $r > 0$ arbitrarily fixed, then we have for any f, $f_1\in D$

$$\|f\| \le \|f-u_0\|+\|u_0\| < r+\|u_0\|, \quad \|f_1\| < r+\|u_0\|$$

and so

$$\alpha(\|f_1\|,\|f\|) \leq \alpha(r+\|u_0\|,\ r+\|u_0\|) = \gamma$$

$$\alpha_1(\|f\|) \leq \alpha_1(r+\|u_0\|) = \gamma_1, \quad \forall f_1, \quad f\in D$$

and (5.33b,c,d,e) follow from (5.63b,c,d,e).

We also remark that (5.63f) follows from (5.63a,b) if $\alpha(\|f_1\|,\|f\|) \leq \hat{\alpha}\ \forall f_1, f\in X$. To see this, we recall that, if $w(t)$ is a strict solution of (5.31) over $[0,t_1]$, then $w(t)$ satisfies the integral equation (5.34) $\forall t\in[0,t_1]$ and so

$$\|w(t)\| \leq M_0\|u_0\|+M_0 \int_0^t [\|F(w(s))-F(\theta_X)\|+\|F(\theta_X)\|]ds$$

where $M_0 = M\exp(\beta t_0)$ if $\beta > 0$ and $M_0 - M$ if $\beta \leq 0$. Hence,

$$\|w(t)\| \leq M_0[\|u_0\|+t_0\|F(\theta_X)\|]+M_0\hat{\alpha} \int_0^t \|w(s)\|ds$$

and Lemma 3.2 gives

$$\|w(t)\| \leq M_0[\|u_0\|+t_0\|F(\theta_X)\|]\exp(M_0\hat{\alpha}t)$$

$$\leq M_0[\|u_0\|+t_0\|F(\theta_X)\|]\exp(M_0\hat{\alpha}t_0) = \eta \quad \forall t\in[0,t_1]\subset[0,t_0].$$

Theorem 5.5. Under assumptions (5.63), the semilinear initial-value problem (5.31) has a unique strict solution u(t) defined over the whole $[0,t_0]$.

Proof. We divide the proof into two parts: (*a*) existence of a unique strict solution of (5.31) over $[0,\hat{t}]$; (*b*) existence of a unique strict solution of (5.31) over $[0,t_0]$.

(*a*) As in the proof of Theorem 5.4, we first consider the 'integral version' (5.34) of system (5.3]) and define the closed sphere $S_1 \subset X_1 = C([0,\hat{t}];X)$ and the non-linear operator Q_1 as follows:

$$S_1 = \{v: v\in X_1; \|v(t)-\phi_1(t)\| \leq r \quad \forall \hat{t}\in[0,t]\} \tag{5.64}$$

$$Q_1(v)(t) = \phi_1(t) + \int_0^t Z(t-s)F(v(s))ds, \quad D(Q_1) = S_1,\ R(Q_1) \subset X_1 \tag{5.65}$$

where $\|\cdot\| = \|\cdot;X\|$, $\phi_1(t) = Z(t)u_0$ and where $r > 0$ is chosen arbitrarily. Note that the centre of S_1 is ϕ_1 whereas the centre of the sphere Σ_0 of Theorem 5.4 was u_0. If v and v_1 belong to S_1, we have from (5.64)

$$\|v(t)\| \le \|v(t)-\phi_1(t)\|+\|\phi_1(t)\| \le r+\|Z(t)u_0\| \le r+\hat{M}\|u_0\|$$

$$\le r+\eta\hat{M}, \quad \|v_1(t)\| \le r+\eta\hat{M}, \qquad \forall t\in[0,\hat{t}]$$

with $\hat{M} = M\exp(\beta\hat{t})$ if $\beta > 0$ and with $\hat{M} = M$ if $\beta \le 0$. As a consequence, we obtain from (5.65)

$$\begin{aligned}\|Q_1(v)(t)-Q_1(v_1)(t)\| &\le M\int_0^t \exp\{\beta(t-s)\}\|F(v(s))-F(v_1(s))\|ds\\ &\le M\int_0^t \exp\{\beta(t-s)\}\alpha(\|v(s)\|\|v_1(s)\|)\|v(s)-v_1(s)\|\\ &\le \hat{\gamma}M\int_0^t \exp\{\beta(t-s)\}\|v(s)-v_1(s)\|ds,\\ &\qquad \forall t\in[0,\hat{t}]\end{aligned}$$

because (5.63b) gives: $\alpha(\|v(s)\|,\|v_1(s)\|) \le \alpha(r+\eta\hat{M},r+\eta\hat{M}) \le \alpha(r_1+\eta\hat{M},r_1+\eta\hat{M}) = \hat{\gamma}$ with $r_1 > r$ (see below for the definition of $\hat{D}$). Moreover,

$$\|Q_1(v)(t)-\phi_1(t)\|$$

$$\begin{aligned}&\le M\int_0^t \exp\{\beta(t-s)\}[\|F(v(s))-F(\theta_X)\|+\|F(\theta_X)\|]ds\\ &\le M\int_0^t \exp\{\beta(t-s)\}\{\alpha(\|v(s)\|,0)\|v(s)\|+\|F(\theta_X)\|\}ds\\ &\le M\int_0^t \exp\{\beta(t-s)\}\{\hat{\gamma}\|v(s)\|+\|F(\theta_X)\|\}ds \qquad \forall t\in[0,\hat{t}].\end{aligned}$$

Hence, as in the proof of Lemma 5.1, we have for any v, $v_1\in S_1$ and for any $t\in[0,\hat{t}]$

$$\|Q_1(v)(t)-Q_1(v_1)(t)\| \le \hat{\gamma}\, M\, \frac{\exp(\beta\hat{t})-1}{\beta}\, \|v-v_1;X_1\|$$

$$\|Q_1(v)(t)-\phi_1(v)\| \le M\,\frac{\exp(\beta\hat{t})-1}{\beta}\,[\hat{\gamma}\|v;X_1\|+\|F(\theta_X)\|]$$

and so

$$\|Q_1(v)-Q_1(v_1);X_1\| \le p(\hat{t})\|v-v_1;X_1\|, \quad \forall v,v_1\in S_1 \tag{5.66}$$

$$\|Q_1(v)-\phi_1;X_1\| \le p(\hat{t})r, \quad \forall v\in S_1 \tag{5.67}$$

with

$$p(\hat{t}) = \frac{M}{r}\{\hat{\gamma}(r+\eta\hat{M})+\|F(\theta_X)\|\}\,\frac{\exp(\beta\hat{t})-1}{\beta} \tag{5.68}$$

If we choose $\hat{t}$ so that $0<p(\hat{t})<1$, then (5.66) and (5.67) show that Q_1 maps the closed sphere S_1 into itself and that Q_1 is strictly contractive over S_1. Hence, the equation

$$u = Q_1(u) \tag{5.69}$$

(i.e. equation (5.34)) has a unique solution $u = u(t)\in S_1 \subset X_1 = C([0,\hat{t}];X)$, provided that $\hat{t}$ is suitably small, because of the contraction mapping theorem (see Example 2.12).
Now if we define the open and convex set

$$\hat{D} = \{f: f\in X; \|f\| < r_1+\eta\hat{M}\}, \quad r_1 > r$$

then for any f, $f_1\in\hat{D}$

$$\alpha(\|f_1\|,\|f\|) \le \alpha(r_1+\eta\hat{M},r_1+\eta\hat{M}) = \hat{\gamma}$$

$$\alpha_1(\|f\|) \le \alpha_1(r_1+\eta\hat{M}) = \hat{\gamma}_1$$

and $u(t)\in\hat{D}$ $\forall t\in[0,\hat{t}]$ because $u\in S_1$ and so $\|u(t)\| \le r+\eta\hat{M} < r_1+\eta\hat{M}$ $\forall t\in[0,\hat{t}]$. Thus, (5.33b,c,d,e) hold with $\hat{\gamma}$, $\hat{\gamma}_1$, $\hat{D}$ instead of γ, γ_1, D and we can use the procedures of Lemmas 5.2-5.5 to prove that $u(t)$ is a strict solution of (5.31) over $[0,\hat{t}]$.

(*b*) If

$$t = t'+\hat{t}, \quad u(t) = u(t'+\hat{t}) = U(t'), \quad u(\hat{t}) = U_0$$

then equation (5.34) can be rewritten as follows

$$u(t'+\hat{t}) = Z(t'+\hat{t})u_0 + \int_0^{\hat{t}} Z(t'+\hat{t}-s)F(u(s))\,ds + \int_{\hat{t}}^{t'+t} Z(t'+\hat{t}-s)F(u(s))\,ds$$

$$u(t'+\hat{t}) = Z(t')\{Z(\hat{t})u_0 + \int_0^{\hat{t}} Z(\hat{t}-s)F(u(s))\,ds\} + \int_0^{t'} Z(t'-s')F(u(s'+\hat{t}))\,ds'$$

with $s' = s-\hat{t}$, and so

$$U(t') = Z(t')U_0 + \int_0^{t'} Z(t'-s')F(U(s'))\,ds' \tag{5.70}$$

because $u(\hat{t}) = U_0$. Note that (5.70) is formally identical with (5.34), with U_0 instead of u_0, and that assumption (5.63f) gives

$$\|U_0\| = \|u(\hat{t})\| \leq \eta \tag{5.71}$$

because $u(t)$ is a strict solution of (5.31) over $[0,\hat{t}]$. Further, define the closed sphere $S_2 \subset X_1$ and the non-linear operator Q_2 as follows

$$S_2 = \{v : v \in X_1; \|v(t')-\phi_2(t')\| \leq r \quad \forall t' \in [0,\hat{t}]\} \tag{5.64'}$$

$$Q_2(v)(t') = \phi_2(t') + \int_0^{t'} Z(t'-s')F(v(s'))\,ds', \quad D(Q_2) = S_2,$$

$$R(Q_2) \subset X_1 \tag{5.65'}$$

where $\phi_2(t') = Z(t')U_0$ (see (5.64) and (5.65)). Then, if $v, v_1 \in S_2$, we have from (5.64') and from (5.71) for any $t' \in [0,\hat{t}]$

$$\|v(t')\| \leq \|v(t')-\phi_2(t')\| + \|\phi_2(t')\| \leq r + \|Z(t')U_0\| \leq r + \hat{M}\|U_0\|$$

$$\leq r + \eta\hat{M}, \quad \|v_1(t')\| \leq r + \eta\hat{M}$$

just as in part (a). By a procedure similar to that leading to (5.66) and to (5.67), we have from (5.65')

$$\|Q_2(v)-Q_2(v_1);X_1\| \le p(\hat{t})\|v-v_1;X_1\| \quad \forall v,v_1\in S_2 \qquad (5.66')$$

$$\|Q_2(v)-\phi_2;X_1\| \le p(\hat{t})r \quad \forall v\in S_2 \qquad (5.67')$$

where $p(\hat{t})$ is *still* given by (5.68). This is basically due to (5.71), i.e. to the assumption (5.63f); in other words, the existence of an *a priori bound* such as (5.63f) makes the form of $p(\hat{t})$ 'universal'. Since $\hat{t}$ was chosen in part (a) so that $0 < p\ \hat{t}) < 1$, inequalities (5.66') and (5.67') ensure that (5.70) has a unique strongly continuous solution $U = U(t')\in S_2$ and so $\|U(t')\| \le r+\eta\hat{M} < r_1+\eta M\ \forall t'\in[0,\hat{t}]$. As a consequence, $u(t) = u(t'+\hat{t}) = U(t'), t'\in[0,\hat{t}]$, i.e. $t\in[\hat{t},2\hat{t}]$, is the unique strongly continuous solution of (5.34) for $[\hat{t},2\hat{t}]$ and it is such that $\|u(t)\| \le r+\eta\hat{M} < r_1+\eta\hat{M}$ $\forall t\in[\hat{t},2\hat{t}]$. We conclude that $u = u(t)$ is the unique strongly continuous solution of (5.34) over the interval $[0,2\hat{t}]$ and that $u(t)\in\hat{D}\ \ \forall t\in[0,2\hat{t}]$ because $\|u(t)\| < r_1+\eta\hat{M}$ both if $t\in[0,\hat{t}]$ and if $t\in[\hat{t},2\hat{t}]$. The procedures of Lemmas 5.2-5.5 (with $\hat{\gamma},\hat{\gamma}_1,\hat{D}$ instead of γ,γ_1,D) can be used again to prove that $u(t)$ is the unique strict solution of (5.31) over $[\hat{t},2\hat{t}]$ and hence over $[0,\ 2\hat{t}]$.

Since the above technique can be iterated any number of times, Theorem 5.5 is completely proved. □

Example 5.7. The case $\mathrm{Re}(F(f),f) \le \lambda\|f\|^2\ \forall f\in X$, *with* X = *a Hilbert space*

Let X be a Hilbert space and assume that $A\in\mathscr{G}(1,\beta;X), u_0\in D(A)$ and that F satisfies (5.63b,c,d). In addition suppose that

$$\mathrm{Re}(F(f),f) \le \lambda\|f\|^2 \ \ \forall f\in X \qquad (5.72)$$

where λ is a real constant. Then, we have from (5.31a) (see Remark (4.4)

$$\left(\frac{\mathrm{d}}{\mathrm{d}t}u(t),u(t)\right) = (Au(t),u(t))+(F(u(t)),u(t))$$

$$\left(u(t),\ \frac{\mathrm{d}}{\mathrm{d}t}\,u(t)\right) = \overline{(Au(t),u(t))} + \overline{(F(u(t)),u(t))}$$

and so

$$\frac{\mathrm{d}}{\mathrm{d}t}\,(u(t),u(t)) = 2\mathrm{Re}\,(Au(t),u(t)) + 2\mathrm{Re}\,(F(u(t)),u(t))$$

$$\frac{\mathrm{d}}{\mathrm{d}t}\,\|u(t)\|^2 \le 2(\beta+\lambda)\|u(t)\|^2$$

because of Theorem 4.3 and (5.72). Then, Lemma 3.2 gives

$$\|u(t)\|^2 \le \|u_0\|^2 \exp[2(\beta+\lambda)t], \quad t \ge 0$$

and (5.63f) is satisfied with $\eta = \|u_0\| \exp\{(\beta+\lambda)t_0\}$ if $\beta+\lambda > 0$ and with $\eta = \|u_0\|$ if $\beta+\lambda \le 0$. Note that $\lim\|u(t)\| = 0$ as $t \to \infty$ if $\beta+\lambda < 0$. □

Example 5.8. $F(f) = (1+f^2)^{-1}, \tilde{F}(f) = f^2[1+f^2]^{-1}$, *with* $D(F) = D(\tilde{F}) = C([a,b])$

If X is the *real* B-space $C([a,b])$ and if

$$F(f) = \frac{1}{1+f^2}, \quad D(F) = X, \quad R(F) \subset X$$

then we have

$$|F(f_1)-F(f)| = \frac{|f^2-f_1^2|}{(1+f_1^2)(1+f^2)} = \frac{|f+f_1|}{(1+f_1^2)(1+f^2)}\,|f_1-f|$$

$$\le \frac{|f|+|f_1|}{(1+f_1^2)(1+f^2)}\,|f_1-f| \le \left(\frac{1}{2}+\frac{1}{2}\right)\,|f_1-f| = |f_1-f|$$

because $|f|/(1+f^2) \le \frac{1}{2}$ and $1/(1+f^2) \le 1$. Hence,

$$|F(f_1)(x)-F(f)(x)| \le |f_1(x)-f(x)| \quad \forall x \in [a,b]$$

and taking the maximum of both sides we obtain

$$\|F(f_1)-F(f)\| \le \|f_1-f\| \, \forall f, f_1 \in X. \tag{5.73}$$

Furthermore,

$$F(f+g)-F(f) + \frac{2f}{(1+f^2)^2}g = \frac{3f^2+2fg-1}{\{1+(f+g)^2\}(1+f^2)^2} g^2$$

and so

$$|F(f+g)(x)-F(f)(x) + \frac{2f(x)}{\{1+f^2(x)\}^2} g(x)|$$

$$\leq [3|f(x)|^2+2|f(x)||g(x)|+1]|g(x)|^2, \quad \forall x\in[a,b].$$

Hence,

$$\|F(f+g)-F(g)+ \frac{2f}{(1+f^2)^2} g\| \leq [3\|f\|^2+2\|f\|\|g\|+1]\|g\|^2$$

and $F(f)$ is Fréchet differentiable with

$$F_f g = -\frac{2f}{(1+f^2)^2} g \quad \forall f,g\in X$$

because

$$\|F(f+g)-F(f) + \frac{2f}{(1+f^2)^2} g\| / \|g\| \leq [3\|f\|^2+2\|f\|\|g\|+1]\|g\| \to 0$$

as $\|g\|\to 0$. Note that

$$|F_f g| = \frac{2|f|}{1+f^2}\frac{1}{1+f^2}|g| \leq 2\times\frac{1}{2}|g| = |g|$$

$$\|F_f g\| \leq \|g\| \quad \forall f,g\in X$$

and so $F_f\in\mathscr{B}(X)$. Finally,

$$F_{f_1}g-F_f g = 2\frac{ff_1(f_1^2+f_1f+f^2)+2ff_1-1}{(1+f^2)^2(1+f_1^2)^2}(f_1-f)g$$

and consequently

$$\|F_{f_1}g-F_f g\| \leq 2[\|f\|\|f_1\|(\|f_1\|^2+\|f\|\|f_1\|+\|f\|^2) +$$

$$2\|f\|\|f_1\|+1]\|f_1-f\|\|g\| \to 0 \tag{5.75}$$

as $\|f_1-f\|\to 0$. Inequalities (5.73), (5.74), and (5.75) show that F satisfies (5.63b,c,d) with $\alpha = 1$, $\alpha_1 = 1$.

Note that if

$$\tilde{F}(f) = \frac{f^2}{1+f^2}, \quad D(\tilde{F}) = X = C([a,b]), \quad R(\tilde{F}) \subset X$$

then

$$\tilde{F}(f) = 1-F(f)$$

and so

$$\|\tilde{F}(f_1)-\tilde{F}(f)\| = \|F(f_1)-F(f)\| \leq \|f_1-f\| \quad \forall\ f,f_1\in X$$

$$\tilde{F}_f g = -F_f g. \qquad \square$$

Example 5.9. $F(f) = (Af)(Bf)$ with A and B belonging to $\mathscr{B}(C([a,b]))$.

If $X = C([a,b])$ in Example 2.11, then the non-linear operator defined by (2.28)

$$F(f) = (Af)(Bf)$$

has domain $D(F) = C([a,b])$, and $A,B\in\mathscr{B}(C([a,b]))$. We have from Example 2.11

$$\|F(f_1)-F(f)\| \leq \|A\|\|B\|(\|f_1\|+\|f\|)\|f_1-f\|$$

where we use the symbol $\|\cdot\|$ for both the norm in $C([a,b])$ and in $\mathscr{B}(C([a,b]))$. Moreover, it is easy to verify that F in Fréchet differentiable with

$$F_f g = (Af)Bg+(Bf)Ag \quad \forall\ f,g\in C([a,b])$$

and so

$$\|F_f g\| \leq \|Af\|\|Bg\|+\|Bf\|\|Ag\| \leq [2\|A\|\|B\|\|f\|]\|g\|$$

$$\|F_{f_1}g-F_f g\| = \|[A(f_1-f)]Bg+[B(f_1-f)]Ag\|$$

$$\leq 2\|A\|\|B\|\|f_1-f\|\|g\| .$$

We conclude that F satisfies (5.63b,c,d) with

$$\alpha(\rho_1,\rho_2) = \|A\|\|B\|(\rho_1+\rho_2), \quad \alpha_1(\rho) = 2\|A\|\|B\|\rho .$$

If in particular

$$Bf = f \quad \forall f\in C([a,b])$$

$$Af = \int_a^b h(x,y)f(y)\,\mathrm{d}y \quad \forall f\in C([a,b])$$

with $h\in C([a,b]\times[a,b])$, then the above results hold with $\|B\| = 1$ and with $\|A\| = \hat{h} = \max\{|h(x,y)|,\ a\leq x,y\leq b\}$. □

Example 5.10. A linearization procedure.

Under the assumptions (5.63), Theorem 5.5 guarantees that the semilinear initial-value problem (5.31) has a unique strict solution $u(t)$ defined over the whole $[0,t_0]$. However, when dealing with concrete problems, it is often useful to consider a *linearized* version of (5.31) which generally gives 'good' results for small values of t. To understand this, note first that the definition (3.47) of the Fréchet derivative of F at $f = u_0$ leads to

$$F(u(t))-F(u_0) = F_{u_0}[u(t)-u_0]+G(u_0,u(t)-u_0) \tag{5.76}$$

with

$$\frac{\|G(u_0,u(t)-u_0);X\|}{\|u(t)-u_0;X\|} \to 0 \quad \text{as} \quad t\to 0+ \tag{5.77}$$

because of (5.31b). As a consequence, system (5.31) becomes

$$\frac{\mathrm{d}}{\mathrm{d}t}u(t) = Au(t)+Bu(t)+[G(u_0,u(t)-u_0)+F(u_0)-Bu_0] \tag{5.78a}$$

$$X\text{-}\lim_{t\to 0+} u(t) = u_0 \tag{5.78b}$$

where $B = F_{u_0} \in \mathscr{B}(X)$. Then the linearized version of (5.78) reads as follows:

$$\frac{\mathrm{d}}{\mathrm{d}t} w(t) = Aw(t)+Bw(t)+[F(u_0)-Bu_0], \quad t>0 \tag{5.79a}$$

$$X\text{-}\lim_{t\to 0+} w(t) = u_0 \tag{5.79b}$$

and the *error* $\Delta(t) = u(t)-w(t)$ is such that

$$\frac{\mathrm{d}}{\mathrm{d}t} \Delta(t) = (A+B)\Delta(t)+G(u_0,u(t)-u_0), \quad t>0 \tag{5.80a}$$

$$X\text{-}\lim_{t\to 0+} \Delta(t) = \theta_X. \tag{5.80b}$$

Thus, Theorem 4.8 with $\hat{Z}(t) = \exp\{t(A+B)\}$ instead of $Z(t)$ and with $g(s) = G(u_0,u(s)-u_0)$ gives

$$\Delta(t) = \int_0^t \hat{Z}(t-s)G(u_0,u(s)-u_0)\mathrm{d}s \tag{5.81}$$

because of (5.80b). However, if $\varepsilon > 0$ and $\varepsilon' > 0$ are given, a $t_1 = t_1(\varepsilon,\varepsilon') \leq t_0$ can be found such that

$$\frac{\|G(u_0,u(t)-u_0);X\|}{\|u(t)-u_0;X\|} < \varepsilon, \quad \|u(t)-u_0;X\| < \varepsilon', \ \forall t\in [0,t_1]$$

owing to (5.77) and to (5.78b). Hence, we obtain from (5.81)

$$\|\Delta(t);X\| < \varepsilon \int_0^t \|\hat{Z}(t-s)\| \|u(s)-u_0;X\|\mathrm{d}s$$

$$< \varepsilon\varepsilon' \int_0^t M \exp\{(\beta+M\|B\|)(t-s)\}\mathrm{d}s, \quad \forall t\in[0,t_1]$$

since $A+B = A+F_{u_0} \in \mathscr{G}(M,\beta+M\|B\|)$ because of Theorem 5.1, with $\|B\| = \|F_{u_0}\| \leq \alpha_1(\|u_0;X\|) \leq \alpha_1(\eta)$, see (5.63b,f).

We conclude that the norm of the error $\Delta(t)$ satisfies the inequality

$$\|\Delta(t);X\| \leq \varepsilon\varepsilon' M \frac{\exp\{(\beta+M\|B\|)t\} -1}{\beta+M\|B\|}, \ \forall t\in [0,t_1]. \quad \square$$

Some 'concrete' examples of semilinear initial-value

problems will be examined later on.

EXERCISES

5.1. Prove that (5.15) can be extended to the whole X. Hint: if $f\in X$ and $\varepsilon > 0$, then an element $u_0 \in D(A+B)$ can be found such that $\| u_0 - f;X \| < \varepsilon$. Furthermore, $\hat{Z}(t)f = \hat{Z}(t)u_0 + \hat{Z}(t)[f-u_0] = \tilde{Z}(t)u_0 + \hat{Z}(t)[f-u_0] = \tilde{Z}(t)f + \tilde{Z}(t)[u_0-f] + \hat{Z}(t)[f-u_0]$, $\| \hat{Z}(t)f - \tilde{Z}(t)f;X \| \leq \ldots$.

5.2. Prove that $\exp\{t(A+B)\} = \exp(bt)\exp(At)$ if $A\in\mathcal{G}(M,\beta;X)$ and $B = bI$. Hint: use (5.21) and (5.22).

5.3. Show that $B(t) = \alpha(t)B_0$ satisfies (i), (ii), and (iii) of Theorem 5.2 if $B_0\in\mathcal{B}(X)$ and $\alpha(t)$ is a continuously differentiable real function.

5.4. Prove that $\|B(t)\| \leq \hat{b}\quad \forall t\in[t_1,t_2]$ if $\{B(t), t\in[t_1,t_2]\}\in\mathcal{B}(X)$ and if, for each $f\in X$, $\phi(t) = B(t)f$ is continuous in $t\in[t_1,t_2]$ in the sense of $\|\cdot;X\|$. Hint: note first that for each $f\in X$ $\|B(t)f;X\|$ is continuous in the classical sense $\forall t\in[t_1,t_2]$ (see Exercise 4.1) and so it is bounded: $\|B(t)f;X\| \leq b(f)\quad \forall t\in[t_1,t_2]$ where the bound b usually depend on f. Then, use the *uniform-boundedness* theorem: if the family of linear operators $\{K_\lambda, \lambda\in\Lambda\}$ is contained in $\mathcal{B}(X)$ and if for each $f\in X$ there is a 'constant' $c = c(f)$ such that $\sup\{\|K_\lambda f;X\|, \lambda\in\Lambda\} \leq c(f)$, then a constant $\hat{c}$ exists such that $\sup\{\|K_\lambda\|, \lambda\in\Lambda\} \leq \hat{c}$.

5.5. Prove that B is A-bounded if $B\in\mathcal{B}(X)$.

5.6. If $X = L^{s,2}(R^1)$ and if the heat-diffusion operator A and the convection operator B are defined as follows

$$Af = \mathcal{F}^{-1}[-ky^2\hat{f}(y)], \quad D(A) = L^{s+2,2}(R^1) \subset X, \quad R(A) \subset X$$

$$Bf = \mathcal{F}^{-1}[-ivy\hat{f}(y)], \quad D(B) = L^{s+1,2}(R^1) \subset X, \quad R(B) \subset X$$

(see Example 2.15), then B is A-bounded with

$$\|Bf\|_{s,2} \leq \frac{vn}{\sqrt{2}}\|f\|_{s,2} + \frac{v}{nk\sqrt{2}}\|Af\|_{s,2} \quad \forall f\in D(A) \quad n = 1,2,\ldots .$$

Hint:

$$(\|Bf\|_{s,2})^2 = v^2 \int_{-\infty}^{+\infty} (1+y^2)^s y^2 |\hat{f}(y)|^2 dy$$

$$\le \frac{v^2}{2} \int_{-\infty}^{+\infty} (1+y^2)^s (n^2 + \frac{1}{n^2} y^4) |\hat{f}(y)|^2 dy$$

since $2y^2 \le n^2+y^4/n^2 \forall n = 1,2,\ldots$ because $(n-y^2/n)^2 \ge 0$.

5.7. Prove that each of the $u_1^{(n)}(t)$ defined by (5.44) belongs to $C([0,\hat{t}];X)$.

5.8. Under the assumptions of Lemma 5.4, show that

$$X\text{-}\lim_{h\to 0+} \frac{1}{h} \int_0^h Z(h-\hat{s})F(u(\hat{s}))d\hat{s} = F(u_0)$$

Hint: if $s = h-\hat{s}$, then the above integral becomes

$$\frac{1}{h} \int_0^h Z(s)F(u(h-s))ds = \frac{1}{h} \int_0^h Z(s)F(u(s))ds +$$

$$\frac{1}{h} \int_0^h Z(s)\{F(u(h-s))-F(u(s))\}ds$$

where

$$X - \lim_{h\to 0+} \frac{1}{h} \int_0^h Z(s)F(u(s))ds = \left\{\frac{d}{dt} \int_0^t Z(s)F(u(s))ds\right\}_{t=0} = Z(0)F(u(0))$$

and

$$\|\frac{1}{h} \int_0^h Z(s)\{F(u(h-s)-u(s)\}ds;X\| \le \frac{\hat{M}_Y}{h} \int_0^h \|u(h-s)-u(s);X\|ds .$$

5.9. Prove that the non-linear operator F defined by (5.32) satisfies (5.72) with $\lambda = \mu/2$. Hint: note that X is now the *real* Hilbert space $L^2(a,b)$, and see Example 2.9.

6

SEQUENCES OF SEMIGROUPS

6.1. SEQUENCES OF SEMIGROUPS $\exp(tA_j) \in \mathcal{B}(X)$

The unique strict solution $u = u(t)$ of the homogeneous initial-value problem (4.3) with $A \in \mathcal{G}(M,\beta;X)$ and with $u_0 \in D(A)$ has the form: $u(t) = \exp(tA)u_0$, $t \geq 0$, because of Theorem 4.7. However, as we have remarked at the beginning of Example 4.10 on the discretization of the time-like variable, an explicit expression of the semigroup $\{\exp(tA), t \geq 0\}$ is usually difficult to derive. Time discretization is an approximation technique especially suitable for numerical purposes (see Example 4.10), whereas the method described below is important from both a theoretical and a numerical viewpoint. The basic idea of such a method is to 'approximate' the semigroup $\exp(tA)$ by means of a sequence of semigroups $\exp(tA_j)$, $j = 1,2,\ldots$ whose structure is simpler than that of $\exp(tA)$ because their generators A_j, $j = 1,2,\ldots$, are in some sense simpler than A.

In the following lemma, we deriva a formula that will be used in the proof of Theorem 6.1.

Lemma 6.1. If A and B belong to $\mathcal{G}\rho(M,\beta;X)$ and if $z \in \rho(A) \cap \rho(B)$ (i.e. z belongs to the resolvent set of A and to the resolvent set of B, see section 2.7), then

$$R(z,B)\{\exp(tA)-\exp(tB)\}R(z,A)f \tag{6.1}$$

$$= \int_0^t \exp\{(t-s)B\}\{R(z,A)-R(z,B)\}\exp(sA)f\,ds \quad \forall f \in X, t \geq 0$$

where $R(z,A) = (zI-A)^{-1}$ and $R(z,B) = (zI-B)^{-1}$.

Proof. If $z \in \rho(A)$, $R(z,A)f \in D(A)$ $\forall f \in X$ because $D(R(z,A)) = X$ and $R(R(z,A)) = D(A)$ (see Remark 2.9), and so we have from (4.36e) with $R(z,A)f$ instead of f:

$$\frac{d}{ds}\,[\exp(sA)\{R(z,A)f\}] = \exp(sA)A\{R(z,A)f\}$$

$$= \exp(sA)\{zR(z,A)-I\}f, \quad z\epsilon\rho(A), \quad f\epsilon X, \quad s \geq 0 \tag{6.2}$$

because $AR(z,A)f = zR(z,A)f-f$ (see Exercise 2.14). In an analogous way, we obtain for any $g\epsilon X$ and for $0 \leq s \leq t$:

$$\frac{d}{ds}[\exp\{(t-s)B\}R(z,B)g] = -\exp\{(t-s)B\}\{zR(z,B)-I\}g, \quad z\epsilon\rho(B). \tag{6.3}$$

Hence, if $z\epsilon\rho(A) \cap \rho(B)$, (6.2) and (6.3) give

$$\frac{d}{ds}[\exp\{(t-s)B\}R(z,B)\exp(sA)R(z,A)f]$$

$$= -\exp\{(t-s)B\}\{zR(z,B)-I\}\{\exp(sA)R(z,A)f\}+$$

$$[\exp\{(t-s)B\}R(z,B)]\exp(sA)\{zR(z,A)-I\}f$$

and so

$$\frac{d}{ds}[\exp\{(t-s)B\}R(z,B)\exp(sA)R(z,A)f]$$

$$= \exp\{(t-s)B\}\{R(z,A)-R(z,B)\}\exp(sA)f, \quad f\epsilon X, \quad z\epsilon\rho(A) \cap \rho(B),$$

$$0 \leq s \leq t \tag{6.4}$$

where we took into account that $R(z,A)$ commutes with $\exp(sA)$ because of (4.36c): $R(z,A)\exp(sA)f = \exp(sA)R(z,A)f$ $\forall f\epsilon X, s \geq 0$, $z\epsilon\rho(A)$ (see Exercise 6.1). Integrating both sides of (6.4) from 0 to t we obtain (6.1) □

Note that, if z is real and larger than β, then $z\epsilon\rho(A) \cap \rho(B)$ because both A and B belong to $\mathscr{G}(M,\beta;X)$.

Theorem 6.1. If $A\epsilon\mathscr{G}(M,\beta;X)$, $A_j\epsilon\mathscr{G}(M,\beta;X)$ $\forall j = 1,2,\ldots,$ *and if*

$$X\text{-}\lim_{j\to\infty} R(z,A_j)f = R(z,A)f \quad \forall f\epsilon X \tag{6.5}$$

for at least one $z > \beta$ *then*

$$X\text{-}\lim_{j\to\infty} \exp(tA_j)f = \exp(tA)f, \quad \forall f\epsilon X, \ t \geq 0. \tag{6.6}$$

Proof. Since any $z > \beta$ belongs to both $\rho(A)$ and $\rho(B)$, we obtain from (6.1) with $B = A_j$

$$\|R(z,A_j)\{\exp(tA)-\exp(tA_j)\}R(z,A)g\|$$

$$\leq M \int_0^t \exp\{\beta(t-s)\}\|\{R(z,A)-R(z,A_j)\}\exp(sA)g\|ds \tag{6.7}$$

where $\|\cdot\|$ is the norm in X and where we used (4.36a) for the semigroup generated by A_j. Note that the integrand approaches zero as $j\to\infty$ for each $s\in[0,t]$ because of assumption (6.5) with $f = \exp(sA)g$. On the other hand, the operators A and A_j belong to $\mathscr{G}(M,\beta;X)$ and so

$$\exp\{\beta(t-s)\}\|\{R(z,A)-R(z,A_j)\}\exp(sA)g\|$$

$$\leq \exp\{\beta(t-s)\}\{\|R(z,A)\exp(sA)g\|+\|R(z,A_j)\exp(sA)g\|\}$$

$$\leq 2\frac{M}{z-\beta}M\exp(\beta t)\|g\|, \quad z>\beta, \quad 0\leq s\leq t$$

and the integrand on the right-hand side of (6.7) is also bounded $\forall s\in[0,t]$, $\forall j = 1,2,\ldots$. As a consequence, we obtain from (6.7)

$$\lim_{j\to\infty}\|R(z,A_j)\{\exp(tA)-\exp(tA_j)\}R(z,A)g\| = 0 \tag{6.8}$$

$\forall g\in X, t\geq 0$, because of the bounded-convergence theorem (Royden 1963, p.200; see also the footnote on p. 203). However, $\hat{f} = R(z,A)g\in D(A)$ and so (6.8) gives

$$\lim_{j\to\infty}\|R(z,A_j)\{\exp(tA)-\exp(tA_j)\}\hat{f}\| = 0 \quad \forall\ \hat{f}\in D(A), t\geq 0 \tag{6.9a}$$

Furthermore, if $f\in X$ and $t\geq 0$, we have from (6.5)

$$\|\exp(tA_j)\{R(z,A)-R(z,A_j)\}f\|$$

$$\leq M\exp(\beta t)\|R(z,A)f-R(z,A_j)f\|\to 0 \text{ as } j\to\infty \tag{6.9b}$$

$$\|\{R(z,A)-R(z,A_j)\}\{\exp(tA)f\}\|\to 0 \quad \text{as } j\to\infty . \tag{6.9c}$$

Finally, using (6.9a,b,c) we obtain

$$\|\{\exp(tA_j)-\exp(tA)\}R(z,A)\hat{f}\| = \|\exp(tA_j)\{R(z,A)-R(z,A_j)\}\hat{f}+$$

$$R(z,A_j)\{\exp(tA_j)-\exp(tA)\}\hat{f}+\{R(z,A_j)-R(z,A)\}\exp(tA)\hat{f}\|$$

$$\leq \|\exp(tA_j)\{R(z,A)-R(z,A_j)\}\hat{f}\|+\|R(z,A_j)\{\exp(tA_j)-\exp(tA)\}\hat{f}\|+$$

$$\|\{R(z,A_j)-R(z,A)\}\exp(tA)\hat{f}\| \to 0 \quad \text{as} \quad j\to\infty$$

i.e.

$$\lim_{j\to\infty} \|\{\exp(tA_j)-\exp(tA)\}R(z,A)\hat{f}\| = 0, \quad \forall\ \hat{f}\in D(A), t \geq 0. \tag{6.10}$$

However, $\phi = R(z,A)\hat{f}\in D(A^2)$ and so (6.10) becomes

$$\lim_{j\to\infty} \|\{\exp(tA_j)-\exp(tA)\}\phi\| = 0, \quad \forall\ \phi\in D(A^2), t \geq 0. \tag{6.11}$$

Since $D(A^2)$ is dense in X(see Exercise 4.8), relation (6.11) can be extended to the whole X in the usual way (for instance, see the discussion after (4.28)) and (6.6) is proved. □

Remark 6.1.

It can be shown that the limit (6.6) holds uniformly with respect to t in any finite interval $[0,\hat{t}]$:

$$\|\exp(tA_j)f-\exp(tA)f\| < \varepsilon \quad \forall\, t\in[0,\hat{t}],\ \forall j > j_0 \tag{6.12}$$

with $j_0 = j_0(\varepsilon,\hat{t},f)$ (Kato 1966, p.502). □

The assumption (6.5) is sometimes difficult to check because it involves the resolvent operators $R(z,A)$ and $R(z,A_j)$. However, if

$$A\in\mathscr{G}(M,\beta;X), \quad A_j\in\mathscr{G}(M,\beta;X) \quad \forall j = 1,2,\ldots \tag{6.13a}$$

$$D(A) \subset D(A_j) \quad \forall\ j = 1,2,\ldots \tag{6.13b}$$

$$X\text{-}\lim_{j\to\infty} A_j\hat{f} = A\hat{f} \quad \forall \hat{f}\in D(A) \tag{6.13c}$$

then (6.5) is certainly satisfied. To see this, note that $R(z,A)\in\mathscr{B}(X)$, $R(z,A_j)\in\mathscr{B}(X)$ $\forall z > \beta$ and $\hat{f} = R(z,A)f\in D(A) \subset D(A_j)$ $\forall f\in X$. Then,

$$\begin{aligned}\|R(z,A_j)f-R(z,A)f\| &= \|R(z,A_j)\{(zI-A)-(zI-A_j)\}R(z,A)f\| \\ &= \|R(z,A_j)[A_j-A]R(z,A)f\| \\ &\le \frac{M}{z-\beta}\|[A_j-A]\hat{f}\|, \quad z>\beta\end{aligned}$$

and (6.5) follows from (6.13c).

Example 6.1. $\exp(tA_j)f \to \exp(tA)f$ *with* $A_j\in\mathscr{B}(X)$

Assume that $A\in\mathscr{G}(1,0;X)$ and consider the sequence $\{A_j,\ j = 1,2,\ldots\}$:

$$A_j = A\left(I - \frac{1}{j}A\right)^{-1} = j\,AR(j,A), \quad j = 1,2,\ldots . \tag{6.14}$$

Definition (6.14) gives (see Exercise 2.14)

$$A_j = j[AR(j,A)] = j[jR(j,A)-I]$$

$$D(A_j) = D(R(j,A)) \cap D(I) = X \cap X = X$$

$$\|A_j\| \le j[j\|R(j,A)\| + \|I\|] \le j\left[j\frac{1}{j}+1\right] = 2j, \quad j = 1,2,\ldots \tag{6.15}$$

because $A\in\mathscr{G}(1,0;X)$ and so $j\in\rho(A)$ with $\|R(j,A)\| \le 1/j$ (we use the symbol $\|\cdot\|$ for both the norms in X and in $\mathscr{B}(X)$). Since the norm of A_j is bounded by $2j$ and $D(A_j) = X$, we conclude that $A_j\in\mathscr{B}(X)$ and that (6.13b) is satisfied. Moreover, if $\hat{f}\in D(A)$ we obtain from (6.14)

$$A_j\hat{f}-A\hat{f} = jAR(j,A)\hat{f}-A\hat{f} = \{jR(j,A)-I\}A\hat{f} = \{V(1/j)-I\}A\hat{f}$$

where $V(1/j)$ is given by (4.23) with $s = 1/j$, and (6.13c) is satisfied because of (4.24d). Finally, since $\|jR(j,A)\| \le j(1/j) = 1$, we have from (2.62) of Example 2.8 (with $jR(j,A)$

instead of A and with $(1+z/j)$ instead of z):

$$\begin{aligned}\|R(z,A_j)\| &= \|\{zI-j[jR(j,A)-I]\}^{-1}\| \\ &= \|\{(z+j)I-j^2R(j,A)\}^{-1}\| \\ &= \frac{1}{j}\left\|\left\{\left(\frac{z}{j}+1\right)I-jR(j,A)\right\}^{-1}\right\| \le \frac{1}{j}\,\frac{1}{|1+z/j|-\|jR(j,A)\|} \\ &\le \frac{1}{j}\,\frac{1}{|1+z/j|-1}\end{aligned}$$

provided that $|1+z/j| > 1$. Hence, if $z > 0$, we obtain

$$\|R(z,A_j)\| \le \frac{1}{z+j-j} = \frac{1}{z}$$

and $A_j \in \mathscr{G}(1,0;X)$. Thus, the three assumptions (6.13) are satisfied and consequently (6.6) holds, uniformly in $t\in[0,\hat{t}]$. Note that the *explicit* expression of $\exp(tA_j)$ is given by

$$\exp(tA_j) = I + \frac{t}{1!}A_j + \frac{t^2}{2!}A_j^2 + \ldots$$

because $A_j \in \mathscr{B}(X)$; however, since A does not necessarily belong to $\mathscr{B}(X)$, the sequence $\{A_j, j = 1,2,\ldots\}$ is *not* uniformly bounded with respect to j (see (6.15)). □

Consider now the homogeneous initial-value problem (4.3) with $A\in\mathscr{G}(1,0;X)$, $u_0\in D(A)$, and the following sequence of 'approximating' problems:

$$\left.\begin{aligned}\frac{d}{dt}u_j(t) &= A_j u_j(t), \quad t > 0 \\ X - \lim_{t\to 0+} u_j(t) &= u_0, \quad j = 1,2,\ldots\end{aligned}\right\} \qquad (6.16)$$

where the sequence $\{A_j, j = 1,2,\ldots\}$ satisfies the assumptions of Theorem 6.1 and where $u_0\in D(A_j)$ $\forall j = 1,2,\ldots$. Then the strict solution of each of (6.16) has the form

$$u_j(t) = \exp(tA_j)u_0, \quad t \ge 0$$

and so, if $u(t) = \exp(tA)u_0$ is the solution of (4.3), we have

$$\lim_{j\to\infty} \| u(t)-u_j(t) \| = 0 \tag{6.17}$$

uniformly in $t\in[0,\hat{t}]$, because of Theorem 6.1 and of Remark 6.1. Relation (6.17) shows that the solution of (6.16) with j large enough is a 'good approximation' of the solution of (4.3) (see also Exercise 6.2).

6.2. SEQUENCES OF BANACH SPACES

In Theorem 6.1 the operator A and the A_j's have domains contained in the *same* B-space X; in other words, the 'original' system (4.3) and the 'approximating' systems (6.16) are all initial-value problems in X: $u(t)\in X$, $u_j(t)\in X$ $\forall t\geq 0$. However, it is often necessary to derive the approximating operators A_j by discretizing some of the space-like variables appearing in the formal expression of A (see Example 6.5). This procedure usually leads to a sequence of B-spaces $\{X_j, j = 1, 2,\ldots\}$ with $D(A_j)\subset X_j \neq X$ and the jth approximating system is an initial-value problem in the space X_j (see (6.26)). To be more specific, let X and X_j, $j = 1,2,\ldots$, be B-spaces and assume that, for each j, an operator P_j exists such that

$$P_j\in\mathscr{B}(X,X_j) \quad \text{with} \quad \| P_jf;X_j\| \leq \| f;X\| \quad \forall f\in X \tag{6.18a}$$

$$\lim_{j\to\infty} \| P_jf;X_j\| = \| f;X\| \quad \forall f\in X. \tag{6.18b}$$

If the assumptions (6.18a,b) are satisfied, then $\{X_j, j = 1,2,\ldots\}$ is said to be *a sequence of B-spaces approximating X* (Trotter 1958). Note that (6.18b) suggests that $P_jf\in X_j$ is in some sense a 'representation' of the element $f\in X$ in the approximating space X_j.

Example 6.2. A sequence of B spaces approximating $C([a,b])$

Let X be the (real) B-space $C([a,b])$ and assume that X_j is the space of all ordered j-tuples of real numbers, with norm

$$\| f_j;X_j\| = \max\{|f_j^{(i)}|, i = 1,2,\ldots,j\}, \quad f_j = \begin{pmatrix} f_j^{(1)} \\ \cdots \\ f_j^{(j)} \end{pmatrix} \tag{6.19}$$

(see Example 1.1). Further, for any given $j = 2,3,\ldots$, the *space-step* δ_j, the co-ordinates x_i, and the operator P_j are defined as follows:

$$\delta_j = \frac{b-a}{j-1}\,; \quad x_i = a+(i-1)\delta_j, \quad i = 1,2,\ldots,j \tag{6.20}$$

$$P_j f = \begin{pmatrix} f(x_1) \\ \cdots \\ f(x_j) \end{pmatrix} \quad \forall f = f(x)\in X = C([a,b]). \tag{6.21}$$

Note that the components of $P_j f$, the representation of f in the space X_j, are the values of $f(x)$ at x_i, $i = 1,2,\ldots,j$. We have from (6.19) and from (6.21)

$$\|P_j f;X_j\| = \max\{|f(x_i)|,\ i = 1,2,\ldots,j\}$$

$$\leq \max\{|f(x)|, x\in[a,b]\} = \|f;X\|$$

and (6.18a) is satisfied; on passing to the limit as $j\to\infty$, we also obtain

$$\lim_{j\to\infty} \|P_j f;X_j\| \leq \|f;X\| \quad \forall f\in X \quad . \tag{6.22}$$

On the other hand, since $|f(x)|$ is continuous on $[a,b]$,

$$\|f;X\| = \max\{|f(x)|,\ x\in[a,b]\} = |f(\hat{x})| = \lim_{x\to\hat{x}} |f(x)|$$

where $\hat{x}$ is a suitable value in $[a,b]$, and so

$$|f(\hat{x})|-\varepsilon < |f(x)| < |f(\hat{x})|+\varepsilon \quad \forall x\in(\hat{x}-\delta_\varepsilon,\hat{x}+\delta_\varepsilon) \ .$$

However, if j_ε is such that $\delta_{j_\varepsilon} = (b-a)/(j_\varepsilon-1) < 2\delta_\varepsilon$, then some of the co-ordinates x_i defined by (6.20) with $j > j_\varepsilon$ belong to $(\hat{x}-\delta_\varepsilon,\hat{x}+\delta_\varepsilon)$. Thus, if for each $j > j_\varepsilon$ we indicate by $\hat{x}^{(j)}$ the smallest of the x_i's that belong to $(\hat{x}-\delta_\varepsilon,\hat{x}+\delta_\varepsilon)$, we have

$$|f(\hat{x})|-\varepsilon < |f(\hat{x}^{(j)})| \leq \max\{|f(x_i)|,\ i=1,2,\ldots,j\} = \|P_j f;X_j\|, \quad j > j_\varepsilon$$

and so

$$\lim_{j\to\infty} \|P_j f;X_j\| \geq |f(\hat{x})|-\varepsilon = \|f;X\|-\varepsilon .$$

Since $\varepsilon > 0$ is arbitrary, we conclude that

$$\lim_{j\to\infty} \|P_j f;X_j\| \geq \|f;X\| \quad \forall f\in X \tag{6.23}$$

and (6.18b) follows from (6.22) and (6.23). Hence, $\{X_j, j = 2,3,\ldots\}$ is a sequence of B-spaces approximating $X = C([a,b])$. □

Example 6.3. A sequence of B spaces approximating $L^1(a,b)$

Let X be the (real) B-space $L^1(a,b)$ with $-\infty < a < b < +\infty$ and assume that X_j is the space of all ordered j-tuples of real numbers with norm

$$\|f_j;X_j\| = \sum_{i=1}^{j} |f_j^{(i)}| , \quad f_j = \begin{pmatrix} f_j^{(1)} \\ \cdots\cdot \\ f_j^{(j)} \end{pmatrix} . \tag{6.24}$$

Also, for any given $j = 1,2,\ldots$, define the *space-step* δ_j, the co-ordinates x_i, and the operator P_j as follows

$$\left.\begin{aligned} &\delta_j = \frac{b-a}{j} \;; \quad x_i = a+i\,\delta_j, \quad i = 0,1,2,\ldots j \\ &P_j f = f_j = \begin{pmatrix} f_j^{(1)} \\ \cdots\cdot \\ f_j^{(j)} \end{pmatrix} \quad \forall f\in X = L^1(a,b) \\ &f_j^{(i)} = \int_i f(x)\,dx = \int_{x_{i-1}}^{x_i} f(x)\,dx, \quad i = 1,2,\ldots,j . \end{aligned}\right\} \tag{6.25}$$

Note that the ith component of $P_j f$, the representation of f in X_j, is the average of $f(x)$ over (x_{i-1},x_i) multiplied by δ_j. We have from (6.24) and from (6.25)

$$\|P_j f;X_j\| = \sum_{i=1}^{j} \left|\int_i f(x)\,dx\right| \leq \sum_{i=1}^{j} \int_i |f(x)|\,dx$$

$$= \int_a^b |f(x)|\,dx = \|f;X\|$$

(see Example 1.4), and (6.18a) is satisfied. As far as (6.18b) is concerned, assume that $g = g(x)$ is a polynomial with real coefficients and, say, of degree m : $g(x) = \alpha_0 x^m + \alpha_1 x^{m-1} + \ldots + \alpha_m$. Then, the equation $g(x) = 0$ has m solutions and so $g(x)$ changes its sign *at most* m times when x runs over the interval $[a,b]$. As a consequence, for each $j \geq m$, $g(x)$ changes its sign at most in m of the j subintervals $[x_{i-1}, x_i]$. Thus, we have

$$0 \leq \|g;X\| - \|P_j g;X_j\| = \int_a^b |g(x)|\,dx - \sum_{i=1}^{j} \{ | \int_i g(x)\,dx | \}$$

$$= \sum_{i=1}^{j} \{ \int_i |g(x)|\,dx - | \int_i g(x)\,dx | \}$$

$$= \sum_i{}' \{ \int_i |g(x)|\,dx - | \int_i g(x)\,dx | \} + \sum_i{}'' \{ \int_i |g(x)|\,dx - | \int_i g(x)\,dx | \}$$

$$= \sum_i{}' \{ \int_i |g(x)|\,dx - | \int_i g(x)\,dx \}$$

where $\sum_i{}'$ indicates a sum extended over the (at most m) subintervals over which $g(x)$ takes some negative values, and so

$$0 \leq \|g;X\| - \|P_j g;X_j\| \leq 2 \sum_i{}' \{ \int_i |g(x)|\,dx \}$$

$$\leq 2m\,\delta_j \max\{|g(x)|, x \in [a,b]\} \to 0 \quad \text{as } j \to \infty$$

because $\delta_j = (b-a)/j \to 0$ as $j \to \infty$. We conclude that (6.18b) is satisfied for any f belonging to the family of all polynomials with real coefficients. However, such a family is dense in the real B-space $X = L^1(a,b)$ and so (6.18b) holds for any $f \in X$ (see Exercise 6.5) and $\{X_j, j = 1,2,\ldots\}$ is a sequence of B-spaces approximating $X = L^1(a,b)$. □

6.3. SEQUENCES OF SEMIGROUPS $\exp(tA_j) \in \mathscr{B}(X_j)$

Consider the initial-value problem (4.3) with $A \in \mathscr{G}(M,\beta;X)$, $u_0 \in D(A)$, and assume that the approximating operators A_j's have domain and range in X_j, where $\{X_j, j = 1,2,\ldots\}$ is a sequence of B-spaces approximating X. Correspondingly, the approximating initial-value problems have the form $(j = 1,2,\ldots)$

$$\frac{d}{dt} u_j(t) = A_j u_j(t), t > 0; \quad X_j - \lim_{t \to 0+} u_j(t) = u_{j,0} \qquad (6.26)$$

where $u_j(t)$ is now a function from $[0,+\infty)$ into the B-space X_j and $u_{j,0}$ is a suitable element of $D(A_j) \subset X_j$.

The following theorem is a generalization of Theorem 6.1 and makes clear why (6.26) is a sequence of problems *approximating* (4.3); of course, if $X_j = X$ and $P_j = I$ $\forall j = 1,2,\ldots$, then the results of Theorems 6.1 and 6.2 coincide.

Theorem 6.2. If $A \in \mathscr{G}(M,\beta;X)$, $A_j \in \mathscr{G}(M,\beta;X_j)$ $\forall j = 1,2,\ldots$ *and if*

$$\lim_{j\to\infty} \| R(z,A_j)P_j f - P_j R(z,A)f;X_j \| = 0 \quad \forall f \in X \tag{6.27}$$

for at least one $z > \beta$, *then*

$$\lim_{j\to\infty} \| \exp(tA_j)P_j f - P_j \exp(tA)f;X_j \| = 0 \quad \forall f \in X, t \geq 0. \quad \square \tag{6.28}$$

The proof of this theorem is similar to that of Theorem 6.1 and will not be given here (see Trotter 1958 for the details; see also Exercise 6.6). Note that, since $P_j f$ and $P_j \exp(tA)f$ are the representations of f and of $\exp(tA)f$ in X_j, relation (6.28) shows how, for j large enough, the representation of $\exp(tA)f$ in X_j is close to the element $g_j = \exp(tA_j)P_j f$, namely to the image under the approximating semigroup $\exp(tA_j)$ of the representation of f in X_j.

Going back to (6.26), assume that $P_j u_0 \in D(A_j)$ $\forall j = 1,2,\ldots$, and let $u_{j,0} = P_j u_0$; then, if $u(t) = \exp(tA)u_0$ is the solution of system (4.3) and $u_j(t) = \exp(tA_j)u_{j,0} = \exp(tA_j)P_j u_0$ is the solution of (6.26) with $u_{j,0} = P_j u_0$, we obtain from (6.28)

$$\lim_{j\to\infty} \| u_j(t) - P_j u(t);X_j \| = 0, \quad \forall t \geq 0 \tag{6.29}$$

i.e. $P_j u(t)$, the representation in X_j of the solution of (4.3) is close to the solution $u_j(t)$ of the approximating problem (6.26), provided that j is large enough.

Remark 6.2.

It can be proved that the limit (6.28) holds uniformly with respect to t in any finite interval $[0,\hat{t}]$. Thus

$$\|u_j(t)-P_ju(t);X_j\| < \varepsilon \quad \forall t\in[0,\hat{t}] \tag{6.30}$$

for any $j > j_0 = j_0\ (\varepsilon,\hat{t},u_0)$. □

We observe that relation (6.27) is certainly satisfied if in particular

$$A\in\mathscr{G}(M,\beta;X);\ A_j\in\mathscr{G}(M,\beta;X_j) \quad \forall j = 1,2\ldots, \tag{6.31a}$$

$$P_j[D(A)] \subset D(A_j),\ \forall j = 1,2,\ldots,$$

$$(\text{i.e. } P_j\hat{f}\in D(A_j)\forall\hat{f}\in D(A)) \tag{6.31b}$$

$$\lim_{j\to\infty} \|A_jP_j\hat{f}-P_jA\hat{f};X_j\| = 0 \quad \forall\,\hat{f}\in D(A). \tag{6.31c}$$

In fact, if $z > \beta$, $R(z,A)\in\mathscr{B}(X)$ and $R(z,A_j)\in\mathscr{B}(X_j)$ with $\|R(z,A_j)g_j;X_j\| \le M(z-\beta)^{-1}\|g_j;X_j\|\ \forall\, g_j\in X_j$ because of (6.31a) and $\hat{f} = R(z,A)f\in D(A)$ with $P_j\hat{f} = P_jR(z,A)f\in D(A_j)\ \forall\, f\in X$ because of (6.31b). Thus,

$$\|R(z,A_j)P_jf-P_jR(z,A)f;X_j\| = \|R(z,A_j)[P_j\{zI-A\}-\{zI-A_j\}P_j]R(z,A)f;$$

$$X_j\| = \|R(z,A_j)[A_jP_j-P_jA]\hat{f};X_j\|$$

$$\le \frac{M}{z-\beta}\|A_jP_j\hat{f}-P_jA\hat{f};X_j\|, \quad z > \beta,$$

and (6.31c) implies (6.27).

The following perturbation theorem is often useful for applications.

Theorem 6.3. Let A and $\{A_j, j = 1,2,\ldots\}$ satisfy the assumptions of Theorem 6.2 and assume that $B\in\mathscr{B}(X)$, $B_j\in\mathscr{B}(X_j)$ $\forall j = 1,2,\ldots$, with

$$\|B_jg_j;X_j\| \le b\|g_j;X_j\| \quad \forall\, g_j\in X_j,\ j = 1,2,\ldots \tag{6.32}$$

and with

$$\lim_{j\to\infty} \|B_jP_jf-P_jBf;X_j\| = 0 \quad \forall f\in X. \tag{6.33}$$

Then,

$$\lim_{j\to\infty} \| \exp[t(A_j+B_j)]P_jf - P_j \exp[t(A+B)]f; X_j \| = 0 \tag{6.34}$$

$$\forall f \in X, \quad t \geq 0.$$

Proof. First note that since $\| P_jBf; X_j \| \to \| Bf; X \|$ as $j\to\infty$ because of (6.18b) and since assumption (6.33) implies that

$$\lim_{j\to\infty} \left| \| B_jP_jf; X_j \| - \| P_jBf; X_j \| \right| = 0$$

we have

$$\lim_{j\to\infty} \| B_jP_jf; X_j \| = \lim_{j\to\infty} \| P_jBf; X_j \| = \| Bf; X \| \quad \forall f \in X.$$

However (6.32) gives

$$\| B_jP_jf; X_j \| \leq b \| P_jf; X_j \| \quad \forall f \in X, j = 1, 2, \ldots$$

and on passing to the limit as $j\to\infty$:

$$\| Bf; X \| \leq b \lim_{j\to\infty} \| P_jf; X_j \| = b \| f; X \| \quad \forall f \in X$$

where we again used (6.18b). Then, $A+B \in \mathcal{G}(M, \beta+bM; X)$, $A_j+B_j \in \mathcal{G}(M, \beta+bM; X_j)$ $\forall j = 1, 2, \ldots$, because of Theorem 5.1 and, for each $z > \beta+bM$, we have (see Example 2.20)

$$R(z, A+B)f = (zI-A-B)^{-1}f = [\{I-B(zI-A)^{-1}\}(zI-A)]^{-1}f$$

$$= R(z, A)(I-G)^{-1}f, \quad \forall f \in X$$

$$R(z, A_j+B_j)f_j = R(z, A_j)(I-G_j)^{-1}f_j, \quad \forall f_j \in X_j$$

with

$$G = B(zI-A)^{-1} = BR(z, A) \in \mathcal{B}(X)$$

$$G_j = B_jR(z, A_j) \in \mathcal{B}(X_j).$$

Note that if $z > \beta+bM$ (i.e. if $1 > bM/(z-\beta) = q$) we have

$$\|Gf;X\| \le b\,\frac{M}{z-\beta}\,\|f;X\| = q\|f;X\| \quad \forall f\in X$$

$$\|G_jf_j;X_j\| \le b\,\frac{M}{z-\beta}\,\|f_j;X_j\| = q\|f_j;X_j\| \quad \forall f_j\in X_j$$

and so $(I-G)^{-1}\in\mathscr{B}(X)$, $(I-G_j)^{-1}\in\mathscr{B}(X_j)$ with $\|(I-G)^{-1}f;X\| \le \|f;X\|/(1-q)$, $\|(I-G_j)^{-1}f_j;X_j\| \le \|f_j;X_j\|/(1-q)$ because $q<1$ (see Example 2.16 with $z = 1$ and with G and G_j instead of A). On the other hand, we have for any $\phi\in X$

$$\|G_jP_j\phi-P_jG\phi;X_j\| = \|B_jR(z,A_j)P_j\phi-P_jBR(z,A)\phi;X_j\|$$

$$= \|B_j[R(z,A_j)P_j\phi-P_jR(z,A)\phi] + [B_jP_j-P_jB]R(z,A)\phi;X_j\|$$

$$\le b\|R(z,A_j)P_j\phi-P_jR(z,A)\phi;X_j\| + \|B_jP_j\hat{\phi}-P_jB\hat{\phi};X_j\|$$

where $\hat{\phi} = R(z,A)\phi$ and where we used (6.32). Hence,

$$\lim_{j\to\infty}\|G_jP_j\phi-P_jG\phi;X_j\| = 0 \quad \forall\,\phi\in X \tag{6.35}$$

because of (6.27) and (6.33). Moreover,

$$\|(I-G_j)^{-1}P_jg-P_j(I-G)^{-1}g;X_j\|$$

$$= \|(I-G_j)^{-1}[P_j(I-G)-(I-G_j)P_j](I-G)^{-1}g;X_j\|$$

$$= \|(I-G_j)^{-1}[G_jP_j\phi-P_jG\phi];X_j\|$$

$$\le \frac{1}{1-q}\,\|G_jP_j\phi-P_jG\phi;X_j\|$$

with $\phi = (I-G)^{-1}g$, and (6.35) gives

$$\lim_{j\to\infty}\|(I-G_j)^{-1}P_jg-P_j(I-G)^{-1}g;X_j\| = 0 \quad \forall\, g\in X. \tag{6.36}$$

Finally, since for any $z > \beta+bM$ and for any $f\in X$

$$R(z,A_j+B_j)P_jf-P_jR(z,A+B)f = R(z,A_j)(I-G_j)^{-1}P_jf-P_jR(z,A)(I-G)^{-1}f$$

$$= R(z,A_j)[(I-G_j)^{-1}P_jf-P_j(I-G)^{-1}f] +$$

$$[R(z,A_j)P_j-P_jR(z,A)](I-G)^{-1}f$$

(6.27) and (6.36) lead to

$$\| R(z,A_j+B_j)P_jf-P_jR(z,A+B)f;X_j\| \le \frac{M}{z-\beta} \| (I-G_j)^{-1}P_jf-$$

$$P_j(I-G)^{-1}f;X_j\| + \| R(z,A_j)P_j\gamma-P_jR(z,A)\gamma;X_j\| \to 0 \quad \text{as } j\to\infty$$

with $\gamma = (I-G)^{-1}f$. We conclude that $\hat{A} = A+B$ and $\hat{A}_j = A_j+B_j$ satisfy the assumptions of Theorem 6.2 and so relation (6.34) is proved. □

Example 6.4. Galerkin method

Let X be a Hilbert space over the field $\underline{K}$, with inner product $(\cdot,\cdot)$, and assume that a sequence $\{\phi^{(k)}, k = 1,2,...\} \subset X$ exists, with

$$(\phi^{(k)},\phi^{(h)}) = 0 \text{ if } k \neq h;\ (\phi^{(k)},\phi^{(k)}) = \|\phi^{(k)};X\|^2 = 1 \quad (6.37)$$

and with the property that each $f\in X$ may be put into the form

$$f = X - \lim_{h\to\infty} \sum_{h=1}^{r} c_h\,\phi^{(h)} = \sum_{h=1}^{\infty} c_h\,\phi^{(h)} \quad (6.38)$$

where the c_h's are suitable coefficients depending on the given $f\in X$. We have from (6.37) and from (6.38)

$$(f,\phi^{(k)}) = \sum_{h=1}^{\infty} c_h(\phi^{(h)},\phi^{(k)}) = c_k(\phi^{(k)},\phi^{(k)}) = c_k \quad (6.39)$$

that gives the value of the *Fourier coefficient* c_k.

If in particular $X = R^n$ (see Example 1.1), the *finite* family

$$\phi^{(1)} = \begin{pmatrix}1\\0\\\cdots\\0\end{pmatrix},\quad \phi^{(2)} = \begin{pmatrix}0\\1\\\cdots\\0\end{pmatrix},\quad \ldots,\quad \phi^{(n)} = \begin{pmatrix}0\\0\\\cdots\\1\end{pmatrix}$$

satisfies (6.37) and (6.38) and so a Hilbert-space, for which the representation (6.38) is valid, is in some sense a generalization of R^n.

If X_j is the *subset* of X composed of all the linear combinations of the elements $\{\phi^{(1)},\phi^{(2)},\ldots,\phi^{(j)}\} \subset \{\phi^{(k)}, k = 1,2,\ldots\}$

$$X_j = \{g : g = \sum_{i=1}^{j} \alpha_i \phi^{(i)} \ \forall\, \alpha_1,\alpha_2,\ldots,\alpha_j \in \underline{K}\} \tag{6.40}$$

then it is easy to verify that X_j is a j-dimensional† closed linear subset of X. Hence, X_j is itself a Hilbert space with inner product $(\cdot,\cdot)_j = (\cdot,\cdot)$ and with norm $\|\cdot;X_j\| = \|\cdot;X\| = \sqrt{(\cdot,\cdot)}$ (see Example 1.5). In what follows, we shall use the symbols $(\cdot,\cdot)$ and $\|\cdot\|$ for the inner product and the norm in both X and X_j.

Now if the operator P_j is defined by the relation

$$P_j f = \sum_{h=1}^{j} (f,\phi^{(h)})\phi^{(h)} \quad \forall\, f = \sum_{h=1}^{\infty} (f,\phi^{(h)})\phi^{(h)} \tag{6.41}$$

i.e. if the image of f under P_j is the sum of the first j terms of the series (6.38) that represents f, then we have that $D(P_j) = X$ and $R(P_j) \subset X_j$. Moreover,

$$\begin{aligned}\|P_j f;X_j\|^2 &= \|P_j f\|^2 = (P_j f, P_j f)\\ &= \Big(\sum_{k=1}^{j} (f,\phi^{(k)})\phi^{(k)}, \sum_{h=1}^{j} (f,\phi^{(h)})\phi^{(h)}\Big)\\ &= \sum_{k,h=1}^{j} \{(f,\phi^{(k)})\overline{(f,\phi^{(h)})}\}(\phi^{(k)},\phi^{(h)})\\ &= \sum_{k=1}^{j} |(f,\phi^{(k)})|^2\end{aligned}$$

---

† The elements $f_1,f_2,\ldots,f_m$ of the vector space Y over the field $\underline{K}$ are said to be *linearly independent* if $\alpha_1 f_1+\alpha_2 f_2+\ldots+\alpha_m f_m = \theta_Y$ *only* if $\alpha_1 = \alpha_2 = \ldots = \alpha_m = 0$; otherwise, $f_1,f_2,\ldots,f_m$ are *linearly dependent*. A linear subset $Y_0 \subset Y$ is said to be *m-dimensional* ($\dim[Y_0] = m$) if m is the largest number of linearly independent elements that exist in Y_0. Hence, if $\dim[Y_0] = m$ and $\{f_1,\ldots,f_m\}$ are linearly independent elements of Y_0, then any $f \in Y_0$ can be put into the form: $f = c_1 f_1+\ldots+c_m f_m$, where the coefficients c_i's depend on the f considered.

where we used (1.7a,b) and (6.37), and (6.18a) is satisfied because

$$\|f;X\|^2 = \|f\|^2 = (f,f) = \Big(\sum_{k=1}^{\infty} (f,\phi^{(k)})\phi^{(k)}, \sum_{h=1}^{\infty} (f,\phi^{(h)})\phi^{(h)}\Big)$$
$$= \sum_{k=1}^{\infty} (f,\phi^{(k)})^2 \geq \sum_{k=1}^{j} (f,\phi^{(k)})^2 = \|P_j f\|^2 \quad \forall f \in X.$$

As far as (6.18b) is concerned, we have from (6.41)

$$f - P_j f = \sum_{h=j+1}^{\infty} (f,\phi^{(h)})\phi^{(h)}$$

$$\|f-P_j f\|^2 = (f-P_j f, f-P_j f) = \sum_{k=j+1}^{\infty} |(f,\phi^{(k)})|^2 \to 0 \text{ as } j\to\infty$$

because the series $\Sigma_{k=1}^{\infty}|(f,\phi^{(k)})|^2$ is convergent (its sum is $\|f\|^2$), and so

$$0 \leq \lim_{j\to\infty} \big|\|f\| - \|P_j f\|\big| \leq \lim_{j\to\infty} \|P_j f - f\| = 0.$$

Hence, (6.18b) is also satisfied and $\{X_j, j = 1,2,\ldots\}$ is a sequence of Hilbert spaces approximating X. Note that the operator P_j has the properties

$$P_j f_j = f_j \ \forall f_j \in X_j; \qquad P_j^2 f = P_j f \ \ \forall f \in X \tag{6.42}$$

as is easy to verify directly from the definition (6.41).

Now, given $A \in \mathcal{G}(1,0;X)$, assume that the sequence of Hilbert spaces $\{X_j, j = 1,2,\ldots\}$ is also such that

$$P_j \hat{f} \in D(A) \ \forall \hat{f} \in D(A); \qquad P_j A\hat{f} = AP_j \hat{f} \ \forall \hat{f} \in D(A) \tag{6.43}$$

and define the approximating operators A_j as follows:

$$D(A_j) = P_j[D(A)] = \{f_j : f_j = P_j \hat{f} \ \forall f \in D(A)\} \tag{6.44a}$$

$$A_j f_j = P_j A\hat{f} \quad \forall f_j \in D(A_j), \tag{6.44b}$$

where $\hat{f}$ is chosen in $D(A)$ so that $f_j = P_j \hat{f}$. The operator A_j is properly defined by (6.44a,b) since, given any $f_j \in D(A_j)$, at least one $\hat{f} \in D(A)$ exists for which $f_j = P_j \hat{f}$ because of (6.44a); furthermore, if $\hat{f}$ and $\hat{g}$ are two elements of $D(A)$ that have the property $f_j = P_j\hat{f} = P_j\hat{g}$, then (6.43) gives $P_jA\hat{f} = AP_j\hat{f} = AP_j\hat{g} = P_jA\hat{g}$ and A_jf_j is uniquely defined even if f_j is the image under P_j of more than one element of $D(A)$.

The operator A_j belongs to $\mathscr{G}(1,0;X_j)$; to see this, we first note that $D(A_j)$ is dense in X_j because $D(A)$ is dense in X (see Exercise 6.7). Secondly, we consider the equation

$$(zI-A_j)f_j = g_j \quad g_j \in X_j, \quad z > 0 \tag{6.45}$$

where the unknown f_j must be sought in $D(A_j)$. If we put $\hat{f} = R(z,A)g_j$, then $\hat{f} \in D(A)$, $f_j = P_j\hat{f} \in D(A_j)$ and we have

$$(zI-A)\hat{f} = g_j, \quad P_j(zI-A)\hat{f} = P_jg_j, \quad (zI-A_j)f_j = g_j$$

where we used (6.44b) and took into account that $P_jg_j = g_j$ because $g_j \in X_j$. Hence, $f_j = P_j\hat{f} = P_jR(z,A)g_j$ belongs to $D(A_j)$, satisfies eqn (6.45), and is uniquely determined by g_j; moreover, we have

$$\|f_j\| = \|P_jR(z,A)g_j\| \le \|R(z,A)g_j\| \le \frac{1}{z}\|g_j\|, \quad z > 0$$

because of (6.18b) and because of the assumption $A \in \mathscr{G}(1,0;X)$, and so $R(z,A_j) \in \mathscr{B}(X_j)$ $\forall z > 0$ with

$$R(z,A_j)g_j = P_jR(z,A)g_j \ \forall g_j \in X_j, \ z > 0; \quad \|R(z,A_j)\| \le \frac{1}{z} \ \forall z > 0.$$

To complete the proof that $A_j \in \mathscr{G}(1,0;X_j)$ we have to show that $A_j \in \mathscr{C}(X_j)$, but such a property follows from the fact that $-R(z,A_j) = -(zI-A_j)^{-1} \in \mathscr{B}(X_j)$ (see Remark 2.9).

Since (6.31b) and (6.31c) follow directly from (6.44a) and from (6.44b), we conclude that all the assumptions (6.31) are satisfied and so (6.28) and (6.29) hold. Note that here the approximating systems (6.26) are initial-value problems in the *finite-dimensional* spaces X_j that are closed linear *subsets* of the 'original' spaces X (Galerkin method). We also remark that the above approximation method works particularly well if the operator A has the sequence of eigenvalues $\{z_1, z_2, \ldots\}$ with eigenfunctions $\{\phi^{(1)}, \phi^{(2)}, \ldots\}$ satisfying conditions (6.37) and

(6.38) (see Exercise 6.8). □

Example 6.5. Discretization of the operator $- v\mathrm{d}/\mathrm{d}x$

If X_0 is the closed linear subset of the (real) B-space $C([a,b])$

$$X_0 = \{f : f = f(x) \in C([a,b]); f(a) = 0\} \tag{6.46}$$

then X_0 is itself a (real) B-space with norm $\| f; X_0 \| = \max\{|f(x)|, x \in [a,b]\}$. The convection operator $A = -v\mathrm{d}/\mathrm{d}x$ can be defined in X_0 as follows

$$Af = -vf', \quad D(A) = \{f : f \in X_0; f' \in X_0\} \tag{6.47}$$

where f' is a classical derivative and v is a positive constant, and $A \in \mathscr{G}(1,0;X_0)$. In fact, it is easy to verify that $R(z,A)g$ is still given by (2.77):

$$R(z,A)g = \frac{1}{v} \int_a^x \exp\left\{- \frac{z}{v}(x-y)\right\} g(y)\,\mathrm{d}y \quad \forall\, g \in X_0 \tag{6.48}$$

where the integral is now a classical Riemann integral. Thus, if $z > 0$

$$\begin{aligned} |R(z,A)g| &\leq \frac{1}{v} \| g; X_0 \| \int_a^x \exp\left\{-\frac{z}{v}(x-y)\right\} \mathrm{d}y \\ &= \frac{1}{z} \| g; X_0 \| \left[1 - \exp\left\{-\frac{z}{v}(x-a)\right\}\right] \\ &\leq \frac{1}{z} \| g; X_0 \|, \end{aligned}$$

$$\begin{aligned} \| R(z,A)g; X_0 \| &= \max\{|R(z,A)g|, x \in [a,b]\} \\ &\leq \frac{1}{z} \| g; X_0 \| \quad \forall g \in X_0,\ z > 0 \end{aligned}$$

and (4.17c) is satisfied (with $\beta = 0$); then, (4.17a) follows as usual from the fact that $-R(z,A) = -(zI-A)^{-1} \in \mathscr{B}(X_0) \subset \mathscr{C}(X_0)$ $\forall z > 0$. As far as (4.17b) is concerned, assume that $g \in X_0$ is given and let

$$f_\eta(x) = \frac{\alpha}{\eta} \frac{1}{v} \int_a^x \exp\left\{- \frac{\alpha}{v\eta}(x-y)\right\} g(y)\,\mathrm{d}y = \frac{\alpha}{\eta} R\left(\frac{\alpha}{\eta}, A\right) g \tag{6.49}$$

where η is a positive (small) parameter and α is a positive constant (which is needed to adjust dimension). Note that f_η belongs to the range of $R(\alpha/\eta,A)$, i.e. to $D(A)$: $f_\eta \in D(A)\ \forall\, \eta > 0$. Further, it can be shown that $\| f_\eta - g;\ X_0 \| = \max\{ |f_\eta(x) - g(x)|, x \in [a,b]\} < \varepsilon$ provided that $0 < \eta < \eta_0 = \eta_0(\varepsilon, g)$. This can be done by writing the difference $f_\eta(x) - g(x)$ as follows:

$$f_\eta(x) - g(x) = \left\{ \frac{\alpha}{v\eta} \int_{x-(v/\alpha)\sqrt{\eta}}^{x} \Phi_\eta(x-y) g(x)\,dy - g(x) \right\} +$$

$$\frac{\alpha}{v\eta} \int_{x-(v/\alpha)\sqrt{\eta}}^{x} \Phi_\eta(x-y)\{g(y) - g(x)\}dy + \frac{\alpha}{v\eta} \int_{a}^{x-(v/\alpha)\eta} \Phi_\eta(x-y) g(y)\,dy$$

if $x \in [a + (v/\alpha)\sqrt{\eta}, b]$, and

$$f_\eta(x) - g(x) = \left\{ \frac{\alpha}{v\eta} \int_{a}^{x} \Phi_\eta(x-y) g(x)\,dy - g(x) \right\} +$$

$$\frac{\alpha}{v\eta} \int_{a}^{x} \Phi_\eta(x-y)\{g(y) - g(x)\}dy \quad \text{if } x \in [a, a + (v/\alpha)\sqrt{\eta}]$$

where

$$\Phi_\eta(x-y) = \exp\left\{ -\frac{\alpha}{v\eta}(x-y) \right\},$$

and by taking into account that $g = g(x)$ is uniformly continuous over $[a,b]$. Hence, f_η belongs to $D(A)$ and it is such that $\| f - g; X_0 \| < \varepsilon$ if $0 < \eta < \eta_0$. This means that $D(A)$ is dense in X_0 and concludes the proof that $A \in \mathscr{G}(1,0;X_0)$.

Now assume that X_j is the space of all ordered j-tuples of real numbers with norm given by (6.19) and with $f_j^{(1)} = 0$:

$$X_j = \left\{ f_j : f_j = \begin{pmatrix} f_j^{(1)} \\ f_j^{(2)} \\ \cdots \\ f_j^{(j)} \end{pmatrix},\ f_j^{(1)} = 0 \right\}; \quad \| f_j; X_j \| = \max\{ |f_j^{(i)}|,\ i = 1,2,\ldots\} \tag{6.50}$$

and let the operator P_j be defined as follows

$$P_j f = \begin{pmatrix} 0 \\ f(x_2) \\ \cdots \\ f(x_j) \end{pmatrix} \forall f = f(x) \in X_0 \tag{6.51}$$

where $x_i = a+(i-1)\delta_j$, $i = 1,2,\ldots,j$, and $\delta_j = (b-a)/(j-1)$ (see Example 6.2 and recall that X_0 is the closed linear subset of $X = C([a,b])$, composed of all $f = f(x) \in C([a,b])$ such that $f(x_1) = 0$). The proof that $\{X_j, j = 2,3,\ldots\}$ is a sequence of B-spaces approximating X_0 is similar to that of Example 6.2 and is left to the reader.

We have from (6.47) and from (6.51):

$$P_j Af = -vP_j f' = -v \begin{pmatrix} 0 \\ f'(x_2) \\ f'(x_3) \\ \cdots \\ f'(x_j) \end{pmatrix} \quad \forall f \in D(A); \tag{6.52}$$

relation (6.52) suggests that the approximating operator A_j be defined as

$$A_j f_j = -\frac{v}{\delta_j} \begin{pmatrix} 0 \\ f_j^{(2)} \\ f_j^{(3)} - f_j^{(2)} \\ \cdots\cdots\cdots \\ f_j^{(j)} - f_j^{(j-1)} \end{pmatrix} \quad \forall f_j \in D(A_j) = X_j . \tag{6.53}$$

Note that the ith components of $P_j Af$ and of $A_j f_j$ with $f_j = P_j f$ are respectively $-vf'(x_i)$ and $-v\{f(x_i)-f(x_{i-1})\}/\delta_j$ $= -v\{f(x_i)-f(x_{i-1})\}/(x_i-x_{i-1})$.

The operator A_j belongs to $\mathcal{G}(1,0;X_j)$; in fact, it follows from (6.53) that $A_j \in \mathcal{B}(X_j)$ with $\|A_j f_j;X_j\| \le 2(v/\delta_j)\|f_j;X_j\| \, \forall f_j \in X_j$ and so $A_j \in \mathcal{C}(X_j)$ because of (2.34) and $D(A_j)$ is dense in X_j because $D(A_j) = X_j$. Furthermore, since the equation

$$(zI-A_j)f_j = g_j, \quad g_j \in X_j, \quad z > 0$$

can be put into the form

$$zf_j^{(2)}+(v/\delta_j)f_j^{(2)} = g_j^{(2)}, \; zf_j^{(3)}+(v/\delta_j)[f_j^{(3)}-f_j^{(2)}] = g_j^{(3)}, \ldots,$$

$$zf_j^{(j)}+(v/\delta_j)[f_j^{(j)}-f_j^{(j-1)}] = g_j^{(j)} \;,$$

we obtain for $i = 2,3,\ldots,j$:

$$f_j^{(i)} = \lambda[g_j^{(i)}+\{(v/\delta_j)\lambda\}g_j^{(i-1)}+\ldots+\{(v/\delta_j)\lambda\}^{i-2}g_j^{(2)}]$$

with $\lambda = (z+v/\delta_j)^{-1}$. Hence, f_j is uniquely determined by g_j, with

$$|f_j^{(i)}| \le \lambda\|g_j;X_j\| \sum_{h=0}^{i-2} \{(v/\delta_j)\lambda\}^h$$

$$\le \lambda\|g_j;X_j\| \sum_{h=0}^{\infty} \{(v/\delta_j)\lambda\}^h = \lambda\|g_j;X_j\| \frac{1}{1-(v/\delta_j)\lambda} = \frac{1}{z}\|g_j;X_j\| \quad \forall z>0$$

because $(v/\delta_j)\lambda < 1 \;\forall z > 0$. Thus,

$$\|f_j;X_j\| = \|R(z,A_j)g_j;X_j\| = \max\{|f_j^{(i)}|, \; i = 2,3,\ldots,j\}$$

$$\le \frac{1}{z}\|g_j;X_j\| \quad \forall z>0, \quad g_j\in X_j$$

and this completes the proof that $A_j\in\mathscr{G}(1,0;X_j)$ (see (4.17a,b,c)). Finally, we have from (6.47), (6.52), and (6.53)

$$P_j[D(A)] \subset X_j = D(A_j)$$

because $P_jf\in X_j \;\forall f\in X_0 \supset D(A)$, and also

$$P_jAf-A_jP_jf = \frac{v}{\delta_j}\begin{pmatrix} 0 \\ f(x_2)-\delta_jf'(x_2) \\ f(x_3)-f(x_2)-\delta_jf'(x_3) \\ \cdots\cdots\cdots\cdots \\ f(x_j)-f(x_{j-1})-\delta_jf'(x_j) \end{pmatrix} \quad \forall f\in D(A) \;.$$

However, if $f\in D(A)$, $f'(x)$ is continuous and hence uniformly continuous over $[a,b]$ and so $|f'(x)-f'(y)| < \varepsilon \;\forall x,y\in[a,b]$ with $|x-y| < \delta_\varepsilon$. Thus, if j_ε is such that $(b-a)/(j_\varepsilon-1) < \delta_\varepsilon$, then $x_i-x_{i-1} = \delta_j = (b-a)/(j-1) < \delta_\varepsilon \;\forall j \ge j_\varepsilon$ and consequently $|f'(x_{i+1})-f'(y)| < \varepsilon \;\forall y\in[x_i,x_{i+1}]$, $i = 1,2,\ldots,j$, with $j\ge j_\varepsilon$. It follows that for each $j \geq j_\varepsilon$

$$|f(x_{i+1})-f(x_i)-\delta_j f'(x_{i+1})| = |\int_{x_i}^{x_{i+1}} f'(y)\,dy-\delta_j f'(x_{i+1})|$$

$$= |\int_{x_i}^{x_{i+1}} [f'(y)-f'(x_{i+1})]\,dy \leq \int_{x_i}^{x_{i+1}} |f'(y)-f'(x_{i+1})|\,dy$$

$$< \varepsilon[x_{i+1}-x_i] = \varepsilon\delta_j$$

$\forall i = 1,2,\ldots,j$. Hence, $\forall j \geq j_\varepsilon$, we obtain

$$\|P_j Af-A_j P_j f;X_j\| = \frac{v}{\delta_j}\max\{|f(x_{i+1})-f(x_i)-\delta_j f'(x_{i+1})|,\ i = 1,\ldots,j-1\} < v\varepsilon .$$

Thus, the conditions (6.31) are satisfied and (6.28) holds.

We remark that the above procedure is a simple way of 'discretizing' the operator $-v\mathrm{d}/\mathrm{d}x$. □

EXERCISES

6.1 If $A\in\mathcal{G}(M,\beta;X)$ and $z\in\rho(A)$, show that the following commutation formula holds:

$$\exp(sA)R(z,A)f = R(z,A)\exp(sA)f \quad \forall\, s \geq 0,\ f\in X.$$

Hint: $(zI-A)\exp(sA)\hat{f} = \exp(sA)(zI-A)\hat{f}\ \forall\, s \geq 0,\ \hat{f}\in D(A)$, because of (4.36c).

6.2. If (a) $A\in\mathcal{G}(M,\beta;X)$, (b) $u_0\in D(A)$, (c) $g = g(t)\in C^1([0,\hat{t}];X)$ (i.e. $g(t)$ is continuously differentiable over $[0,\hat{t}]$ in the sense of $\|\cdot;X\|$), then the non-homogeneous initial-value problem (4.2) has a unique strict solution $u = u(t)$ given by (4.49)$\forall t\in[0,\hat{t}]$. Consider also the following sequence of non-homogeneous problems

$$\frac{\mathrm{d}}{\mathrm{d}t}u_j(t) = A_j u_j(t)+g_j(t), t>0;\ X\text{-}\lim_{t\to 0+} u_j(t) = u_{j,0},\quad j = 1,2,\ldots$$

and assume that (a_1) $A_j\in\mathcal{G}(M,\beta;X)$, $j = 1,2,\ldots$; (b_1) $u_{j,0}\in D(A_j)$, $j = 1,2,\ldots$; (c_1) $g_j = g_j(t)\in C^1([0,\hat{t}];X)$; (d_1) X-lim $\exp(tA_j)f =$

$\exp(tA)f$ as $j\to\infty$, uniformly with respect to $t\in[0,\hat{t}]$; (e_1) X-lim $g_j(t) = g(t)$ as $j\to\infty$ uniformly with respect to $t\in[0,\hat{t}]$; (f_1) X-lim $u_{j,0} = u_0$ as $j\to\infty$. Show that X, X- lim $u_j(t) = u(t)$ as $j\to\infty$ uniformly in $t\in[0,\hat{t}]$.

Hint: write $u(t)$ and $u_j(t)$ into the form (4.49) and then use Theorem 3.3. Note that (d_1) and (e_1) imply that $\exp(tA_j)f$ and $g_j(t)$ converge respectively to $\exp(tA)f$ and to $g(t)$ in the sense of the norm of $C([0,\hat{t}];X)$.

6.3. Let A and $\{A_j, j = 1,2,\ldots\}$ satisfy the assumptions of Theorem 6.1 and suppose that the operators B and $\{B_j, j = 1,2,\ldots\}$ belong to $\mathscr{B}(X)$ with $\|B_j\| \le b$ $\forall j = 1,2,\ldots$, and with X-lim $B_jf = Bf$ as $j\to\infty$ $\forall f\in X$. Prove that X-lim $\exp[t(A_j+B_j)]f = \exp[t(A+B)]f$ as $j\to\infty$, uniformly with respect to $t\in[0,\hat{t}]$.

Hint: see the proof of Theorem 6.3 with $X_j = X$, $P_j = I$ $\forall j = 1,2,\ldots$.

6.4. Assume that $w(t)$ is a strongly differentiable function from (a,b) into the B-space X and that (6.18a,b) are satisfied. Prove that $d[P_jw(t)]/dt = P_j\, dw(t)/dt$, where the derivative on the left-hand side is in the sense of $\|\cdot;X_j\|$ and the other one is in the sense of $\|\cdot;X\|$.

6.5. Prove that (6.18b) follows from (6.18a) and from the assumption $\lim\|P_jg;X_j\| = \|g;X\|$ as $j\to\infty$, $\forall g\in X_0$, where X_0 is a dense subset of X.

Hint: if $f\in X$ and $\varepsilon > 0$, a suitable $g\in X_0$ exists such that $|\|f;X\| - \|g;X\|| \le \|f-g;X\| < \varepsilon/2$, i.e. such that $0 \le \|g;X\| < \|f;X\| + \varepsilon/2$. Further, $0 \le \|P_jf;X_j\| \le \|P_jg;X_j\| + \|P_j[f-g];X_j\| < \|P_jg;X_j\| + \varepsilon/2$ and so $0 \le \lim_{j\to\infty} \|P_jf;X_j\| \le \|g;X\| + \varepsilon/2 < \|f;X\| + \varepsilon$ ... .

6.6. Assume that $\{X_j, j=1,2,\ldots\}$ is a sequence of B-spaces approximating X and that $A\in\mathscr{G}(M,\beta;X)$, $A_j\in\mathscr{G}(M,\beta;X_j)$ $\forall j = 1,2,\ldots$. Prove that formula (6.1) must be modified as follows:

$$R(z,A_j)\{P_j \exp(tA) - \exp(tA_j)P_j\}R(z,A)f$$

$$= \int_0^t \exp\{(t-s)A_j\}\{P_jR(z,A) - R(z,A_j)P_j\}\exp(sA)f\,ds, \quad f\in X, \quad z > \beta.$$

6.7. Prove that $D(A_j)$ is dense in X_j, where $D(A_j)$ is defined by (6.44a). Hint: if $\varepsilon > 0$ and $f_j \in X_j \subset X$, an element $\hat{f} \in D(A)$ exists such that $\| f_j - \hat{f} \| < \varepsilon$ because $D(A)$ in dense in X; but $\hat{g}_j = P_j \hat{f} \in D(A_j)$, $f_j = P_j f_j$ and so $\| f_j - \hat{g}_j \| = \| P_j [f_j - \hat{f}] \| \leq \ldots$.

6.8. Discuss the remark at the end of Example 6.4; in particular, show that

$$A\hat{f} = A \sum_{k=1}^{\infty} c_k \phi^{(k)} = \sum_{k=1}^{\infty} c_k z_k \phi^{(k)} \;\forall\; \hat{f} \in D(A) \qquad (A \in \mathscr{C}(X))$$

if the eigenvalues are bounded: $|z_k| \leq \alpha_0$ $\forall k = 1,2,\ldots$.

7

SPECTRAL REPRESENTATION OF CLOSED OPERATORS AND OF SEMIGROUPS

7.1. INTRODUCTION

We begin with an elementary example that illustrates the main purpose of this chapter. If A is the matrix operator

$$Af = \begin{pmatrix} a_{11} \cdots a_{1n} \\ \cdots\cdots \\ a_{n1} \cdots a_{nn} \end{pmatrix} \begin{pmatrix} f_1 \\ \cdot\cdot \\ f_n \end{pmatrix}, \qquad f = \begin{pmatrix} f_1 \\ \cdot\cdot \\ f_n \end{pmatrix} \in \mathbb{C}^n$$

where the a_{ij}'s are given complex numbers, then $A \in \mathscr{B}(\mathbb{C}^n)$ (see Example 2.2). Here, $\mathbb{C}^n$ is the Hilbert space of all ordered n-tuples of complex numbers with inner product and norm

$$(f,g) = \sum_{j=1}^{n} f_j \bar{g}_j, \quad \|f\| = \sqrt{(f,f)} = (\sum_{j=1}^{n} |f_j|^2)^{1/2}.$$

It is known from the elementary theory of linear systems that the equation $(zI-A)f = 0$ has non-trivial solutions if and only if the parameter z satisfies the algebraic equation of degree n $\det\{(zI-A)\} = 0$. Thus, if z_j is such that $\det\{(z_jI-A)\} = 0$, an element $\phi^{(j)} \in \mathbb{C}^n$ exists for which

$$(z_jI-A)\phi^{(j)} = 0, \quad \|\phi^{(j)}\| = 1, \quad j = 1,2,\ldots,n \tag{7.1}$$

and so z_j is an eigenvalue of A and $\phi^{(j)}$ is a corresponding normalized eigenfunction. Note that if $(z_jI-A)f^{(j)} = 0$ with $0 < \|f^{(j)}\| \neq 1$ (i.e. if $f^{(j)}$ is an eigenfunction that is not normalized), then $\phi^{(j)} = f^{(j)}/\|f^{(j)}\|$ satisfies (7.1) and $\|\phi^{(j)}\| = 1$. Now assume that $z_i \neq z_j \; \forall i \neq j$; then the eigenfunctions $\{\phi^{(1)}, \phi^{(2)}, \ldots, \phi^{(n)}\}$ are linearly independent (see the footnote on p. 234 and Exercise 7.1) and each $f \in \mathbb{C}^n$ can be put into the form

$$f = c_1\phi^{(1)} + c_2\phi^{(2)} + \ldots + c_n\phi^{(n)} \tag{7.2}$$

where the c_j's are suitable coefficients depending on the f

considered. We have from (7.2)

$$Af = c_1 A\phi^{(1)} + \ldots + c_n A\phi^{(n)} = c_1 z_1 \phi^{(1)} + \ldots + c_n z_n \phi^{(n)} \quad (7.3)$$

$$A^k f = c_1 z_1^k \phi^{(1)} + \ldots + c_n z_n^k \phi^{(n)}$$

and so

$$\exp(tA)f = f + \sum_{k=1}^{\infty} \frac{t^k}{k!} A^k f = [c_1\phi^{(1)} + \ldots + c_n\phi^{(n)}] + \sum_{k=1}^{\infty} \frac{t^k}{k!} [c_1 z_1^k \phi^{(k)} + \ldots + c_n z_n^k \phi^{(n)}]$$

$$\exp(tA)f = c_1 \exp(z_1 t)\phi^{(1)} + \ldots + c_n \exp(z_n t)\phi^{(n)}. \quad (7.4)$$

Relations (7.3) and (7.4) show how a *complete* knowledge of the structure of the spectrum of $A \in \mathscr{B}(\mathbb{C}^n)$ leads to simple expressions for Af and for $\exp(tA)f$. The following sections are devoted to investigating how a *partial* knowledge of the structure of the spectrum of an operator $A \in \mathscr{G}(M, \beta; X)$ leads to formulas similar in some sense to (7.3) and to (7.4).

7.2. PROJECTIONS

In this section we briefly examine some properties of projection operators that will be used in what follows.

An operator P is said to be a *projection* if

$$(a)\ P \in \mathscr{B}(X); \quad (b)\ P^2 f = Pf\ \forall f \in X. \quad (7.5)$$

Note that $Q = I - P$ is also a projection because $\|Q\| \le 1 + \|P\|$, $Q^2 f = (I-P)(I-P)f = (I-2P+P^2)f = (I-2P+P)f = Qf$. Moreover, $PQf = QPf = 0\ \forall f \in X$ because $PQf = QPf = (P-P^2)f = (P-P)f = 0$. Here and in what follows we use the symbol $\|\cdot\|$ to indicate the norm in both X and $\mathscr{B}(X)$ (see Theorem 2.1). If

$$M_1 = P[X] = \{g : g = Pf\ \forall f \in X\}$$
$$\quad (7.6)$$
$$M_2 = Q[X] = \{g : g = Qf\ \forall f \in X\}$$

i.e. if M_1 is the range of P and M_2 is the range of Q, then we have the following theorem.

Theorem 7.1. (a) $M_1 = \{g: g\in X;\ g = Pg\}$, $M_2 = \{g: g\in X;\ g = Qg\}$ *(b)* M_1 *and* M_2 *are closed linear subsets (closed subspaces) of the B-space* X*; (c) each* $f\in X$ *can be written as* $f = f_1+f_2$ *with* $f_1 = Pf\in M_1$, $f_2 = Qf\in M_2$, *and such a decomposition of* f *is unique.*

Proof. (*a*) To show that M_1 is the set of all $g\in X$ such that $g = Pg$, we first consider an element $f\in\{g: g\in X; g = Pg\}$. Then, $f = Pf$ and $f\in M_1$ because f is the image under P of $f\in X$ (see (7.6)); as a consequence $\{g: g\in X; g = Pg\} \subset M_1$. Conversely, if $g\in M_1$, then $g = Pf$ for some $f\in X$ because of (7.6) and $Pg = P^2f = Pf = g$ because of (7.5b). Hence, $g\in\{g: g\in X; g = Pg\}$ and $M_1 \subset \{g: g\in X; g = Pg\}$. We conclude that $M_1 = \{g: g\in X; g = Pg\}$ and, in an analogous way, $M_2 = \{g: g\in X; g = Qg\}$.

(*b*) The complement of M_1 with respect to X, $M_1^c = \{f: f\in X; f\notin M_1\}$ is open in X (see Example 1.5), because, if $f_0\in M_1^c$, the open sphere $S(f_0;r) = \{f: f\in X; \|f-f_0\| < r\}$ is contained in M_1^c if $r = \|f_0-Pf_0\|/[2(1+\|P\|)]$ (note that $r>0$ because $f_0\in M_1^c$ and so $f_0 \neq Pf_0$). In fact, we have for any $g\in S(f_0;r)$

$$0 < \|f_0-Pf_0\| = \|(f_0-g)+(g-Pg)+P[g-f_0]\| \leq \|f_0-g\| + \|g-Pg\| +$$

$$\|P\|\,\|g-f_0\| = (1+\|P\|)\|f_0-g\| + \|g-Pg\| \leq (1+\|P\|)r + \|g-Pg\|$$

$$= \frac{\|f_0-Pf_0\|}{2} + \|g-Pg\|$$

and so $0 < \|f_0-Pf_0\|/2 \leq \|g-Pg\|$, i.e. $g \neq Pg$. Hence, $g\notin M_1$ and $g\in M_1^c$. We conclude that $S(f_0;r) \subset M_1^c$ and that M_1^c is open. Of course, the same result holds for M_2^c.

(*c*) Obviously, $f = Pf+(I-P)f = f_1+f_2$ with $f_1 = Pf\in M_1$ and with $f_2 = (I-P)f = Qf\in M_2$. Such a decomposition is unique since, if $f = \phi_1+\phi_2$ with $\phi_1 = P\phi_1\in M_1$ and $\phi_2 = Q\phi_2\in M_2$, then $Pf = P\phi_1+P\phi_2 = P\phi_1$ because $P\phi_2 = PQ\phi_2 = 0$. Hence, $f_1 = \phi_1$ and, in an analogous way, $f_2 = \phi_2$. □

We say that X is the *direct sum* of the closed subspaces M_1 and M_2 and we write $X = M_1 \oplus M_2$ because *each* $f \in X$ can be *uniquely* expressed as

$$f = f_1 + f_2, \quad f_1 = Pf \in M_1, \quad f_2 = Qf \in M_2 \,. \tag{7.7}$$

In general, if the family of operators $\{P_1, \ldots, P_m\}$ is such that

$$\left.\begin{array}{l} (a)\ P_j \in \mathscr{B}(X);\ (b)\ P_j^2 f = P_j f\ \forall f \in X; \\ (c)\ P_i P_j f = 0\ \forall i \neq j,\ \forall f \in X;\ (d)\ \sum_{j=1}^{m} P_j f = f\ \forall f \in X \end{array}\right\} \tag{7.8}$$

then the results of Theorem 7.1 hold for $M_j = R(P_j)$ $\forall j = 1, 2, \ldots, m$, $X = M_1 \oplus M_2 \oplus \ldots \oplus M_m$, and each $f \in X$ can be uniquely expressed as

$$f = P_1 f + P_2 f + \ldots + P_m f, \quad P_j f \in M_j \,. \tag{7.9}$$

Example 7.1. Projections in R^n

If $X = R^n$, then $\{\phi^{(1)}, \phi^{(2)}, \ldots, \phi^{(n)}\}$ with

$$\phi^{(1)} = \begin{pmatrix} 1 \\ 0 \\ \cdots \\ 0 \end{pmatrix}, \quad \phi^{(2)} = \begin{pmatrix} 0 \\ 1 \\ \cdots \\ 0 \end{pmatrix}, \ \ldots, \ \phi^{(n)} = \begin{pmatrix} 0 \\ 0 \\ \cdots \\ 1 \end{pmatrix}$$

is a family of linearly independent elements of R^n and so each $f \in R^n$ can be put into the form

$$f = (f, \phi^{(1)})\phi^{(1)} + \ldots + (f, \phi^{(n)})\phi^{(n)} \tag{7.10}$$

because $\dim[R^n] = n$ (see Example 6.4 and the footnote on p.234. The operator

$$P_h f = (f, \phi^{(h)})\phi^{(h)}, \quad f \in D(P_h) = R^n \tag{7.11}$$

is a projection because

$$\|P_h f\|^2 = (P_h f, P_h f) = |(f, \phi^{(h)})|^2 \|\phi^{(h)}\|^2 = |(f, \phi^{(h)})|^2$$

$$\leq \sum_{h=1}^{n} |(f,\phi^{(h)})|^2 = \|f\|^2$$

and $P_h \in \mathscr{B}(R^n)$ with $\|P_h\| \leq 1$. Furthermore,

$$P_h^2 f = P_h[(f,\phi^{(h)})\phi^{(h)}] = (f,\phi^{(h)})[P_h\phi^{(h)}]$$

$$= (f,\phi^{(h)})[(\phi^{(h)},\phi^{(h)})\phi^{(h)}] = (f,\phi^{(h)})\phi^{(h)} = P_h f$$

since $(\phi^{(h)},\phi^{(h)}) = \|\phi^{(h)}\|^2 = 1$. Note that (7.10) can be written as $f = P_1 f + \ldots + P_n f$, and so $P_1 + \ldots + P_n = I$. Finally, we have from (7.11)

$$P_j P_h f = P_j[(f,\phi^{(h)})\phi^{(h)}] = (f,\phi^{(h)})[P_j\phi^{(h)}]$$

$$= (f,\phi^{(h)})[(\phi^{(h)},\phi^{(j)})\phi^{(j)}] = 0 \quad \forall j \neq h$$

because $(\phi^{(h)},\phi^{(j)}) = 0$ if $j \neq h$. Hence, $\{P_1,\ldots,P_n\}$ is a family of projections that satisfies (7.8) with $m = n$. □

Example 7.2. Projections on a subspace of a Hilbert space.

Let X be the Hilbert space of Example 6.4 and assume that $Y = X_m$, $P = P_m$ (see (6.40) and (6.41) with $j = m$), where m is a given positive integer. Then, P is a projection operator and $R(P) = P[X] = Y$. □

Example 7.3. A projection operator in L^1

Let $X = L^1(\Omega)$ with $\Omega = (a,b)\times(a_1,b_1)$, $-\infty < a_1 < b_1 < +\infty$, and define the operator

$$Pf = (b_1-a_1)^{-1}\int_{a_1}^{b_1} f(x,y')\,dy', \quad f \in D(P) = X \tag{7.12}$$

that arises from monoenergetic neutron transport theory (Bell and Glasstone 1970, Chapter 2). Then, $P \in \mathscr{B}(X)$ because

$$|Pf| \leq (b_1-a_1)^{-1}\int_{a_1}^{b_1} |f(x,y')|\,dy', \ \|Pf;X\|$$

$$= \int_a^b dx \int_{a_1}^{b_1} |Pf| \, dy \le \int_a^b dx \int_{a_1}^{b_1} |f(x,y')| \, dy' = \|f;X\| \quad \forall f \in X;$$

moreover

$$P^2 f = P[(b_1 - a_1)^{-1} \int_{a_1}^{b_1} f(x,y')dy]$$

$$= (b_1 - a_1)^{-1} \int_{a_1}^{b_1} dy''[(b_1 - a_1)^{-1} \int_{a_1}^{b_1} f(x,y')dy'] = (b_1 - a_1)^{-1} \int_{a_1}^{b_1} f(x,y')dy' = Pf$$

and P is a projection. □

7.3. ISOLATED POINTS OF THE SPECTRUM OF $A \in \mathscr{C}(X)$

Assume that the spectrum $\sigma(A)$ of an operator $A \in \mathscr{C}(X)$ satisfies the following conditions: (i) $\sigma(A)$ is composed of a single point $\sigma_i(A) = \{z_1\}$ and of a part $\sigma_0(A)$, i.e. $\sigma(A) = \{z_1\} \cup \sigma_0(A)$; (ii) $\{z_1\}$ is an *isolated* point of $\sigma(A)$. Condition (ii) means that an open ball

$$S_1 = S(z_1;r_0) = \{z : z \in \mathbb{C}; \ |z - z_1| < r_0\}$$

exists, such that each $z \in S_1$ and different from z_1 is contained in the resolvent set $\rho(A)$ (see section 2.7). Then, the operator P_1 is defined as follows

$$P_1 = \frac{1}{2\pi i} \int_\Gamma R(\lambda,A) d\lambda, \qquad D(P_1) = X \tag{7.13}$$

where $i = \sqrt{(-1)}$ and Γ is a *regular closed curve* contained in $\rho(A)$ and enclosing an open set containing the point $\{z_1\}$ (see section 3.5). By the locution 'regular closed curve', we mean that $\Gamma \subset \mathbb{C}$ is defined by an equation $\lambda = \lambda(s)$, $s \in [0,1]$, such that (a) $\lambda(s) = x(s) + iy(s)$ where $x(s)$ and $y(s)$ are continuously differentiable functions of $s \in [0,1]$, with $x(0) = x(1)$, $y(0) = y(1)$, (b) $x(s) + iy(s) \ne x(s_1) + iy(s_1)$ $\forall s, s_1 \in (0,1)$ with $s \ne s_1$, (c) when s increases from 0 to 1, the point $\lambda(s) \in \mathbb{C}$ describes Γ counterclockwise. For instance, Γ might be the circle of radius $r < r_0$ and with centre z_1: $\lambda(s) = z_1 + r \exp(2\pi i s) = z_1 + r[\cos(2\pi s) + i \sin(2\pi s)], s \in [0,1]$. Note that the definition (7.13) does *not* depend on the curve

Γ chosen to evaluate the integral because

$$\int_{\Gamma} R(\lambda,A)\,\mathrm{d}\lambda = \int_{\Gamma_1} R(\lambda_1,A)\,\mathrm{d}\lambda_1 \tag{7.14}$$

where Γ and Γ_1 are both regular closed curves contained in $\rho(A)$ and enclosing the point $\{z_1\}$. Relation (7.14) follows from the fact that $R(\lambda,A)$ is a holomorphic function of $\lambda\in\rho(A)$ (see Example 3.9) and will not be proved here[†]. We also remark that the integral on the right-hand side of (7.13) is not identically zero because Γ encloses the point $\{z_1\}$ that does not belong to $\rho(A)$ and so (3.40) does not hold. We have the following theorem.

Theorem 7.2. The operator P_1 *has the properties* (*a*) $P_1\in\mathscr{B}(X)$, $P_1 = P_1^2$, *i.e.* P_1 *is a projection;* (*b*) $P_1R(z,A)g = R(z,A)P_1g$ $\forall g\in X, z\in\rho(A)$; (*c*) $P_1f\in D(A^j)$ $\forall f\in X$, $j = 1,2,\ldots$; (*d*) $P_1A\hat{f} = AP_1\hat{f}$ $\forall\hat{f}\in D(A)$.

Proof. (*a*) If Γ is the circle of equation $\lambda(s) = z_1 + r\exp(2\pi\mathrm{i}s), s\in[0,1]$ with $0 < r < r_0$, we have from (7.13) and from (3.38)

$$P_1 = \frac{1}{2\pi\mathrm{i}}\int_0^1 R(\lambda(s),A)\{2\pi\mathrm{i}\; r\exp(2\pi\mathrm{i}s)\}\mathrm{d}s$$

$$\|P_1\| \le r\int_0^1 \|R(\lambda(s)),A)\|\,\mathrm{d}s < \infty \tag{7.15}$$

because $R(\lambda(s),A)$ is a holomorphic function from $\rho(A)$ into $\mathscr{B}(X)$ (see Example 3.9), and so $\|R(\lambda(s),A)\|$ is continuous in $s\in[0,1]$. Since the domain of P_1 coincides with the whole

---

[†] By using Cauchy's integral formula (3.40) (which holds under assumptions weaker than (i)-(iv) of section 3.5) it can be shown that

$$\int_{\Gamma} w(\lambda)\,\mathrm{d}\lambda = \int_{\Gamma_1} w(\lambda')\,\mathrm{d}\lambda'$$

where $w(\lambda)$ is holomorphic in the open set $\Omega\subset\mathbb{C}$ and Γ and Γ_1 are regular closed curves contained in Ω and enclosing the point $\{z_1\}$ at which $w(\lambda)$ is not holomorphic. In particular, the above relation holds if $w(\lambda) = R(\lambda,A)$ and $\Omega = \{\lambda:\lambda\in S_1;\lambda\neq z_1\}$ (see for instance Taylor 1958, Chapter 5).

space X because $D(R(\lambda,A)) = X$, inequality (7.15) shows that $P_1 \in \mathscr{B}(X)$. As far as the relation $P_1^2 = P_1$ is concerned, the definition (7.13) and the discussion after (7.14) give

$$P_1^2 = P_1\left\{\frac{1}{2\pi i}\int_\Gamma R(\lambda,A)d\lambda\right\} = \frac{1}{2\pi i}\int_{\Gamma_1} R(\lambda_1,A)\left[\frac{1}{2\pi i}\int_\Gamma R(\lambda,A)d\lambda\right]d\lambda_1$$

where Γ_1 is, say, the circle $\lambda_1(s) = z_1+r_1\exp(2\pi i s_1)$, $s_1\in[0,1]$, with $r < r_1 < r_0$. Hence,

$$\begin{aligned} P_1^2 &= -\frac{1}{4\pi^2}\int_{\Gamma_1} d\lambda_1 \int_\Gamma R(\lambda_1,A)R(\lambda,A)d\lambda \\ &= \frac{1}{4\pi^2}\int_{\Gamma_1} d\lambda_1 \int_\Gamma \frac{R(\lambda_1,A)-R(\lambda,A)}{\lambda_1-\lambda}\,d\lambda \\ &= \frac{1}{4\pi^2}\int_{\Gamma_1} R(\lambda_1,A)\left[\int_\Gamma \frac{d\lambda}{\lambda_1-\lambda}\right]d\lambda_1 - \frac{1}{4\pi^2}\int_\Gamma R(\lambda,A)\left[\int_{\Gamma_1}\frac{d\lambda_1}{\lambda_1-\lambda}\right]d\lambda \end{aligned}$$

because $R(\lambda_1,A)-R(\lambda,A) = -(\lambda_1-\lambda)R(\lambda_1,A)R(\lambda,A)$ (see Exercise 2.15). Note that $\lambda_1-\lambda \neq 0$ since $\lambda\in\Gamma = \{z: z\in\mathbb{C};\ |z-z_1| = r\}$ and $\lambda_1 = \Gamma_1 = \{z: z\in\mathbb{C};\ |z-z_1| = r_1\}$ with $r < r_1$. Using (3.40) we obtain

$$\int_\Gamma \frac{d\lambda}{\lambda_1-\lambda} = 0$$

because λ_1 is a point outside Γ and so $w(\lambda) = 1/(\lambda_1-\lambda)$ is holomorphic in any open set $\Omega \subset \mathbb{C}$ such that $\Omega \supset \Gamma$ and $\lambda_1 \notin \Omega$. Further, since λ belongs to the interior of Γ_1

$$\begin{aligned} \int_{\Gamma_1}\frac{d\lambda_1}{\lambda_1-\lambda} &= \int_\gamma \frac{d\hat\lambda}{\hat\lambda-\lambda} = \int_0^1 \frac{2\pi i\hat r\exp(2\pi i\hat s)}{\hat r\exp(2\pi i\hat s)}d\hat s \\ &= 2\pi i\int_0^1 d\hat s = 2\pi i \end{aligned}$$

where γ is the (small) circle of radius $\hat r$ and with centre λ, defined by the equation $\hat\lambda(s) = \lambda+\hat r\exp(2\pi i\hat s)$, $\hat s\in[0,1]$ (see the footnote on p. 250). Thus,

$$P_1^2 = -\frac{2\pi i}{4\pi^2}\int_\Gamma R(\lambda,A)d\lambda = P_1$$

and (*a*) is proved.

(*b*) Since $R(z,A) \in \mathscr{B}(X) \subset \mathscr{C}(X)$ and $R(z,A)R(\lambda,A) = R(\lambda,A)R(z,A)$ $\forall \lambda, z \in \rho(A)$ (see Exercise 2.15), we have from (7.13)

$$2\pi i\, P_1[R(z,A)] = \int_\Gamma R(\lambda,A)\{R(z,A)\}d\lambda = \int_\Gamma R(z,A)R(\lambda,A)d\lambda$$

$$= R(z,A)[\int_\Gamma R(\lambda,A)d\lambda] = 2\pi i\, R(z,A)P_1 .$$

(*c*), (*d*) As in the proof of (*a*)

$$P_1 f = r\int_0^1 R(\lambda(s),A)f\, \exp(2\pi i s)ds$$

where $R(\lambda(s),A)f$ belongs to $D(A)$ and is strongly continuous $\forall s \in [0,1]$. Hence, $P_1 f \in D(A)$ because of (3.25),

$$AP_1 f = r\int_0^1 AR(\lambda(s),A)f\, \exp(2\pi i s)ds$$

and so $AP_1\hat{f} = P_1 A\hat{f}$ because $AR(\lambda(s),A)\hat{f} = R(\lambda(s),A)A\hat{f}$ $\forall \hat{f} \in D(A), s \in [0,1]$ (see Exercise 2.16). Finally, since $P_1^2 f = P_1 f \in D(A)$ $\forall f \in X$, we have $AP_1 f = AP_1^2 f = AP_1(P_1 f) = P_1 A(P_1 f)$ $= P_1(AP_1 f) = P_1 g$ with $g = AP_1 f$. Hence, $AP_1 f \in D(A)$ because $AP_1 f = P_1 g$, and so $A(AP_1 f) = A^2 P_1 f$ is defined, i.e. $P_1 f \in D(A^2)$. Iterating this procedure, we obtain that $P_1 f \in D(A^j)$ $\forall f \in X$, $j = 1,2,\ldots$ and Theorem 7.2 is completely proved. □

The above results can be generalized to the case in which $\sigma(A) = [\sigma_1(A) \cup \sigma_2(A) \cup \ldots \cup \sigma_n(A)] \cup \sigma_0(A)$ where $\sigma_j(A), j = 1,\ldots,n$, is a *bounded* subset of $\sigma(A)$ completely 'surrounded' by points of $\rho(A)$ (of course, $\sigma_j(A)$ might be an isolated point $\{z_j\}$ of the spectrum). In fact, if Γ_j is a regular closed curve contained in $\rho(A)$ and enclosing an open set containing $\sigma_j(A)$ and if

$$P_j = \frac{1}{2\pi i}\int_{\Gamma_j} R(\lambda_j,A)d\lambda_j, \quad D(P_j) = X, \quad j = 1,2,\ldots,n \tag{7.16}$$

then the results of Theorem 7.2 hold for each of the P_j's. Further, if the Γ_j's lie outside each other, it can be shown that

$$P_j P_h = 0 \quad \text{if } j \neq h \tag{7.17}$$

(see Exercise 7.2). It is also easy to verify that the operator

$$P_0 = I - \sum_{j=1}^{n} P_j \tag{7.18}$$

is a projection as well, and so $X = M_1 \oplus \ldots \oplus M_n \oplus M_0$, with $M_j = R(P_j), j = 0,1,\ldots,n$ (see (7.8),(7.9) with $m = n+1$).

7.4. LAURENT EXPANSION OF $R(z,A)$

As in section 7.3 we assume that $\sigma(A) = \{z_1\} \cup \sigma_0(A)$ and that $\{z_1\}$ is an isolated point of $\sigma(A)$. Then, if $z \in \rho(A)$ and $z \notin \Gamma$, we have from (7.13)

$$2\pi i\ R(z,A)P_1 = \int_\Gamma R(z,A)R(\lambda,A)\,d\lambda = \int_\Gamma \frac{d\lambda}{\lambda - z} R(z,A) - \int_\Gamma \frac{1}{\lambda - z} R(\lambda,A)\,d\lambda$$

because $R(z,A)R(\lambda,A) = (\lambda-z)^{-1}[R(z,A)-R(\lambda,A)]$. Hence,

$$R(z,A)P_1 = -\frac{1}{2\pi i}\int_\Gamma \frac{1}{\lambda-z} R(\lambda,A)\,d\lambda \quad \text{if } z \text{ is } \textit{outside } \Gamma \tag{7.19a}$$

$$R(z,A)P_1 = R(z,A) - \frac{1}{2\pi i}\int_\Gamma \frac{1}{\lambda-z} R(\lambda,A)\,d\lambda \quad \text{if } z \text{ is } \textit{inside } \Gamma \tag{7.19b}$$

since, as in the proof of (*a*) of Theorem 7.2, the integral $\int_\Gamma (\lambda-z)^{-1}d\lambda$ is zero if z is outside Γ and it is equal to $2\pi i$ if z is inside Γ. Note that Γ is *any* regular closed curve contained in $\rho(A)$ and enclosing an open set containing the point $\{z_1\} \in \sigma(A)$ and it is then advisable to choose a 'small' circle with centre z_1 for the curve Γ in (7.19a) and a 'large' circle for the curve Γ in (7.19b).

Now, if z is a *given* complex number belonging to the ball $S_1 = S(z_1;r_0)$ (see the beginning of section 7.3), then $z \in \rho(A)$ and is outside the 'small' circle $\Gamma = \{\lambda : \lambda \in \mathbb{C}; |\lambda - z_1| = r\}$ with $0 < r < |z-z_1|$. Moreover, since $|\lambda-z_1|/|z-z_1| = r/|z-z_1| < 1$ $\forall \lambda \in \Gamma$, we have

$$\frac{1}{\lambda - z} = -\frac{1}{z-z_1}\left[1 - \frac{\lambda - z_1}{z-z_1}\right]^{-1} = -\frac{1}{z-z_1}\sum_{k=0}^{\infty}\left(\frac{\lambda-z_1}{z-z_1}\right)^k$$

and substituting the above expansion into (7.19a) we obtain

$$R(z,A)P_1 = (z-z_1)^{-1}P_1 + \sum_{k=1}^{\infty} (z-z_1)^{-k-1} \frac{1}{2\pi i} \int_{\Gamma} (\lambda-z_1)^k R(\lambda,A)\,d\lambda. \tag{7.20}$$

If we put

$$D_1 = \frac{1}{2\pi i} \int_{\Gamma} (\lambda-z_1)R(\lambda,A)\,d\lambda \tag{7.21}$$

by procedures similar to those of Theorem 7.2, we have (see Exercise 7.3)

(*a*) $D_1 \in \mathscr{B}(X)$

(*b*) $P_1D_1f = D_1P_1f = D_1f \;\forall f \in X$

(*c*) $D_1f \in D(A^k) \;\forall f \in X,\; k = 1,2,\ldots, D_1A\hat{f} = AD_1\hat{f} \;\forall \hat{f} \in D(A)$ (7.22)

(*d*) $D_1f = -(z_1I-A)P_1f \quad \forall f \in X$

$$D_1^k = \frac{1}{2\pi i} \int_{\Gamma} (\lambda-z_1)^k R(\lambda,A)\,d\lambda, \quad k = 1,2,\ldots \tag{7.23}$$

and (7.20) becomes

$$R(z,A)P_1 = (z-z_1)^{-1}P_1 + \sum_{k=1}^{\infty} (z-z_1)^{-k-1} D_1^k . \tag{7.24}$$

The importance of the operators $D_1, D_1^2, \ldots,$ will be explained in section 7.5.

On the other hand, if Γ_1 is the 'large' circle $\{\lambda' : \lambda' \in \mathbb{C};\; |\lambda'-z_1| = r_1\}$ with $|z-z_1| < r_1 < r_0$, then z is inside Γ_1 and, for any $\lambda' \in \Gamma_1$,

$$\frac{1}{\lambda'-z} = \frac{1}{\lambda'-z_1}\left[1 - \frac{z-z_1}{\lambda'-z_1}\right]^{-1} = \frac{1}{\lambda'-z_1} \sum_{k=0}^{\infty} \left(\frac{z-z_1}{\lambda'-z_1}\right)^k$$

because $|z-z_1|/|\lambda'-z_1| = |z-z_1|/r_1 < 1 \;\forall \lambda' \in \Gamma_1$. Substituting the above expansion into (7.19b) with Γ_1 instead of Γ, we have

$$R(z,A)P_1 = R(z,A) - \sum_{k=0}^{\infty} (z-z_1)^k \frac{1}{2\pi i} \int_{\Gamma_1} (\lambda'-z_1)^{-k-1} R(\lambda',A)\,d\lambda' . \tag{7.25}$$

If we put

$$H_1 = \frac{1}{2\pi i} \int_{\Gamma_1} (\lambda'-z_1)^{-1} R(\lambda',A)\,d\lambda' \tag{7.26}$$

by the usual procedures we have (see Exercises 7.3, 7.4, and 7.5)

$$\left.\begin{aligned}
&(a) \quad H_1 \in \mathscr{B}(X);\\
&(b) \quad P_1 H_1 f = H_1 P_1 f = 0 \quad \forall f \in X;\\
&(c) \quad H_1 \hat{f} \in D(A),\ H_1 A\hat{f} = AH_1\hat{f} \quad \forall \hat{f} \in D(A);\\
&(d) \quad H_1(z_1 I - A)\hat{f} = (I-P_1)\hat{f} \quad \forall \hat{f} \in D(A)
\end{aligned}\right\} \tag{7.27}$$

$$H_1^k = \frac{1}{2\pi i} \int_{\Gamma_1} (\lambda'-z_1)^{-k} R(\lambda',A)\,d\lambda', \qquad k = 1,2,\ldots, \tag{7.28}$$

and (7.25) becomes

$$R(z,A)P_1 = R(z,A) - \sum_{k=0}^{\infty} (z-z_1)^k H_1^{k+1} . \tag{7.29}$$

The series on the right-hand side of (7.29) and the operator H_1 will play a fundamental role in section 7.5 (see Theorem 7.5).

We finally obtain from (7.24) and from (7.29)

$$R(z,A) = (z-z_1)^{-1} P_1 + \sum_{k=1}^{\infty} (z-z_1)^{-k-1} D_1^k + \sum_{k=0}^{\infty} (z-z_1)^k H_1^{k+1} \tag{7.30}$$

which is the *Laurent expansion* of $R(z,A)$ relative to the isolated point $\{z_1\} \in \sigma(A)$.

Remark 7.1.

The expansion (7.30) holds for *any* $z \in \{z : z \in \mathbb{C};\, 0 < |z-z_1| < r_0\}$, i.e.

for any z belonging to the ball $S_1 = S(z_1;r_0)$ and different from z_1. In fact, if z is given so that $0 < |z-z_1| < r_0$, the radius of Γ can be chosen arbitrarily small (and hence smaller than $|z-z_1|$) and the radius of Γ_1 can be taken arbitrarily close to r_0 (and hence larger than $|z-z_1|$). □

7.5. ISOLATED EIGENVALUES

If z_1 is an eigenvalue of the closed operator A, the linear subset of the B-space X

$$N(z_1I-A) = \{f: f\in D(A); (z_1I-A)f = 0\} \tag{7.31}$$

composed of all the $f\in D(A)$ that satisfy the equation $(z_1I-A)f = 0$ is the *geometric eigenspace* for the eigenvalue z_1 and $m_g = \dim[N(z_1I-A)]$ is the *geometric multiplicity* of z_1. Note that, since $z_1\in P_\sigma(A)$, an element $g\in D(A)$ exists such that $(z_1I-A)g = 0$ with $\|g;X\| \neq 0$ and so $N(z_1I-A)$ contains at least all the f proportional to g including the zero element of X.

Moreover, if for instance $m_g = 2$ and $\phi^{(1)}$ and $\phi^{(2)}$ are linearly independent eigenfunctions for the eigenvalue z_1, then $f\in N(z_1I-A)$ if and only if $f = c_1\phi^{(1)}+c_2\phi^{(2)}$ (see the footnote on p.234).

Now if we assume that the eigenvalue z_1 is also an *isolated* point of $\sigma(A)$, then the expansion (7.30) holds, the range of the operator P_1, $R(P_1) = M_1 = \{g: g = P_1f, f\in X\}$ is the *algebraic eigenspace* for the isolated eigenvalue z_1, and $m_a = \dim[M_1]$ is the *algebraic multiplicity* of z_1. We have

$$N(z_1I-A) \subset M_1, \quad m_g = \dim[N(z_1I-A)] \leq \dim[M_1] = m_a \tag{7.32}$$

because, if $\hat{f}\in N(z_1I-A)$, then $(z_1I-A)\hat{f} = 0$ and (7.27d) gives $\hat{f}-P_1\hat{f} = (I-P_1)\hat{f} = H_1(z_1I-A)\hat{f} = 0$, i.e. $\hat{f} = P_1\hat{f}\in M_1$ because of Theorem 7.1. Thus, $\hat{f}\in N(z_1I-A)$ implies $\hat{f}\in M_1$ and (7.32) is proved.

The case of an isolated eigenvalue with *finite* algebraic multiplicity, $m_a < \infty$, is of particular interest as is shown by the following theorem.

Theorem 7.3. If $z_1 \in P_\sigma(A)$ *is an isolated point of* $\sigma(A)$ *and its algebraic multiplicity* m_a *is finite, then (a)* $D_1^k f = 0$ $f \in X$, $k \geq m_a$, *(b)* $M_1 = R(P_1) = \{f : f \in D(A^m); (z_1 I - A)^m f = 0;\ m = m_a\}$.

Proof. (*a*) We only sketch the proof of (*a*) (for more details see Kato 1966, pp.181, 39, 22; Taylor 1958, p.306): since the first series on the right-hand side of (7.30) is convergent in the sense of $\|\cdot\| = \|\cdot;\mathscr{B}(X)\|$, the spectral radius of D_1 is zero, $\mathrm{spr}[D_1] = \lim_{k\to\infty} \|D_1^k\|^{1/k} = 0$. Hence, also $\mathrm{spr}[\hat{D}_1] = 0$ where $\hat{D}_1$ is the restriction of D_1 to the m_a-dimensional subspace $M_1 : \hat{D}_1 g = D_1 g\ \forall g \in M_1$, and so we have that $\hat{D}_1^m g = 0\ \forall g \in M$, with $m \geq m_a$ (Kato 1966, Problem 3.10, p.22 and Problem 5.6, p.38). However, if $f \in X$, then $P_1 f \in M_1$ and (7.22b) gives

$$D_1^m f = D_1^m P_1 f = \hat{D}_1^m P_1 f = 0 \quad \forall\, m \geq m_a.$$

(*b*) If $g \in M_1$, then $g = P_1 g$ and (*a*) gives: $0 = D_1^m g = [-(z_1 I - A)]^m g$ with $m \geq m_a$ (see also (i) of Exercise 7.4). Hence, $(z_1 I - A)^m g = 0$ with $m = m_a$ and so $g \in M_1$ implies

$$g \in \Lambda = \{f : f \in D(A^m); (z_1 I - A)^m f = 0; m = m_a\},$$

i.e. $M_1 \subset \Lambda$. On the other hand, if $g \in \Lambda$, then

$$(I - P_1) g = (I - P_1)^m g = H_1^m (z_1 I - A)^m g = 0$$

with $m = m_a$ and so $g = P_1 g \in M_1$. Hence, $g \in \Lambda$ implies $g \in M_1$, i.e. we also have $\Lambda \subset M_1$ and (*b*) is proved. □

Note that the first series on the right-hand side of (7.30) becomes a finite sum of m_a terms if the isolated eigenvalue z_1 has algebraic multiplicity $m_a < \infty$. In particular, if $m_a = 1$, the isolated eigenvalue z_1 is said to be *simple*, $m_g = 1$ because of (7.32), and Theorem 7.3 shows that $D_1 f = 0\ \forall f \in X$ and that

$$M_1 = \{f : f \in D(A); (z_1 I - A) f = 0\} = N(z_1 I - A).$$

Finally, the following theorem partially inverts the

results of Theorem 7.3.

Theorem 7.4. If z_1 is an isolated point of $\sigma(A)$ and $D_1^m f = 0\ \forall f \in X$ with $D_1^{m-1} f \not\equiv 0$, then

$$(a_1) \quad z_1 \in P_\sigma(A)$$

$$(b_1) \quad M_1 = R(P_1) = \{f : f \in D(A^m);\ (z_1 I - A)^m f = 0\}.$$

Proof. (a_1) Since $D_1^m f = 0\ \forall f \in X$ and $D_1^{m-1} f \not\equiv 0$, an element $g \in X$ exists such that $D_1^m g = 0$, $D_1^{m-1} g \neq 0$ and so (ii) of Exercise 7.4 gives

$$0 = D_1^m g = -(z_1 I - A) D_1^{m-1} g.$$

Hence, $f_0 = D_1^{m-1} g$ satisfies the equation $(z_1 I - A) f_0 = 0$ and $\| f_0; X \| \neq 0$. We conclude that $z_1 \in P_\sigma(A)$ and that f_0 is a corresponding eigenfunction.

(b_1) See the proof of (b) of Theorem 7.3. □

If in particular $D_1 f = 0\ \forall f \in X$, Theorem 7.4 shows that $z_1 \in P_\sigma(A)$ and that $M_1 = N(z_1 I - A)$. Then, the eigenvalue z_1 is said to be *semisimple*. Note that, if $m_a = m_g = 1$, z_1 is a simple eigenvalue and $D_1 f = 0\ \forall f \in X$, and so a simple eigenvalue is also semisimple. The converse statement is not necessarily true.

Example 7.4. Spectral properties of an operator in $\mathbb{C}^2$

Let X be the Hilbert space $\mathbb{C}^2$ and assume that the matrix operator A is defined as follows:

$$Af = \begin{pmatrix} a_{11}, & a_{12} \\ a_{21}, & a_{22} \end{pmatrix} \begin{pmatrix} f_1 \\ f_2 \end{pmatrix}, \quad f = \begin{pmatrix} f_1 \\ f_2 \end{pmatrix} \in D(A) = X = \mathbb{C}^2 \tag{7.33}$$

(see the example in section 7.1 with $n = 2$). The equation $(zI - A) f = g$, i.e. the linear system

$$(z-a_{11})f_1-a_{12}f_2 = g_1, \quad -a_{21}f_1+(z-a_{22})f_2 = g_2$$

has the solution f uniquely determined by g:

$$f = \begin{pmatrix} f_1 \\ f_2 \end{pmatrix} = \frac{1}{\det\{(zI-A)\}} \begin{pmatrix} z-a_{22}, & a_{12} \\ a_{21}, & z-a_{11} \end{pmatrix} \begin{pmatrix} g_1 \\ g_2 \end{pmatrix} = R(z,A)g \qquad (7.34)$$

if $\det\{(zI-A)\} = (z-a_{11})(z-a_{22})-a_{12}a_{21} \neq 0$, (Cramer's rule), and so $\rho(A) = \{z:z\in\mathbb{C};\ \det\{(zI-A)\} \neq 0\}$. If z_1 and z_2 are the solutions of the equation $\det\{(zI-A)\} = 0$

$$z_1 = \tfrac{1}{2}\{(a_{11}+a_{22})-\sqrt{\Delta}\}, \qquad z_2 = \tfrac{1}{2}\{(a_{11}+a_{22})+\sqrt{\Delta}\}$$

where $\sqrt{\Delta}$ is the principal square root of

$$\Delta = (a_{11}-a_{22})^2+4a_{12}a_{21},$$

then the equation $(zI-A)f = 0$ with $z = z_1$ or with $z = z_2$ has non-trivial solutions and so

$$P_\sigma(A) = \{z_1,z_2\} = \sigma(A).$$

(a) If $\Delta \neq 0$, then $z_1 \neq z_2$ and

$$\phi^{(1)} = \begin{pmatrix} a_{12} \\ z_1-a_{11} \end{pmatrix}, \qquad \phi^{(2)} = \begin{pmatrix} z_2-a_{22} \\ a_{21} \end{pmatrix}$$

are such that $(z_jI-A)\phi^{(j)} = 0$, $\|\phi^{(j)};X\| \neq 0$, $j = 1,2$, i.e. $\phi^{(1)}$ and $\phi^{(2)}$ are eigenfunctions for z_1 and for z_2 respectively. Furthermore, $\phi^{(1)}$ and $\phi^{(2)}$ are linearly independent elements of $X = \mathbb{C}^2$ (see Exercise 7.1) and each $f\in X$ can be put into the form

$$f = \begin{pmatrix} f_1 \\ f_2 \end{pmatrix} = c_1(f)\phi^{(1)} + c_2(f)\phi^{(2)} \qquad (7.35a)$$

where the coefficients c_1 and c_2 are uniquely determined by f:

$$c_1(f) = \frac{-a_{21}f_1+(z_2-a_{22})f_2}{(z_1-a_{11})(z_2-a_{22})-a_{12}a_{21}}$$

$$c_2(f) = \frac{(z_1-a_{11})f_1-a_{12}f_2}{(z_1-a_{11})(z_2-a_{22})-a_{12}a_{21}} \tag{7.35b}$$

as is easy to verify by equating each of the components of f to the corresponding components of the right-hand side of (7.35a). Note that $N(z_jI-A) = \{f:f = c\phi^{(j)}\forall c\in\mathbb{C}\}$ because $(z_jI-A)f = 0$ if and only if $f = c\phi^{(j)}$ and so $\dim[N(z_jI-A)] = 1$, $j = 1,2$. On the other hand, since $\det\{(zI-A)\} = (z-z_1)(z-z_2)$ with $z_1 \neq z_2$, we have

$$\frac{b}{\det\{(zI-A)\}} = \frac{b}{z_1-z_2}\left[\frac{1}{z-z_1} - \frac{1}{z-z_2}\right], \qquad (b = a_{12} \text{ or } b = a_{21})$$

$$\frac{z-a}{\det\{(zI-A)\}} = \frac{1}{z_1-z_2}\left[\frac{z_1-a}{z-z_1} - \frac{z_2-a}{z-z_2}\right], \qquad (a = a_{11} \text{ or } a = a_{22})$$

and substituting into (7.34)

$$R(z,A) = (z-z_1)^{-1}[(z_1-z_2)^{-1}R_1] + (z-z_2)^{-1}[(z_2-z_1)^{-1}R_2]$$

with

$$R_1 = \begin{pmatrix} z_1-a_{22}, & a_{12} \\ a_{21}, & z_1-a_{11} \end{pmatrix}$$

$$R_2 = \begin{pmatrix} z_2-a_{22}, & a_{12} \\ a_{21}, & z_2-a_{11} \end{pmatrix}.$$

However, if $|z-z_1| < |z_1-z_2|$, then $|z-z_1|/|z_1-z_2| < 1$ and $(z-z_2)^{-1}$ can be written

$$\frac{1}{z-z_2} = \frac{1}{z-z_1+z_1-z_2} = \frac{1}{z_1-z_2}\,\frac{1}{1+(z-z_1)/(z_1-z_2)}$$

$$= \frac{1}{z_1-z_2} \sum_{k=0}^{\infty} (-1)^k \left(\frac{z-z_1}{z_1-z_2}\right)^k$$

and so we obtain

$$R(z,A) = (z-z_1)^{-1}[(z_1-z_2)^{-1}R_1] + \sum_{k=0}^{\infty} (z-z_1)^k[-(z_2-z_1)^{-k-2}R_2] \ . \quad (7.36)$$

Hence, by comparing (7.36) with (7.30)

$$P_1 = (z_1-z_2)^{-1}R_1 \ , \quad D_1 = 0 \ , \quad H_1 = -(z_2-z_1)^{-2}R_2 \quad (7.37a)$$

and in an analogous way

$$P_2 = (z_2-z_1)^{-1}R_2 \ , \quad D_2 = 0 \ , \quad H_2 = -(z_1-z_2)^{-2}R_1 \quad (7.37b)$$

where P_2 and D_2 are defined by (7.13) and by (7.21) with z_2 instead of z_1 and with $\Gamma = \{\lambda:\lambda\in\mathbb{C};|\lambda-z_2| = r\}$, and H_2 is given by (7.26) with z_2 instead of z_1 and with

$$\Gamma_1 = \{\lambda':\lambda'\in\mathbb{C};|\lambda'-z_2| = r_1\}.$$

Of course, (7.37a,b) can also be derived from definitions (7.13), (7.21), and (7.26) (see Exercise 7.6). It follows from (7.37a,b) and from Theorem 7.4 that $M_j = N(z_jI-A), j = 1,2$ and so

$$\dim[M_j] = \dim[N(z_jI-A)] = 1,$$

i.e. z_1 and z_2 are simple eigenvalues of A. In fact it is easy to verify directly from (7.35) and from (7.37) that $P_jf = c_j(f)\phi^{(j)}$, $j = 1,2$.

(*b*) If $\Delta = 0$, then $z_1 = z_2$, $\det\{(zI-A)\} = (z-z_1)^2$ and we have

$$\frac{z-a}{\det\{(zI-A)\}} = \frac{z-z_1+z_1-a}{(z-z_1)^2} = \frac{1}{z-z_1} + \frac{z_1-a}{(z-z_1)^2} \quad (a = a_{11} \text{ or } a = a_{22}) \ .$$

Substituting into (7.34), we obtain

$$R(z,A) = (z-z_1)^{-1}I + (z-z_1)^{-2}R_1$$

and so

$$P_1 = I, \quad D_1 = R_1, \quad H_1 = 0. \quad \square$$

Example 7.5. Spectral properties of an operator in l^1.

If $X = l^1$ and A is the operator defined in Example 2.17 then

$$P_\sigma(A) = \sigma(A) = \{z : z = z_n = -(n-1), n = 1,2,\ldots\}$$

$$\rho(A) = \{z : z\in\mathbb{C}; z \neq z_n \,\forall n = 1,2,\ldots\}$$

and

$$R(z,A) = \begin{pmatrix} z^{-1}, & 0 & 0, & 0, & \ldots \\ 0, & (z+1)^{-1}, & 0, & 0, & \ldots \\ 0, & 0, & (z+2)^{-1}, & 0, & \ldots \\ \ldots & \ldots & \ldots & \ldots & \ldots \end{pmatrix}, \quad z\in\rho(A) \tag{7.38}$$

see (2.69) and (2.73). Furthermore, $\dim[N(z_nI-A)] = 1$ because $N(z_nI-A) = \{g : g = c\phi^{(n)} \,\forall c\in\mathbb{C}\}$ where $\phi^{(n)}$ is the element of l^1 whose components $\phi_h^{(n)}$ are such that $\phi_h^{(n)} = 0 \,\forall h \neq n$, $\phi_n^{(n)} = 1$.

If Γ_n is the circle of radius $\varepsilon < 1 = |z_n - z_{n-1}| = |z_n - z_{n+1}|$ and with centre $z_n = -(n-1)$, defined by the equation

$$\lambda(s) = z_n + \varepsilon \exp(2\pi i s), \quad s\in[0,1]$$

then Γ_n satisfies the conditions listed after (7.13) and we have

$$\int_{\Gamma_n} \frac{d\lambda}{[\lambda+(j-1)]} = 0 \text{ if } j \neq n, \quad \int_{\Gamma_n} \frac{d\lambda}{[\lambda+(n-1)]} = 2\pi i$$

because $z_j = -(j-1)$ is outside Γ_n if $j \neq n$ whereas $z_n = -(n-1)$ is the centre of Γ_n (see (*a*) of Theorem 7.2). Then (7.38) gives

$$P_n = \frac{1}{2\pi i}\int_{\Gamma_n} R(\lambda,A)\,d\lambda = \begin{pmatrix} p_{11}, & p_{12}, & \ldots \\ p_{21}, & p_{22}, & \ldots \\ \ldots & \ldots & \ldots \end{pmatrix}$$

with $p_{jh} = 0$ if $j \neq h$, $p_{jj} = 0$ if $j \neq n$ and with $p_{nn} = 1$. Hence, the element $g = P_nf$ is such that $g_h = 0$ if $h \neq n$ and $g_n = f_n$:

$$P_n f = P_n \begin{pmatrix} f_1 \\ f_2 \\ \cdots \end{pmatrix} = f_n \phi^{(n)}$$

and so $M_n = R(P_n) = \{g : g = f_n \phi^{(n)} \;\forall f \in X\} \subset N(z_n I - A)$. We conclude that $M_n = N(z_n I - A)$ because of (7.32) and $\dim[M_n] = 1$, i.e. all the eigenvalues z_n are simple. Note that, for instance, we have directly from (7.38)

$$R(z,A) = (z - z_2)^{-1} \begin{pmatrix} 0, & 0, & 0, & \cdots \\ 0, & 1, & 0, & \cdots \\ 0, & 0, & 0, & \cdots \\ \cdot & \cdot & \cdot & \cdot \end{pmatrix} + \begin{pmatrix} z^{-1}, & 0, & 0, & \cdots \\ 0, & 0, & 0, & \cdots \\ 0, & 0, & (z+2)^{-1}, & \cdots \\ \cdot & \cdot & \cdot & \cdot \end{pmatrix}$$

and the second matrix is finite at $z = z_2 = -1$. Hence,

$$P_2 = \begin{pmatrix} 0, & 0, & 0, & \cdots \\ 0, & 1, & 0, & \cdots \\ 0, & 0, & 0, & \cdots \\ \cdot & \cdot & \cdot & \cdot \end{pmatrix}, \quad D_2 = 0, \quad H_2 = \begin{pmatrix} -1, & 0, & 0, & \cdots \\ 0, & 0, & 0, & \cdots \\ 0, & 0, & -1, & \cdots \\ \cdot & \cdot & \cdot & \cdot \end{pmatrix}. \quad \square$$

Example 7.6. Spectral properties of the heat-diffusion operator

If $X = L^2(a,b)$ with $0 < \delta = b - a < \infty$ and A is the heat-diffusion operator defined in Example 2.19, then

$$P_\sigma(A) = \sigma(A) = \{z : z = z_n = -k\pi^2 n^2/\delta^2,\; n = 1,2,\ldots\}$$

$$\rho(A) = \{z : z \in \mathbb{C};\, z \neq z_n,\; n = 1,2,\ldots\}$$

and

$$N(z_n I - A) = \{f : f = c\phi^{(n)} \;\forall c \in \mathbb{C}\}, \qquad \dim[N(z_n I - A)] = 1,$$

where $\phi^{(n)}$ is given by (2.86) with $c_1^{(n)} = 1$. Moreover, if $z \in \rho(A)$,

$$R(z,A)g = \frac{1}{2k\mu\Delta(\mu)} \left[\int_a^x \chi(x;y;\mu) g(y)\,dy + \int_x^b \chi(y,x;\mu) g(y)\,dy \right] \qquad (7.39)$$

where

$$\chi(x,y;\mu) = [\exp\{\mu(b-x)\}-\exp\{-\mu(b-x)\}][\exp\{\mu(y-a)\}-\exp\{-\mu(y-a)\}]$$

$$\Delta(\mu) = \exp(\delta\mu)-\exp(-\delta\mu) = \exp(-\delta\mu)\{\exp(2\delta\mu)-1\}, \qquad \delta = b-a$$

with $\mu = \surd(z/k)$ and with $\Delta(\mu) = 0$, $\forall\, \mu = \mu_n = \surd(z_n/k)$, $n = 1,2,\ldots$ (see Example 2.19).

The procedure that follows is a typical 'shortcut' to evaluate integrals of functions of a complex variable and, as such, should be handled with care. Of course, the final results are correct because each step can be made rigorous. Let D_1 be defined by (7.21) with $z_1 = -k\pi^2/\delta^2$ and with $\Gamma = \Gamma_1 = \{\lambda:\lambda\in\mathbb{C};\ |\lambda-z_1| = \varepsilon\}$ where ε is a 'small' radius. Then, each $\lambda = \Gamma_1$ is 'close' to z_1 and so (7.39) gives

$$(\lambda-z_1)R(\lambda,A)g \simeq \frac{\lambda-z_1}{\Delta(\mu)}\,\frac{1}{2k\mu_1}\,\{\int_a^x \chi(x,y;\mu_1)g(y)\,dy + \int_x^b \chi(y,x;\mu_1)g(y)\,dy\}$$

where now $\mu = \surd(\lambda/k)$. Note that we have substituted $\mu = \mu_1 = \surd(z_1/k)$ in all the parts of $(\lambda-z_1)R(\lambda,A)g$ which are 'well behaved' as $\mu\to\mu_1$ because $\mu \simeq \surd(z_1/k)$ if $\lambda = k\mu^2\in\Gamma_1$. The crucial part is the factor $(\lambda-z_1)/\Delta(\mu)$ which has the form 0/0 as $\lambda\to z_1$. We have

$$\frac{\lambda-z_1}{\Delta(\mu)} = \frac{k(\mu^2-\mu_1^2)\exp(\delta\mu)}{\exp(2\delta\mu)-1} = \frac{k(\mu^2-\mu_1^2)\exp(\delta\mu)}{\exp\{2\delta(\mu-\mu_1)\}-1}$$

because $\exp(2\delta\mu_1) = 1$ (see the discussion after (2.84)), and so $\exp(-2\delta\mu_1) = 1$. Hence, if $\mu \simeq \mu_1$:

$$\frac{\lambda-z_1}{\Delta(\mu)} = \frac{k(\mu+\mu_1)(\mu-\mu_1)\exp(\delta\mu)}{2\delta(\mu-\mu_1)/1!+\{2\delta(\mu-\mu_1)\}^2/2!+\ldots}$$

$$= \frac{k(\mu+\mu_1)\exp(\delta\mu)}{2\delta+(2\delta)^2(\mu-\mu_1)/2!+\ldots} \simeq \frac{2k\mu_1\exp(\delta\mu_1)}{2\delta}$$

and consequently

$$D_1 g = \frac{1}{2\pi i} \int_{\Gamma_1} (\lambda - z_1) R(\lambda, A) g d\lambda = \frac{\exp(\delta\mu_1)}{4\pi i\delta} \left[\int_a^x \chi(x,y;\mu_1) g(y) dy + \right.$$

$$\left. \int_x^b \chi(y,x;\mu_1) g(y) dy \right] \int_{\Gamma_1} d\lambda = 0$$

because

$$\int_{\Gamma_1} d\lambda = 2\pi i\varepsilon \int_0^1 \exp(2\pi i s) ds = \varepsilon\{\exp(2\pi i) - 1\} = 0.$$

Since $D_1 = 0$, we conclude that

$$M_1 = R(P_1) = N(z_1 I - A)$$

$$\dim[M_1] = \dim[N(z_1 I - A)] = 1$$

and that z_1 is a simple eigenvalue. □

Remark 7.2.

The above results are relatively easy to derive because the explicit expressions of the resolvent operators are known. However, it is often possible to evaluate how P_1, D_1, and H_1 behave possibly on some subset $X_0 \subset X$ if some suitable properties of $R(\lambda, A)g$ are known for λ close to z_1 and for $g \in X_0$. □

7.6. SPECTRAL REPRESENTATION OF A AND OF $\exp(tA)$

As in sections 7.3 and 7.4 we assume that the eigenvalue z_1 is an isolated point of the spectrum of the closed operator A. Then, we have from (7.22d)

$$AP_1 f = z_1 P_1 f + D_1 f \quad \forall f \in X$$

where $P_1 f \in D(A)$ $\forall f \in X$ because of (*c*) of Theorem 7.2, and so

$$A\hat{f} = z_1 P_1 \hat{f} + D_1 \hat{f} + AP_0 \hat{f} \quad \forall \hat{f} \in D(A) \tag{7.40a}$$

because $A\hat{f} = AP_1\hat{f} + A(I-P_1)\hat{f} = AP_1\hat{f} + AP_0\hat{f}$ with $P_0 = I - P_1$. Relation (7.40a) is the *spectral representation* of A relative to the isolated eigenvalue z_1 and shows that A can be written as the sum of the operator $(z_1P_1+D_1)\in\mathscr{B}(X)$ and of (AP_0). To prove that AP_0 is in some sense simpler than A, we introduce the restriction of A to $M_0 = R(P_0)$ (see section 2.1):

$$A_0g = Ag \quad \forall g\in D(A_0) = D(A) \cap M_0 \tag{7.41}$$

(we recall that M_0, the range of the projection P_0, is a closed linear subset of X and so it is a B-space with norm $\|\cdot;X\|$, see Example 1.5; moreover, $M_0 = \{g:g\in X;g = P_0g\}$, see Theorem 7.1). It follows from (7.40a) and from (7.41) that

$$A\hat{f} = (z_1P_1+D_1)\hat{f}+A_0P_0\hat{f} \quad \forall \hat{f}\in D(A) \tag{7.40b}$$

because $P_0\hat{f} = (I-P_1)\hat{f}$ belongs to both $D(A)$ and M_0 for any $\hat{f}\in D(A)$. Furthermore, the following theorem is obtained.

Theorem 7.5. (*a*) $A_0\in\mathscr{C}(M_0)$; (*b*) $\rho(A_0) = \rho(A) \cup \{z_1\}$.

Proof. (*a*) If $\{f_n, n = 1,2,\ldots\}$ is a sequence contained in the B-space M_0, with $\{f_n\} \subset D(A_0) = D(A) \cap M_0$, $X\text{-lim } f_n = f$ and $X\text{-lim } A_0f_n = \phi$ as $n\to\infty$, then $\{f_n\} \subset D(A)$ and $X\text{-lim } Af_n = \phi$ because $A_0f_n = Af_n$. Since $A\in\mathscr{C}(X)$, Theorem 2.2 shows $f\in D(A)$ and $Af = \phi$. As a consequence, $f\in D(A_0)$ and $A_0f = \phi$, i.e. $A_0\in\mathscr{C}(M_0)$ again because of Theorem 2.2.

(*b*) The operator

$$R_0(z) = R(z,A)P_0 \text{ if } z\in\rho(A), \quad R_0(z) = H_1 \text{ if } z = z_1 \tag{7.42}$$

is holomorphic at any $z\in\rho(A) \cup \{z_1\}$ because (7.29) gives

$$R_0(z) = \sum_{k=0}^{\infty} (z-z_1)^k H_1^{k+1} \quad \text{if } z\in\rho(A)$$

and the above series is holomorphic even at $z = z_1$ (and its value at $z = z_1$ is just H_1). Note that $R_0(z)$ maps M_0 into itself because of (*b*) of Theorem 7.2 and because of (7.27b) and so it can be considered as an operator from the B-space M_0 into itself. Then, consider the equation

$$(zI-A_0)f = g, \quad g\in M_0 \tag{7.43}$$

where the unknown f must be sought in $D(A_0) = D(A) \cap M_0$.
If $z\in\rho(A)$, we have from (7.41) and from (7.42)

$$(zI-A_0)R_0(z)g = (zI-A_0)R(z,A)P_0g = (zI-A_0)P_0R(z,A)g$$

$$= (zI-A)P_0R(z,A)g = (zI-A)R(z,A)P_0g = P_0g = g$$

and if $z = z_1$

$$(z_1I-A_0)R_0(z_1)g = (z_1I-A_0)H_1g = (z_1I-A_0)H_1P_0g = (z_1I-A_0)P_0H_1g$$

$$= z_1P_0H_1g-A_0P_0H_1g = (z_1I-A)H_1P_0g = (I-P_1)P_0g = P_0^2g = P_0g = g$$

because of (7.27b,d). Hence, $f = R_0(z)g$ is the solution of (7.43) for any $z\in\rho(A) \cup \{z_1\}$, i.e.

$$R_0(z)g = R(z,A_0)g \quad \forall z\in\rho(A) \cup \{z_1\}, \qquad g\in M_0$$

and so $\rho(A_0) = \rho(A) \cup \{z_1\}$. □

Now if we assume that $A\in\mathscr{G}(M,\beta;X)$ and that $\{\exp(tA), t \geq 0\}$ is the semigroup generated by A, then $(AP_0) = A-(z_1P_1+D_1)\in\mathscr{G}(M,\beta+M\|z_1P_1+D_1\|;X)$ because of (a) of Theorem 5.1 and the semigroup $\{\exp[t(AP_0)], t\geq 0\}$ is defined. However, the operators P_1, D_1, and (AP_0) commute with each other because of (d) of Theorem 7.2 and of (7.22) and so (5.22) and (7.40a) give

$$\exp(tA) = \exp(tz_1P_1)\exp(tD_1)\exp\{t(AP_0)\}, \qquad t\geq 0 \tag{7.44}$$

where

$$\exp(tz_1P_1) = \sum_{k=0}^{\infty}\frac{(z_1t)^k}{k!}P_1^k = I + \sum_{k=1}^{\infty}\frac{(z_1t)^k}{k!}P_1 = (I-P_1) + \exp(z_1t)P_1$$

$$\exp(tD_1) = \sum_{k=0}^{\infty}\frac{t^k}{k!}D_1^k = I + \sum_{k=1}^{\infty}\frac{t^k}{k!}D_1^k$$

because P_1, $D_1 \in \mathscr{B}(X)$ and $P_1^k = P_1 \; \forall k = 1,2,\ldots$. Hence,

$$\exp(tz_1P_1)\exp(tD_1)$$

$$= (I-P_1)[I + \sum_{k=1}^{\infty} \frac{t^k}{k!} D_1^k] + \exp(z_1t)P_1[I + \sum_{k=1}^{\infty} \frac{t^k}{k!} D_1^k]$$

$$= (I-P_1)I + \exp(z_1t)\exp(tD_1)P_1$$

$$= P_0 + \exp(z_1t)\exp(tD_1)P_1$$

since $P_1D_1 = D_1P_1 = D_1$ because of (7.22b) and so $(I-P_1)D_1 = 0$. Thus, (7.44) becomes

$$\exp(tA) = P_0 \exp\{t(AP_0)\} + \exp(z_1t)\exp(tD_1)P_1 \exp\{t(AP_0)\}. \tag{7.45}$$

Relation (7.45) can be further simplified by using the following lemma.

Lemma 7.1. If $\hat{A} \in \mathscr{G}(M,\beta;X)$, $\hat{B} \in \mathscr{B}(X)$ and if $\hat{B}\hat{A}f = 0 \; \forall f \in D(\hat{A})$, then $\hat{B}\exp(t\hat{A})f = \hat{B}f \; \forall f \in X, t \geq 0$.

Proof. The unique strict solution of the initial-value problem

$$\frac{dv}{dt} = \hat{A}v(t), \quad t > 0; \qquad X\text{-}\lim_{t\to 0+} v(t) = v_0 \in D(\hat{A})$$

has the form $v(t) = \exp(t\hat{A})v_0$ because $\hat{A} \in \mathscr{G}(M,\beta;X)$. However, $\hat{B} \in \mathscr{B}(X)$ and so

$$\frac{d}{dt}\hat{B}v(t) = \hat{B}\frac{dv(t)}{dt} = \hat{B}\hat{A}v(t) = 0$$

i.e. $\hat{B}v(t) = \hat{B}v(s) \; \forall t,s > 0$ and also $\hat{B}v(t) = \hat{B}v(0) = \hat{B}v_0$ because $v(t)$ is continuous as $t \to 0+$. Thus,

$$\hat{B}\exp(t\hat{A})v_0 = \hat{B}v_0 \quad \forall v_0 \in D(\hat{A})$$

and Lemma 7.1 is proved because $D(\hat{A})$ is dense in X and the above relation can be extended to the whole X in the usual way. □

Lemma 7.1 with $\hat{A} = AP_0$, $\hat{B} = P_1$ and (7.45) give

$$\exp(tA) = P_0\exp\{t(AP_0)\}+\exp(z_1t)\exp(tD_1)P_1$$

and also

$$\exp(tA) = \exp(z_1t)\exp(tD_1)P_1+\exp\{t(AP_0)\}P_0 \tag{7.46a}$$

since P_0 commutes with (AP_0) and so

$$P_0 \exp\{t(AP_0)\} = \exp\{t(AP_0)\}P_0$$

because of (5.25). Relation (7.46a) is the *spectral representation* of the semigroup $\exp(tA)$ relative to the isolated eigenvalue z_1 of A. The properties of the operator A_0 can be used to show that the last term on the right-hand side of (7.46a) has a structure simpler than that of $\exp(tA)$.

Theorem 7.6. If $A\in\mathcal{G}(M,\beta;X)$, *then* (*a*) $A_0\in\mathcal{G}(M\|P_0\|,\beta;M_0)$ *with* $M_0 = R(P_0)$, (*b*) $\exp\{t(AP_0)\}g = \exp(tA_0)g \ \forall g\in M_0, t \geq 0$.

Proof. (*a*) $A_0\in\mathcal{C}(M_0)$ because of (*a*) of Theorem 7.6 and $D(A_0) = D(A) \cap M_0$ is dense in M_0 because $D(A)$ is dense in X (see Exercise 7.8). Furthermore, if $z > \beta$, then $z\in\rho(A) \subset \rho(A) \cup \{z_1\} = \rho(A_0)$ and (7.42) gives

$$R(z,A_0) = R_0(z) = R(z,A)P_0$$

$$\{R(z,A_0)\}^k = \{R(z,A)P_0\}^k = \{R(z,A)\}^kP_0$$

because $R(z,A)P_0 = P_0R(z,A)$ and $P_0^k = P_0 \ \forall k = 1,2,\ldots$. Hence,

$$\|\{R(z,A_0)\}^k\| \leq \|\{R(z,A)\}^k\| \ \|P_0\| \leq \frac{M\|P_0\|}{(z-\beta)^k}, \quad z > \beta, \quad k = 1,2,\ldots$$

and $A_0\in\mathcal{G}(M\|P_0\|,\beta;M_0)$.

(*b*) The unique strict solution of the initial-value problem in the space M_0

$$\frac{du}{dt} = A_0 u(t), \quad t > 0; \quad X\text{-}\lim_{t\to 0+} u(t) = u_0 \in D(A_0) \tag{7.47}$$

has the form $u(t) = \exp(tA_0)u_0, t \geq 0$, because $A_0 \in \mathcal{G}(M\|P_0\|, \beta; M_0)$, and $u(t) \in D(A_0) = D(A) \cap M_0 \subset M_0$, i.e. $u(t) = P_0 u(t)$ $\forall t \geq 0$. Thus, (7.47) can be interpreted as an initial-value problem in the space X:

$$\frac{du}{dt} = AP_0 u(t), t > 0; \quad X\text{-}\lim_{t\to 0+} u(t) = u_0 \in D(A_0) \subset D(AP_0) = D(A)$$

because of definition (7.41) and so $u(t) = \exp\{t(AP_0)\}u_0$. We conclude that

$$\exp\{t(AP_0)\}u_0 = \exp(tA_0)u_0 \ \forall u_0 \in D(A_0), t \geq 0$$

and (*b*) is proved because $D(A_0)$ is dense in M_0 and the above relation can be extended to the whole M_0. □

By using the results of Theorem 7.6 equation (7.46a) becomes

$$\exp(tA)f = \exp(z_1 t)\exp(tD_1)P_1 f + \exp(tA_0)P_0 f \tag{7.46b}$$

$\forall f \in X, t \geq 0$, because $P_0 f \in M_0$ $\forall f \in X$.

Remark 7.3.

It is often possible to prove that the operator A_0 belongs to a class $\mathcal{G}(M', \beta'; M_0)$ with $\beta' < \text{Re } z_1$ and that $D_1 = 0$ (see for instance Examples 7.8 and 7.9 and section 8.2). Then,

$$\frac{\|\exp(tA_0)\|}{|\exp(z_1 t)|} \leq M' \exp\{(\beta' - \text{Re } z_1)t\} \to 0$$

as $t \to +\infty$, i.e. the second term on the right-hand side of (7.46b) becomes negligible in norm with respect to the first as $t \to +\infty$. □

Finally, if $A \in \mathcal{G}(M,\beta;X)$ and $\sigma(A) = \{z_1\} \cup \ldots \cup \{z_n\} \cup \sigma_0(A)$ where each of the z_j's, $j = 1,\ldots,n$, is an eigenvalue of A and an isolated point of $\sigma(A)$, then (7.46b) can be generalized as follows

$$\exp(tA)f = \sum_{j=1}^{n} \exp(z_j t)\exp(tD_j)P_j f + \exp(tA_0)P_0 f \tag{7.48}$$

$\forall f \in X, t \geq 0$, where $P_0 = I-(P_1+\ldots+P_n)$ and where A_0 is still the restriction of A to the closed subspace $M_0 = R(P_0)$.

Example 7.7. Spectral representations of $\exp(tA)$ *with* $A \in \mathcal{B}(\mathbb{C}^2)$.

If $X = \mathbb{C}^2$, A is the matrix operator of Example 7.4 and $z_1 \neq z_2$, then $D_1 = D_2 = 0$ and $P_j f = c_j(f)\phi^{(j)}$, $j = 1,2$. Hence, (7.48) gives

$$\exp(tA)f = c_1(f)\exp(z_1 t)\phi^{(1)} + c_2(f)\exp(z_2 t)\phi^{(2)}$$

where c_1 and c_2 are defined by (7.35b). Note that $P_0 f = (I-P_1-P_2)f = 0 \ \forall f \in X$ since $f-P_1 f-P_2 f = f-c_1(f)\phi^{(1)}-c_2(f)\phi^{(2)} = 0$ because of (7.35a). □

Example 7.8. Spectral representation of $\exp(tA)$ *with* $A \in \mathcal{G}(1,0,l^1)$

If $X = l^1$ and A is the matrix operator of Example 7.5, then $z_1 = 0$ is an isolated eigenvalue of A, with $D_1 = 0$ and with

$$P_1 = \begin{pmatrix} 1, & 0, & 0, & \ldots \\ 0, & 0, & 0, & \ldots \\ 0, & 0, & 0, & \ldots \\ \ldots & \ldots & \ldots & \ldots \end{pmatrix}, \quad P_0 = I-P_1 = \begin{pmatrix} 0, & 0, & 0, & \ldots \\ 0, & 1, & 0, & \ldots \\ 0, & 0, & 1, & \ldots \\ \ldots & \ldots & \ldots & \ldots \end{pmatrix}$$

(see the discussion for z_2 at the end of Example 7.5). Hence, (7.46b) gives

$$\exp(tA)f = \exp(z_1 t)P_1 f + \exp(tA_0)P_0 f \quad \forall f \in X, \quad t \geq 0 \tag{7.49}$$

with

$$P_1 f = P_1 \begin{bmatrix} f_1 \\ f_2 \\ f_3 \\ \cdots \end{bmatrix} = f_1 \begin{bmatrix} 1 \\ 0 \\ 0 \\ \cdots \end{bmatrix} = f_1 \phi^{(1)} , \quad P_0 f = \begin{bmatrix} 0 \\ f_2 \\ f_3 \\ \cdots \end{bmatrix}$$

where $\phi^{(1)}$ is the normalized eigenfunction for the eigenvalue z_1. Since $M_0 = R(P_0)$ is composed of all elements $g \in l^1$ whose first component is zero, the equation

$$(zI - A_0) f = g, \quad g \in l^1$$

leads to the system

$$(z+1) f_2 = g_2, \; (z+2) f_3 = g_3, \ldots, (z+n-1) f_n = g_n, \ldots$$

because $A_0 f = Af \; \forall f \in M_0$. Hence, if $z \neq -1,-2,\ldots$

$$f_2 = (z+1)^{-1} g_2, \; f_3 = (z+2)^{-1} g_3, \ldots, f_n = (z+n-1)^{-1} g_n, \ldots$$

and so the resolvent $R(z,A_0)$ is a diagonal matrix whose elements $r_{hk}(z)$, $h,k = 1,2,\ldots$, are defined by

$$r_{hk}(z) = 0 \text{ if } h \neq k \text{ or if } h = k = 1, r_{hh}(z) = (z+h-1)^{-1},$$

$\forall \; z \in \rho(A_0) = \{z : z \in \mathbb{C}; z \neq -(n-1), n = 2,3,\ldots\} = \rho(A) \cup \{z_1\}$. Moreover, if $z > -1$

$$\| R(z,A_0) g; X \| = \sum_{h=2}^{\infty} (z+h-1)^{-1} |g_h| \leq (z+1)^{-1} \sum_{h=2}^{\infty} |g_h|$$

$$= (z+1)^{-1} \| g; X \|$$

and $A_0 \in \mathcal{G}(1,-1;M_0)$, that improves ($a$) of Theorem 7.6 which states that $A_0 \in \mathcal{G}(\|P_0\|, 0; M_0)$ (here, $A \in \mathcal{G}(1,0;X)$). We conclude that $\|\exp(tA_0)\| \leq \exp(-t) \; \forall t \geq 0$ and that

$$\| \exp(tA) f - \exp(z_1 t) P_1 f; X \| = \| \exp(tA) f - P_1 f; X \|$$

$$\leq \exp(-t) \| f; X \| \to 0$$

as $t\to+\infty$ and the second term on the right-hand side of (7.49) becomes negligible in norm with respect to the first as $t\to+\infty$. Note that the above calculations are quite simple because $A, A_0, R(z,A)$, and $R(z,A_0)$ are diagonal matrices (see also Exercise 7.9). □

Example 7.9. Spectral representation of the heat-diffusion operator

If $X = L^2(a,b)$ and A is the heat-diffusion operator of Example 7.6, then $z_1 = -k\pi^2/\delta^2$, $\delta = b-a < \infty$, is an isolated eigenvalue of A with $D_1 = 0$ and so $A\hat{f} = z_1 P_1 \hat{f} + AP_0\hat{f}$ $\forall \hat{f}\in D(A)$ and

$$\exp(tA)f = \exp\left(-\frac{k\pi^2 t}{\delta^2}\right)P_1 f + \exp(tA_0)P_0 f.$$

Note that $M_1 = N(z_1 I - A)$ because $D_1 = 0$ and consequently $P_1 f = c(f)f^{(1)}$ where $f^{(1)}$ is given by (2.86) with $n = 1$. □

Example 7.10. Relationships between $\sigma(A)$ and $\sigma(\exp(tA))$.

The relationships, listed below, between the spectral properties of the semigroup $\{Z(t) = \exp(tA), t\geq 0\}$ and those of the generator $A\in\mathscr{G}(M,\beta;X)$ can be used to show that, under suitable assumptions on $\sigma(A)$ and on $\rho(Z(t))$, the semigroup generated by A_0 is such that $\|\exp(tA_0)\| \leq M'\exp(\beta' t)$ for t large enough, with $\beta' < \beta$.

If $z\in P_\sigma(A)$, then $\exp(zt)\in P_\sigma(Z(t))$ $\forall t > 0$; hence, $\{\mu : \mu\in\mathbb{C}; \mu = \exp(zt), z\in P_\sigma(A)\} \subset P_\sigma(Z(t))$ $\forall t > 0$ and $P_\sigma(Z(t))$ may also contain at most the point $\mu = 0$. *Conversely*, if $0 \neq \mu\in P_\sigma(Z(\hat{t}))$ for some fixed $\hat{t} > 0$ and if $\{z_n, n = 1,2,\ldots\}$ is the set of all the roots of the equation $\exp(z\hat{t}) = \mu$, then at least one of the z_n belongs to $P_\sigma(A)$. Moreover, the geometric eigenspace $N(\mu I - Z(\hat{t}))$ is the direct sum (see section 7.2) of the geometric eigenspaces relative to all the z_n that belong to $P_\sigma(A)$. (7.50a)

If $z \in R_\sigma(A)$ and if none of the points $z_n = z+2n\pi i/t, n = \pm 1, \pm 2, \ldots$, belongs to $P_\sigma(A)$, then $\exp(zt) \in R_\sigma(Z(t))$. *Conversely*, if $0 \neq \mu \in R_\sigma(Z(\hat{t}))$, then at least one of the roots of the equation $\exp(z\hat{t}) = \mu$ belongs to $R_\sigma(A)$ and none can be in $P_\sigma(A)$. (7.50b)

If $z \in C_\sigma(A)$ and if none of the points $z_n = z+2n\pi i/t, n = \pm 1, \pm 2, \ldots$, belongs to $P_\sigma(A) \cup R_\sigma(A)$, then $\exp(zt) \in C_\sigma(Z(t))$. (7.50c)

The spectral classification of $\mu = 0$ with respect to the semigroup $Z(t)$ is the same for all $t > 0$. (7.50d)

Note that there is no 'converse part' in (7.50c) because it may happen that a $\mu \neq 0$ exists such that $\mu \in C_\sigma(Z(\hat{t}))$ and all the roots of the equation $\exp(z\hat{t}) = \mu$ belong to $\rho(A)$. Furthermore, (7.50d) means that if for instance $0 \in P_\sigma(Z(t_0))$ for some $t_0 > 0$, then $0 \in P_\sigma(Z(t))\ \forall t > 0$ (for a proof of (7.50a,b,c,d), see Hille and Phillips 1957).

Now assume that the operator A belongs to $\mathcal{G}(M,\beta;X)$ and has the following properties

(i) $z_1 = z_1' + iz_1'' \in P_\sigma(A)$ and is an isolated point of $\sigma(A)$.

(ii) A real β_1 exists such that $\beta_1 < z_1' < \beta$ and $\{z : z = z'+iz''; \beta_1 < z' < \beta; z \neq z_1\} \subset \rho(A)$.

(iii) A $t_0 > 0$ exists such that, for any $t > t_0$, $\exp(zt) \in \rho(\exp(tA))$ $\forall\, z = z'+iz''$ with $z \neq z_1$ and with $z' > \beta_1$ (note that $\exp(z_1 t) \in P_\sigma(\exp(tA))$ because of (7.50a)).

The set $\{z : z = z'+iz''; z' > \beta_1; z \neq z_1\} \subset \rho(A)$ because of (ii)[†] and so assumptions (i) and (ii) imply that $\Sigma_0 = \{z : z = z'+iz''; z' > \beta_1\} \subset \rho(A_0) = \rho(A) \cup \{z_1\}$. Moreover,

---

[†]The definition of the family $\mathcal{G}(M,\beta;X)$ states that any *real* $z > \beta$ belongs to $\rho(A)$. However, it can be shown that $z = z'+iz'' \in \rho(A)\ \forall z' > \beta$ if $A \in \mathcal{G}(M,\beta;X)$.

$$\exp(zt)\in\rho(Z_0(t)) \quad \forall z\in\Sigma_0, \quad t>t_0 \tag{7.51}$$

i.e. if $t>t_0$, the resolvent set of the semigroup $Z_0(t) = \exp(tA_0)$ contains each $\mu = \exp(zt)$ with $z' = \mathrm{Re}z > \beta_1$. We divide the proof of (7.51) into four parts.

(*a*) Assume that $z = z'+iz''\in\Sigma_0$ and that $\mu = \exp(zt)\in P_\sigma(Z_0(t))$ for some $t>t_0$. Then an integer n exists for which

$$z_n = z+2\pi in/t = z'+i(z''+2\pi n/t)\in P_\sigma(A_0)$$

because of (7.50a). However, $z_n\in\Sigma_0 \subset \rho(A_0)\ \forall n = 0,\pm1,\pm2,\ldots$ because $\mathrm{Re}z_n = z'>\beta_1$, and this is a contradiction. Hence, $\exp(zt)$ cannot belong to $P_\sigma(Z_0(t)), t>t_0$.

(*b*) Assume that $z\in\Sigma_0$ and that $\mu = \exp(zt)\in R_\sigma(Z_0(t))$ for some $t>t_0$; then, at least one of the z_n must belong to $R_\sigma(A)$ because of (7.50b). This is again a contradiction and so $\exp(zt) \notin R\sigma(Z_0(t))$.

(*c*) Assume that $z\in\Sigma_0, z\neq z_1$ and that $\mu = \exp(zt)\in C_\sigma(Z_0(t))$ for some $t>t_0$. Note that (7.50c) cannot be used to get a contradiction as in (*a*) and in (*b*) because (7.50c) does not have a converse part. However, assumption (iii) can be exploited to prove that $\mu\notin C_\sigma(Z_0(t))$. In fact, if $\mu\in C_\sigma(Z_0(t))$, then the equation

$$\{uI-Z_0(t)\}f = g \tag{7.52}$$

has a unique solution

$$f = \{\mu I-Z_0(t)\}^{-1}g\ \forall g\in\Lambda_0 = R(\{\mu I-Z_0(t)\}) = D(\{\mu I-Z_0(t)\}^{-1}) \subset M_0$$

and Λ_0 is dense in M_0 but $\{\mu I-Z_0(t)\}^{-1}$ is not bounded (see (*c*) and (2.54c) of section 2.7). However, $f = P_0f$ because $f\in M_0$ and (7.46b) gives

$$Z_0(t)f = Z_0(t)P_0f = Z(t)P_0f-\exp(z_1t)\exp(tD_1)P_1P_0f = Z(t)P_0f = Z(t)f$$

because $P_1 P_0 f = P_1(I-P_1)f = 0$. Hence, (7.52) becomes

$$\{\mu I - Z(t)\}f = g$$

and

$$\{\mu I - Z(t)\}^{-1} g = \{\mu I - Z_0(t)\}^{-1} g \;\forall g \in \Lambda_0.$$

Thus, $\{\mu I - Z(t)\}^{-1}$ is not a bounded operator on Λ_0 and consequently it cannot be bounded on any $\Lambda \supset \Lambda_0$, on which it might be defined. We conclude that $\mu \notin \rho(Z(t))$ and this contradicts (iii). Hence, $\exp(zt) \in C_\sigma(Z_0(t))$ if $z \in \Sigma_0$ and $z \neq z_1$.

(*d*) Finally, if we assume that $\mu_1 = \exp(z_1 t) \in C_\sigma(Z_0(t))$, then $\{\mu_1 I - Z(t)\}^{-1}$ exists and it is certainly defined on some $\Lambda_0 \subset M_0 \subset X$ (see (*c*)). As a consequence, $\mu_1 \notin P_\sigma(Z(t))$ which is again a contradiction. We conclude that $\exp(zt) \notin \sigma(Z_0(t)), \forall z \in \Sigma_0, t > t_0$ and (7.51) is proved. □

Consider a fixed value $\hat{t}$ of the parameter t with $\hat{t} > t_0$ and let $\mathrm{spr}[Z_0(\hat{t})]$ be the spectral radius of $Z_0(t)$ at $t = \hat{t}$:

$$\mathrm{spr}[Z_0(\hat{t})] = \lim_{k\to\infty} \|\{Z_0(\hat{t})\}^k\|^{1/k}.$$

Then, at least one $\mu_0 \in \sigma(Z_0(\hat{t}))$ exists such that $|\mu_0| = \mathrm{spr}[Z_0(\hat{t})]$ (see Example 2.16). Furthermore, if $\varepsilon > 0$,

$$\mu = \exp(z\hat{t}) \in \rho(Z_0(\hat{t})) \quad \forall z = z' + iz'', \quad z' \geq \beta_1 + \varepsilon \tag{7.53}$$

because of (7.51), and so

$$\mathrm{spr}[Z_0(\hat{t})] < \exp\{(\beta_1+\varepsilon)\hat{t}\} . \tag{7.54}$$

In fact, if $\mathrm{spr}[Z_0(\hat{t})] \geq \exp\{(\beta_1+\varepsilon)\hat{t}\}$, at least one $\mu_0 \in \sigma(Z_0(\hat{t}))$ exists such that $|\mu_0| \geq \exp\{(\beta_1+\varepsilon)\hat{t}\}$ and so, if $\mu_0 = \exp(z_0\hat{t})$ with $z_0 = z_0' + iz_0''$, $|\mu_0| = \exp(z_0'\hat{t})$ and $z_0' \geq \beta_1+\varepsilon$, i.e. $z_0 \in \Sigma_0$ and $\mu_0 \in \rho(Z_0(\hat{t}))$, because of (7.51). Since this is a contradiction because $\mu_0 \in \sigma(Z_0(\hat{t}))$, inequality (7.54) is proved. Now (7.54) implies that

$$\|\{Z_0(\hat{t})\}^{k}\|^{1/k} \le \exp\{(\beta_1+\varepsilon)\hat{t}\} \ \ \forall\, \hat{k} \ge \hat{k} = \hat{k}(\varepsilon,\hat{t})$$

$$\|Z_0(k\hat{t})\| \le \exp\{(\beta_1+\varepsilon)k\hat{t}\} \ \ \forall\, k \ge \hat{k}. \qquad (7.55)$$

However, if $t \ge \hat{k}\hat{t}$, we can put $t = k\hat{t}+\tau$ with $k \ge \hat{k}$ and with $0 \le \tau < \hat{t}$, and (7.55) gives

$$\begin{aligned}\|Z_0(t)\| &= \|Z_0(k\hat{t}+\tau)\| = \|Z_0(k\hat{t})Z_0(\tau)\| \\ &\le \|Z_0(\tau)\| \exp\{(\beta_1+\varepsilon)k\hat{t}\} \\ &= [\|Z_0(\tau)\| \exp\{-(\beta_1+\varepsilon)\tau\}]\exp\{(\beta_1+\varepsilon)t\} \\ &\le \hat{M} \exp\{(\beta_1+\varepsilon)t\}\end{aligned}$$

where $\hat{M} = \max\{\|Z_0(\tau)\| \exp\{-(\beta_1+\varepsilon)\tau\}, \tau \in [0,\hat{t}]\}$. Hence, if we choose ε so that $\beta_1+\varepsilon < z_1' = \mathrm{Re}z_1$, we obtain

$$\frac{\|Z_0(t)\|}{|\exp(z_1)t|} \le \hat{M} \exp\{(\beta_1+\varepsilon-z_1')t\} \to 0 \quad \text{as } t \to +\infty$$

and the second term on the right-hand side of (7.46b) becomes negligible in norm with respect to the first as $t \to +\infty$ (if, for instance, $D_1 = 0$). □

EXERCISES

7.1. Prove that the eigenfunctions $\{\phi^{(1)},\dots,\phi^{(n)}\}$ of section 7.1 are linearly independent if $z_i \ne z_j \ \forall i \ne j$. Hint: Assume that they are not linearly independent, i.e. $\alpha_1\phi^{(1)}+\dots+\alpha_n\phi^{(n)} = 0$ and, say, $\alpha_1 \ne 0$ (see the footnote on p. 234). Then,

$$0 = A[\alpha_1\phi^{(1)}+\dots+\alpha_n\phi^{(n)}] = \alpha_1 z_1\phi^{(1)}+\dots+\alpha_{n-1}z_{n-1}\phi^{(n-1)}+\alpha_n z_n\phi^{(n)}$$

$$0 = z_n[\alpha_1\phi^{(1)}+\dots+\alpha_n\phi^{(n)}] = \alpha_1 z_n\phi^{(1)}+\dots+\alpha_{n-1}z_n\phi^{(n-1)}+\alpha_n z_n\phi^{(n)}$$

and so

$$\alpha_1(z_1-z_n)\phi^{(1)}+\dots+\alpha_{n-1}(z_{n-1}-z_n)\phi^{(n-1)} = 0$$

i.e.,

$$\beta_1\phi^{(1)}+\ldots+\beta_{n-1}\phi^{(n-1)} = 0$$

with $\beta_1 = \alpha_1(z_1-z_n) \neq 0$. Show that iteration of the above procedure leads to $\gamma_1\phi^{(1)} = 0$ with $\gamma_1 \neq 0$ and so... .

7.2. Prove relation (7.17). Hint:

$$R(\lambda_j,A)R(\lambda_h,A) = -(\lambda_j-\lambda_h)^{-1}R(\lambda_j,A)R(\lambda_h,A) \quad \forall\, \lambda_j\in\Gamma_j,\ \lambda_h\in\Gamma_h$$

where Γ_j and Γ_h are chosen so that they do not have any common point. Then, see the proof of (*a*) of Theorem 7.2.

7.3. Prove formulas (7.22b,c,d). Hint: See (*a*) of Theorem 7.2 to prove (7.22b) and use the relation

$$AR(\lambda,A) = \lambda R(\lambda,A)-I$$

of Exercise 2.14 and definition (7.13) to prove (7.22d). Relation (7.22c) follows from (*c*) and (*d*) of Theorem 7.2.

7.4. Prove that

(i) $D_1^k f = [-(z_1I-A)]^k P_1 f \ \forall f\in X, k = 1,2,\ldots$

(ii) $D_1^{k+1} = -(z_1I-A)D_1^k f \ \forall f\in X, k = 1,2,\ldots$.

Hint: Use (7.22d) to prove (i) and then derive (ii) from (i).

7.5. Prove formulas (7.27 b,c,d).

7.6. Derive (7.37a) from the definitions (7.13), (7.21), and (7.26) Hint: Let Γ be the circle $\Gamma_1 = \{\lambda:\lambda\in\mathbb{C};\ |\lambda-z_1| = \varepsilon\}$, defined by the equation $\lambda(s) = z_1+\varepsilon\exp(2\pi is)$, $s\in[0,1]$, $\varepsilon < |z_1-z_2|$.

7.7. Verify that $P_1P_2f = P_2P_1f\ \forall f\in X = \mathbb{C}^2$, with P_1 and P_2 defined by (7.37a,b).

7.8. Prove that $D(A_0) = D(A) \cap M_0$ is dense in M_0 (see (*a*) of Theorem 7.6). Hint: If $g = P_0 g \in M_0$ and $\varepsilon > 0$, a suitable $\hat{f} \in D(A)$ can be found such that $\|\hat{f}-g;X\| < \varepsilon$ since $D(A)$ is dense in X because $A \in \mathcal{G}(M,\beta;X)$. Then, $P_0\hat{f}$ belongs to M_0 and to $D(A)$ and

$$\|P_0\hat{f}-g;X\| = \|P_0[\hat{f}-g];X\| \leq \ldots .$$

7.9. Prove that the semigroup $\{\exp(tA), t \geq 0\}$, generated by the matrix operator A of Examples 2.17, 7.5, and 7.8, is a diagonal matrix whose elements $s_{hk}(t), h,k = 1,2,\ldots,$ are defined by

$$s_{hk}(t) = 0 \text{ if } h \neq k, \qquad s_{hh}(t) = \exp\{-(h-1)t\}$$

Hint:

$$\left\{V\left(\frac{t}{n}\right)\right\}^n f = \left\{\left(\frac{n}{t}\right)R\left(\frac{n}{t},A\right)\right\}^n f \to \exp(tA)f$$

as $n\to\infty$, where $R(n/t,A)$ is given by (7.38) with $z = n/t$.

7.10. Show that $0 \in \rho(\exp(tA))$ if $A \in \mathcal{G}'(M,\beta;X)$. Hint: Since $Z(t) = \exp(tA)$ is a group, the equation $\{0I-Z(t)\}f = g$ can be solved for any given $g \in X$, by using $Z(-t)$.

8
HEAT CONDUCTION IN RIGID BODIES AND SIMILAR PROBLEMS

8.1. INTRODUCTION

Let S be a homogeneous rigid body that occupies the volume Ω bounded by the regular surface $\partial\Omega$, where Ω is an open subset of R^3 with $\mathrm{cl}[\Omega] = \Omega \cup \partial\Omega$, and assume that the 'state' of S is characterized by the temperature $\tau = \tau(x,y,z;t)$, $(x,y,z)\in\Omega, t \geq 0$. Under suitable assumptions (see Carslow and Jaeger 1959), $\tau(x,y,z;t)$ satisfies the equation

$$\frac{\partial}{\partial t}\tau(x,y,z;t)$$

$$= k\,\nabla^2\tau + F(\tau)+\phi(x,y,z;t), \quad t > 0, \ (x,y,z)\in\Omega \tag{8.1a}$$

where $k > 0$ is the *heat-diffusion* (or heat-conduction) coefficient $\nabla^2\tau = \partial^2\tau/\partial x^2+\partial^2\tau/\partial y^2+\partial^2\tau/\partial z^2$, F is a given linear or non-linear function of the temperature, and ϕ is a known function. Equation (8.1a) is usually supplemented with some kind of *boundary condition* such as

$$\tau(x,y,z;t) = 0, \quad (x,y,z)\in\partial\Omega, \quad t > 0, \tag{8.1b}$$

and with an *initial condition*

$$\tau(x,y,z;0) = \tau_0(x,y,z), \quad (x,y,z)\in\Omega \tag{8.1c}$$

where τ_0 is a given function.

Equation (8.1a) is a balance equation that expresses the conservation of thermal energy in the neighbourhood of each point $(x,y,z)\in\Omega$. More specifically, since (under suitable assumptions) $\tau(x,y,z;t)\,d\Omega$ is a measure of the thermal energy contained at time t in the volume element $d\Omega$ centred at $(x,y,z)\in\Omega$, equation (8.1a) shows that the thermal energy in $d\Omega$ at time t changes because of heat conduction $[k\nabla^2\tau]$, and because of temperature-dependent sources or sinks $[F(\tau)]$ and of temperature-independent sources or sinks $[\phi]$, which may arise for instance from chemical reactions or from a flow of electric current. Furthermore, (.1b) shows that the boundary of S is constantly kept at zero temperature, whereas (8.1c) indicates that the temperature at $t = 0$ is known at each point within S (see (0.1) and (0.2) of the Introduction).

Remark 8.1.

System (8.1), possibly with some minor modifications, is the mathematical model of many physical phenomena, provided that the state vector $\tau(x,y,z;t)$ is suitably interpreted.

(*a*) If the volume Ω is filled, say, with water and if $\tau d\Omega$ is the number of molecules of table salt that are within the $d\Omega$ at time t, then (8.1) with $F = 0$ and $\phi = 0$ describes the diffusion phenomenon of salt in water under the assumption that $\tau_0(x,y,z)$ is the initial concentration.

(*b*) If S is a neutron-multiplying medium and $\tau d\Omega$ is the number of neutrons that are in $d\Omega$ at time t, then (8.1a) is the one-group diffusion approximation of neutron transport theory. In this case, $F = k_1\tau$ and the constant k_1 gives the neutron 'multiplying-power' of S.

(*c*) If $\tau d\Omega$ is the number of some kind of bacteria that at time t are in the volume element $d\Omega$ of a culture S, then (8.1a) describes the diffusion and the growth of the bacteria population in S. In this case, $F(\tau)$ represents the processes of multiplication and of death of bacteria, whereas the τ-independent source ϕ takes into account 'immigration' phenomena (i.e. bacteria introduced into S from outside). □

8.2. A LINEAR HEAT-CONDUCTION PROBLEM IN $L^2(a,b)$

If S is a homogeneous slab bounded by the planes $x = a$ and $x = b$ $(0 < \delta = b-a < \infty)$ and if $\tau_0 = \tau_0(x)$, then $\tau = \tau(x;t)$ and systems (8.1) with $\phi \equiv 0$ becomes

$$\frac{\partial}{\partial t}\tau(x;t) = k\frac{\partial^2}{\partial x^2}\tau(x;t)+F(\tau(x;t)) \qquad a < x < b, t > 0 \qquad (8.2a)$$

$$\tau(a;t) = \tau(b;t) = 0, \qquad t > 0 \qquad (8.2b)$$

$$\tau(x;0) = \tau_0(x), \quad a < x < b \;. \tag{8.2c}$$

In this section we assume that

$$F(\tau(x;t)) = k_1\tau(x;t)$$

where k_1 is a given real constant, whereas in the next section we shall consider a non-linear $F(\tau)$ (see also Example 2.13 where $F(\tau) = 0$ and $\phi = \phi_0\exp(i\omega t)$).

Let X be the Hilbert space $L^2(a,b)$ (see Remark 8.2) and define the heat-diffusion operator A as

$$Af = k\,\frac{d^2}{dx^2}f, \quad D(A) = \{f:f\in X;\ \frac{d^2f}{dx^2}\in X;\ f(a) = f(b) = 0\} \tag{8.3}$$

where d^2/dx^2 is a generalized derivative (see Example 2.19). Then, if $\tau(x;t)$ is interpreted as a function $u(t)$ from $[0,+\infty)$ into X, the abstract version of (8.2) reads as follows

$$\frac{d}{dt}\,u(t) = Au(t)+k_1u(t), \quad t > 0;\ X\text{-}\lim_{t\to 0+} u(t) = u_0 \tag{8.4}$$

with $u_0 = \tau_0(x)$. Note that the boundary conditions (.2b) are implicitly taken into account because they are contained in the definition of $D(A)$ and the solution $u(t)$ of (8.4) is sought in $D(A)$.

Now $A\in\mathcal{G}(1,-2k\delta^{-2};X)$ (see Example 4.5) and k_1I commutes with A. Hence $(A+k_1I)$ generates the semigroup $\{\exp(k_1t)\exp(tA), t\geq 0\}$ and the strict solution of the initial-value problem (8.4) has the form

$$u(t) = \exp(k_1t)\exp(tA)u_0, \quad t\geq 0,\ \forall u_0\in D(A). \tag{8.5}$$

Furthermore,

$$\|u(t);X\| = \|u(t)\|_2 \leq \exp\{(k_1-2k\delta^{-2})t\}\|u_0\|_2, \quad t\geq 0 \tag{8.6}$$

and $\|u(t)\|_2 \to 0$ as $t\to+\infty$ if $k_1 < 2k\delta^{-2}$, i.e. if the source $F(\tau) = k_1\tau$ is not 'too strong'. Note that the heat-diffusion operator produces the factor $\exp(-2k\delta^{-2}t)$ in (8.6) and so it represents an energy-dissipating phenomenon (see Remark 4.4.).

Remark 8.2.

Since $\delta = b-a < \infty$ we have

$$\left|\int_a^b \tau(x;t)\,dx\right| \le \int_a^b |\tau|\,dx \le \left[\int_a^b dx \int_a^b |\tau|^2 dx\right]^{1/2}$$
$$\le \sqrt{\delta}\,\exp\{(k_1 - 2k\delta^{-2})t\}\|u_0\|_2 \to 0$$

as $t \to +\infty$ if $k_1 < 2k\delta^{-2}$. This is important from a physical viewpoint because the integral of τ from a to b is a measure of the thermal energy contained in S at time t. □

Inequality (8.6) can be improved if we recall that

$$\left.\begin{aligned} P_\sigma(A) = \sigma(A) &= \{z : z = z_n = -k\pi^2 n^2 \delta^{-2}, n = 1,2,\ldots\} \\ \rho(A) &= \{z : z \in \mathbb{C}; z \neq z_n, n = 1,2,\ldots\} \end{aligned}\right\} \quad (8.7)$$

and that (2.86) with $c_1^{(n)} = 1/\sqrt{(2\delta)}$ defines the eigenfunction $\phi^{(n)}$ for the eigenvalue z_n:

$$\phi^{(n)} = \frac{1}{\sqrt{(2\delta)}}\left\{\exp\left(\frac{\pi n i}{\delta}\,x\right) - \exp\left(\frac{\pi n i}{\delta}\,(2a-x)\right)\right\}, \quad n = 1,2,\ldots \quad (8.8)$$

that is normalized because $\|\phi^{(n)}\|_2 = 1$. The family of eigenfunctions $\{\phi^{(1)}, \phi^{(2)}, \ldots\}$ is said to be *orthonormal* because $(\phi^{(n)}, \phi^{(n)}) = 1$ and

$$(\phi^{(j)}, \phi^{(n)}) = 0 \quad \text{if } j \neq n \quad (8.9)$$

as is easy to verify using (8.8). Furthermore, each $f \in X = L^2(a,b)$ can be written as follows

$$f = X\text{-}\lim_{h\to\infty} \sum_{j=1}^{h} c_j \phi^{(j)} = \sum_{j=1}^{\infty} c_j \phi^{(j)} \quad (8.10)$$

(Titchmarsh 1948), with $c_m = (f, \phi^{(m)})$ because of (8.9):

$$(f,\phi^{(m)}) = \sum_{j=1}^{\infty} c_j(\phi^{(j)},\phi^{(m)}) = c_m(\phi^{(m)},\phi^{(m)}) = c_m.$$

On the other hand, (7.50a) shows that $\eta_n = \exp(z_n t)\in P_\sigma(\exp(tA))$ $\forall t>0$ and that the geometric eigenspace $N(\eta_n I-\exp(tA))$ is the direct sum of the geometric eigenspaces relative to all the roots of the equation $\exp(zt) = \eta_n$ that belong to $P_\sigma(A)$. However, the only root belonging to $P_\sigma(A)$ is $z = z_n$ because $z_n+2m\pi i/t\in\rho(A)$ $\forall m = \pm 1,\pm 2,\ldots$.

Hence, $N(\eta_n I-\exp(tA)) = N(z_n I-A)$, i.e. $\phi^{(n)}$ is an eigenfunction of $\exp(tA)$ corresponding to the eigenvalue $\exp(z_n t)$: $\exp(tA)\phi^{(n)} = \exp(z_n t)\phi^{(n)}$, $\forall t>0$. Finally, since

$$u_0 = X\text{-}\lim_{h\to\infty}\sum_{j=1}^{h}(u_0,\phi^{(j)})\phi^{(j)} = \sum_{j=1}^{\infty}(u_0,\phi^{(j)})\phi^{(j)}$$

and $\exp(tA)\in\mathscr{B}(X)$, we obtain from (8.5)

$$u(t) = \exp(k_1 t)\sum_{j=1}^{\infty}(u_0,\phi^{(j)})\exp(tA)\phi^{(j)}$$

$$= \exp(k_1 t)\sum_{j=1}^{\infty}(u_0,\phi^{(j)})\phi^{(j)}\exp(-k\pi^2\delta^{-2}j^2 t), \quad t\geq 0 \qquad (8.11)$$

that gives $u(t)$ *explicitly* as a function of t. Note that

(*a*) if $k_1 < k\pi^2\delta^{-2}$, $\|u(t)\|_2 \to 0$ as $t\to+\infty$,

(*b*) if $k_1 = k\pi^2\delta^{-2}$, $\|u(t)-(u_0,\phi^{(1)})\phi^{(1)}\|_2 \to 0$ as $t\to+\infty$ because $k_1-4k\pi^2\delta^{-2}<0$,

(*c*) if $k\pi^2\delta^{-2} < k < 4k\pi^2\delta^{-2}$, $\|u(t)-(u_0,\phi^{(1)})\phi^{(1)}\exp\{(k_1-k\pi^2\delta^{-2})t\}\|_2 \to 0$ as $t\to+\infty$,

and so on for $j^2 = 9,16,\ldots$. The above results improve inequality (8.6) and give the asymptotic behaviour of $u(t)$ as $t\to+\infty$.

8.3. A SEMILINEAR HEAT-CONDUCTION PROBLEM

The closed linear subset of the *real* B-space $C([a,b])$:

$$X_0 = \{f : f \in C([a,b]); f(a) = f(b) = 0\} \tag{8.12}$$

is itself a real B-space with norm

$$\| f \| = \| f; X_0 \| = \max\{|f(x)|, x \in [a,b]\}$$

and the heat-diffusion operator can be defined in X_0 as follows:

$$Af = \frac{d^2 f}{dx^2}, \quad D(A) = \{f : f \in X_0; \frac{d^2 f}{dx^2} \in X_0\} \tag{8.13}$$

where d^2/dx^2 is a classical derivative. Note that the boundary conditions (8.2b) are now included in the definition of the space X_0 and that A is densely defined because $D(A)$ contains $C_0^\infty(a,b)$ which is dense in X_0. Also, the choice of X_0 is justified by the fact that it is often important for practical purposes to find an upper bound for the maximum value of the temperature of the slab S.

Using (8.13), the abstract version of system (8.2) becomes

$$\frac{d}{dt} w(t) = Aw(t) + F(w(t)), t > 0; \ X_0 - \lim_{t \to 0+} w(t) = w_0 \tag{8.14}$$

where $w(t) = \tau(x;t)$ is now a function from $[0,+\infty)$ into X_0, $w_0 = \tau_0(x)$, and where we assume that the non-linear term has the form

$$F(f) = k_1 \frac{f^2}{k_2 + f^2}, \quad f \in D(F) = X_0 \tag{8.15}$$

with k_1 real and $k_2 > 0$ (see Example 5.8).

In order to use the results of section 5.5, we prove the following lemmas.

Lemma 8.1. $A \in \mathscr{G}(1,0;X_0)$.

Proof. By procedures similar to those of Example 2.19 and using classical derivatives and Riemann integrals rather than generalized derivatives and Lebesque integrals, it is not difficult to show that $z \in \rho(A)$ if $z > 0$ and that, for any $z > 0$,

$$R(z,A)g = \frac{1}{2k\mu\Delta(\mu)}\{\int_a^x \chi(x,y;\mu)g(y)\,dy + \int_x^b \chi(y,x;\mu)g(y)\,dy\} \quad (8.16)$$

$$\chi(x,y;\mu) = [\exp\{\mu(b-x)\}-\exp\{-\mu(b-x)\}][\exp\{\mu(y-a)\}-\exp\{-\mu(y-a)\}]$$

$$\Delta(\mu) = \exp(\delta\mu)-\exp(-\delta\mu), \quad \delta = b-a, \quad \mu = \surd(z/k)$$

(see Examples 2.19 and 7.6). Note that $f(x) = \{R(z,A)g\}(x)$ is continuous on $[a,b]$ with $f(a) = f(b) = 0$ for any $g\in X_0$ and $z > 0$. Moreover,

$$k\,\frac{d^2f}{dx^2} = zf-g$$

and so $d^2f/dx^2 \in X_0$ because both f and g belong to X_0, i.e. $f = R(z,A)g\in D(A)$. On the other hand, $\mu = \surd(z/k) > 0$ if $z > 0$ and (8.16) gives

$$|R(z,A)g| \le \frac{\|g\|}{2k\mu\Delta(\mu)}\{\int_a^x \chi(x,y;\mu)\,dy + \int_x^b \chi(y,x;\mu)\,dy\}$$

because $\Delta(\mu) > 0$, $\chi(x,y;\mu) \ge 0$ and $\chi(y,x;\mu) \ge 0$ $\forall x,y\in[a,b]$. Thus, evaluating the two integrals we obtain

$$|R(z,A)g| \le \frac{\|g\|}{2k\mu\Delta(\mu)}\,\frac{2}{\mu}\,[\Delta(\mu)+\exp\{-\mu(x-a)\}-\exp\{\mu(x-a)\}+$$

$$\exp\{-\mu(b-x)\}-\exp\{\mu(b-x)\}]$$

$$\le \frac{\|g\|\,\Delta(\mu)}{k\mu^2\Delta(\mu)} = \frac{\|g\|}{z}, \quad z > 0$$

and $R(z,A)\in\mathscr{B}(X_0)$ with

$$\|R(z,A)g\| = \max\{|R(z,A)g|, x\in[a,b]\} \le \frac{\|g\|}{z}, \quad \forall\, g\in X_0, \quad z > 0. \quad (8.17)$$

Finally, $-R(z,A) = -(zI-A)\in\mathscr{B}(X_0)$ for $z > 0$ implies that $-(zI-A)^{-1}\in\mathscr{C}(X_0)$, and so $-(zI-A)\in\mathscr{C}(X_0)$, $A = -(zI-A)+zI\in\mathscr{C}(X_0)$ because of Theorem 2.3. We conclude that $A\in\mathscr{G}(1,0;X_0)$ because A is a densely defined closed operator and its resolvent satisfies (8.17). □

Lemma 8.2. (a) $\|F(f)-F(f_1)\| \leq (|k_1|/\sqrt{k_2})\|f-f_1\|\ \forall f,f_1 \in D(F) = X_0$; (b) *$F(f)$ is Fréchet differentiable at any $f \in X_0$ and its F-derivative*

$$F_f g = \frac{2k_1 k_2 f}{(k_2+f^2)^2}\, g$$

is such that $\|F_f g\| \leq (|k_1|/\sqrt{k_2})\|g\|\ \forall f,g \in X_0$. *Further,* $\|F_f g - F_{f_1} g\| \to 0$ *as* $\|f-f_1\| \to 0$.

Proof. See Example 5.8 and take into account that now $|f|/(k_2+f^2) \leq 1/2\sqrt{k_2}$ and $k_2/(k_2+f^2) \leq 1$. □

Lemma 8.3. If the initial-value problem (8.14) has a strict solution $w(t)$ over $[0,t_1] \subset [0,t_0]$ then $\|w(t)\| \leq \eta = \exp(|k_1|t_0/\sqrt{k_2})\|w_0\|$, $\forall t \in [0,t_1]$.

Proof. If (8.14) has a strict solution $w(t)\ \forall t \in [0,t_1]$, then

$$w(t) = Z(t)w_0 + \int_0^t Z(t-s)F(w(s))\,ds, \quad \forall\, t \in [0,t_1] \tag{8.18}$$

with $Z(t) = \exp(tA)$ (see (5.34)). Hence,

$$\|w(t)\| \leq \|w_0\| + \int_0^t \|F(w(s))\|\,ds \leq \|w_0\| + (|k_1|/\sqrt{k_2}) \int_0^t \|w(s)\|\,ds$$

because of (*a*) of Lemma 8.2 and because $A \in \mathscr{G}(1,0;X_0)$. Using Lemma 3.2 (Gronwall's inequality), we obtain

$$\|w(t)\| \leq \|w_0\| \exp(|k_1|t/\sqrt{k_2}) \leq \eta, \quad \forall t \in [0,t_1]$$

with $\eta = \exp(|k_1|t_0/\sqrt{k_2})\|w_0\|$. □

We conclude that the assumptions (5.63) are satisfied and so the initial-value problem (8.14) has a unique strict solution over the arbitrarily fixed interval $[0,t_0]$ because of Theorem 5.5. Moreover, since X_0 is a subset of the space $C([a,b])$, $\tau(x;t) = w(t)$ is a solution of the 'physical system' (8.2) that is partially differentiable with respect to t uniformly in $x \in [a,b]$ (see Example 3.6).

Finally, we have from the proof of Lemma 8.3:

$$|\tau(x;t)| \le \max\{|\tau(x;t)|, x\in[a,b]\} = \|w(t)\|$$

$$\le \|w_0\| \exp(|k_1|t/\sqrt{k_2})$$

$$= \max\{|\tau_0(x)|, x\in[a,b]\}\exp(|k_1|t/\sqrt{k_2})$$

that gives a *pointwise* upper bound for the temperature in S.

8.4. POSITIVE SOLUTIONS

Suppose that system (8.2) is a mathematical model for the diffusion and growth of a bacteria population in a culture S. Then, $\tau(x;t)$ and $\tau_0(x)$ are densities of bacteria (see (*c*) of Remark 8.1) and as such they should be non-negative functions. However, $\tau_0(x)$ is the initial density and so we may assume that $\tau_0(x) \ge 0 \;\; \forall\, x\in[a,b]$, whereas $\tau(x;t)$, the solution of (8.2), is a function *a priori* unknown and the structure of (8.2) should lead to a non-negative $\tau(x;t)$.

Remark 8.3.

> System (8.2) is a mathematical model of a physical phenomenon and it can only take into account some basic characteristics of the phenomenon under consideration. *Existence* of a *unique* solution of (8.2) is a first indication that the model (8.2) is correct. Similarly, if $\tau(x;t)$ is a density of bacteria or of particles, that (8.2) leads to a *non-negative* $\tau(x;t)$ is another indication in favour of the model. In other words, if (8.2) does not possess one of the above properties, then (8.2) is not a correct mathematical model. □

To investigate whether the initial-value problem (8.14) has non-negative solutions, we introduce the *closed positive cone* X_0^+ of the space X_0[†]

[†]A subset Y_1 of a vector space Y is a *cone* if (i) $f+g\in Y_1 \forall f,g\in Y_1$; (ii) $\alpha f\in Y_1 \;\forall f\in Y_1, \alpha \ge 0$; (iii) either f or $-f$ does not belong to Y_1, with $f \ne \theta_Y$. It is easy to verify that Y_1 is a convex set.

$$X_0^+ = \{f: f\in X_0; f(x) \geq 0 \ \forall x\in [a,b]\} \tag{8.19}$$

and assume that $w_0 \in D(A) \cap X_0^+$ (hence, $\tau_0(x) \geq 0 \ \forall x\in[a,b]$). Then, the unique strict solution $w(t)$ of (8.14) belongs to the positive cone X_0^+ for any $t\in[0,t_0]$. To see this, we first prove two lemmas.

Lemma 8.4. $Z(t) = \exp(tA)$, with A defined by (8.13), maps X_0^+ into itself for any $t \geq 0$: $Z(t)[X_0^+] \subset X_0^+ \ \forall t \geq 0$.

Proof. Since $Z(0)g = g\in X_0^+ \ \forall g\in X_0^+$, we may consider $Z(t)$ with $t > 0$. However, $A\in \mathscr{G}(1,0;X_0)$ and so

$$Z(t)g = X_0\text{-}\lim_{n\to\infty} \left[\left(I - \frac{t}{n} A\right)^{-1}\right]^n g = X_0\text{-}\lim_{n\to\infty} \left[\frac{n}{t} R\left(\frac{n}{t},A\right)\right]^n g$$

$t > 0$ (see sections 4.4 and 4.5). Now $R(n/t,A)g\in X_0^+ \forall g\in X_0^+$ as is easy to verify from (8.16) with $\mu = \surd(n/tk)$ (i.e. with $z = n/t$) and consequently

$$\left\{\frac{n}{t} R\left(\frac{n}{t},A\right)\right\}^2 g = \frac{n}{t} R\left(\frac{n}{t},A\right)\left\{\frac{n}{t} R\left(\frac{n}{t},A\right)g\right\}\in X_0^+$$

and so on. Hence, if $g\in X_0^+$, $\{(n/t)R(n/t,A)\}^n g\in X_0^+$ and $Z(t)g \in X_0^+$ because X_0^+ is a closed subset of X_0 (see Example 1.5). □

Lemma 8.5. Let $F_1(f) = \lambda f + F(f)$ with $\lambda = 0$ if $k_1 \geq 0$ and with $\lambda = |k_1|/2\surd k_2$ if $k_1 < 0$. Then, $F_1(f)\in X_0^+ \ \forall f\in X_0$.

Proof. If $k_1 \geq 0$, it follows directly from (8.12) that $F_1(f) = F(f)\in X_0^+ \ \forall f\in X_0^+$. If $k_1 < 0$, we have for any $f\in X_0^+$

$$F_1(f) = \frac{|k_1|}{2\surd k_2} f - |k_1| \frac{f}{k^2+f^2} \geq \frac{|k_1|}{2\surd k_2} f - \frac{|k_1|}{2\surd k_2} f = 0$$

because $f/(k_2+f^2) \leq 1/2\surd k_2 \ \forall f\in X_0^+$ and so $F_1(f)\in X_0^+ \ \forall f\in X_0^+$. □

Now the strict solution of (8.14) also satisfies the system

$$\frac{\mathrm{d}}{\mathrm{d}t} w(t) = A_1 w(t) + F_1(w(t)), \quad t > 0; \quad X_0\text{-}\lim_{t\to 0+} w(t) = w_0\in D(A) \cap X_0^+ \tag{8.20}$$

with

$$A_1 = A - \lambda I, \quad F_1(w) = \lambda w + F(w).$$

Note that the semigroup $\{Z_1(t), t \geq 0\}$ generated by A_1 is such that $Z_1(t) = \exp(-\lambda t)Z(t)$ because λI commutes with A_1 and so

$$Z_1(t)[X_0^+] = \exp(-\lambda t)Z(t)[X_0^+] \subset X_0^+ .$$

Furthermore, the integral version of (8.20) reads

$$w(t) = Z_1(t)w_0 + \int_0^t Z_1(t-s)F_1(w(s))\,ds \qquad (8.21)$$

and it can be solved by the usual method of successive approximations:

$$\left.\begin{aligned} w^{(0)}(t) &= w_0 \\ w^{(n+1)}(t) &= Z_1(t)w_0 + \int_0^t Z_1(t-s)F_1(w^{(n)}(s))\,ds, n = 0,1,\ldots \end{aligned}\right\} \qquad (8.22)$$

provided that $t \in [0,\hat{t}]$, with $\hat{t}$ suitably small (see section 5.5). Since $w^{(0)}(t) = w_0 \in X_0^+$ by assumption, (8.22) with $n = 0$ gives $w^{(1)}(t) \in X_0^+ \; \forall t \in [0,\hat{t}]$ because $Z_1(t)[X_0^+] \subset X_0^+ \; \forall t \geq 0$ and because of Lemma 8.5. Similarly, $w^{(2)}(t)$, $w^{(3)}(t)$, ... belong to $X_0^+ \; \forall t \in [0,\hat{t}]$ and so

$$w(t) = X_0\text{-}\lim_{n\to\infty} w^{(n)}(t) \in X_0^+ \; \forall t \in [0,\hat{t}]$$

because X_0^+ is a closed subset of X_0. Iterating this procedure as in section 5.5, we conclude that $w(t) \in X_0^+ \; \forall t \in [0,t_0]$.

The above results may be summarized as follows.

Theorem 8.1. The semilinear initial-value problem (8.14) has a unique strict solution $w(t)$ defined over an arbitrarily given interval $[0,t_0]$ if $w_0 \in D(A)$. Furthermore, if $w_0 \in D(A) \cap X_0^+$ then $w(t) \in X_0^+ \; \forall t \in [0,t_0]$.

EXERCISES

8.1. Study system (8.1) with $F(\tau) = k_1\tau$ in the Sobolev space $X = L^{s,2}(R^3)$ and examine the particular case $s > 3/2$. Hint: Define the heat-diffusion operator as follows

$$Af = \mathscr{F}^{-1}[-k|y^2|\mathscr{F}f], \quad D(A) = L^{s+2,2}(R^3) \subset X$$

where $f = f(x_1,x_2,x_3)$, $\mathscr{F}f = (\mathscr{F}f)(y_1,y_2,y_3)$, $|y|2 = y_1^2+y_2^2+y_3^2$ (see Examples 2.15 and 4.5) and show that $A \in \mathscr{G}(1,0;X)$.

8.2. Prove that the subset X_0 defined by (8.12) is closed in $C([a,b])$ and that X_0^+ defined by (8.19) is a closed subset of X_0.

8.3. The following system is a mathematical model for the diffusion and growth of a population that occupies the region $a < x < b$ and is surrounded by a 'desert' area†

$$\left.\begin{aligned} \frac{\partial}{\partial t}\tau(x;t) &= k\frac{\partial^2}{\partial x^2}\tau + h_1[h_2-\tau]\tau, \quad a<x<b, t>0 \\ \tau(a;t) = \tau(b;t) &= 0, \quad t>0; \quad \tau(x;0) = \tau_0(x) \geq 0, \quad a<x<b \end{aligned}\right\} \quad (8.23)$$

with k, h_1, and h positive constants (here, $\tau(x;t)\mathrm{d}x$ is the number of individuals of the given population that are between x and $x+\mathrm{d}x$ at time t). Study the abstract version of system (8.23) in the space X_0 defined in section 8.3. Hint: If A is defined by (8.13) and if

$$A_1 = A+h_1h_2I, \quad D(A_1) = D(A); \quad F(f) = -h_1f^2, \quad D(F) = X_0$$

then (8.23) leads to the initial-value problem

$$\frac{\mathrm{d}}{\mathrm{d}t}u(t) = A_1u(t)+F(u(t)), \quad t>0; \quad X_0\text{-}\lim_{t\to 0+} u(t) = u_0 \quad (8.24)$$

with $u(t) = \tau(x;t)$ and $u_0 = \tau_0(x) \in D(A) \cap X_0^+$. Show that $A_1 \in \mathscr{G}(1,h_1h_2;X_0)$ with $\exp(tA_1) = \exp(h_1h_2t)\exp(tA)\ \forall\, t \geq 0$, that

---

†See for instance, Murray (1978). *Differential-equation models in biology*. Clarendon Press, Oxford.

F satisfies (5.33b,c,d,e) (see Example 2.8), and that consequently (8.24) has a unique strict solution $u(t)$ over $[0,\hat{t}]$ with $\hat{t}$ suitably small. Then, prove that $u(t)\in X_0^+$ by adding and subtracting on the right-hand side of the first of (8.24) the term $h_1\eta u(t)$ with $\eta = \max\{\|u(t)\|, t\in[0,\hat{t}]\}$ and by defining the new operators

$$A_2 = A_1 - h_1\eta I, \quad G(f) = h_1\eta f + F(f) .$$

8.4. Consider the heat-diffusion operator A in $X = L^2(a,b)$, defined by (8.3), and assume that $z = z'+iz''\in\rho(A)$. Prove that
$\|R(z,A)g\|_2 \le \{(z'+2k\delta^{-2})^2+(z'')^2\}^{-1/2}\|g\|_2$ if $z'+2k\delta^{-2}\ge 0$
$\|R(z,A)g\|_2 \le |z''|^{-1}\|g\|_2$ if $z'+2k\delta^{-2}\le 0$.
Hint: As in Example 2.19, we have

$$((zI-A)f,f) = (z'+iz'')\|f\|_2^2+k\|f'\|_2^2$$

i.e.

$$\mathrm{Re}((zI-A)f,f) = z'\|f\|_2^2+k\|f'\|_2^2 \ge (z'+2k\delta^{-2})\|f\|_2^2$$

$$\mathrm{Im}((zI-A)f,f) = z''\|f\|_2^2.$$

However,

$$|((zI-A)f,f)|^2 = [\mathrm{Re}((zI-A)f,f)]^2+[\mathrm{Im}((zI-A)f,f)]^2$$

and so .... .

8.5. Study the abstract version of the following system with *non-zero boundary conditions*:

$$\frac{\partial}{\partial t}\tau_1(x;t) = k\frac{\partial^2}{\partial x^2}\tau_1 + k_1\tau_1, \quad a<x<b, \quad t>0 \tag{8.25a}$$

$$\tau_1(a;t) = \tau_a(t), \quad \tau_1(b;t) = \tau_b(t), \quad t>0 \tag{8.25b}$$

$$\tau_1(x;0) = \tau_{1,0}(x), \quad a<x<b \tag{8.25c}$$

where τ_a, τ_b, and $\tau_{1,0}$ are given functions, and discuss the assumptions that $\tau_a(t)$ and $\tau_b(t)$ should satisfy. Hint: The non-zero boundary conditions (8.25b) can be transformed into a 'source term' as follows. Let

$$\tau_1(x;t) = \tau(x;t) - \mu(x;t) \tag{8.26}$$

where μ is the *known* function

$$\mu(x;t) = \mu_a(x)\tau_a(t) + \mu_b(x)\tau_b(t) \tag{8.27}$$

with $\mu_a(x)$ and $\mu_b(x)$ 'regular' functions of $x \in [a,b]$, such that $\mu_a(a) = 1$, $\mu_a(b) = 0$, $\mu_b(a) = 0$, $\mu_b(b) = 1$ (for instance, we might choose $\mu_a(x) = (b-x)/(b-a)$, $\mu_b(x) = (x-a)/(b-a)$). Substituting (8.26) into (8.25a,b,c), we obtain

$$\frac{\partial}{\partial t}\tau(x;t) = k\frac{\partial^2}{\partial x^2}\tau + k_1\tau + \phi, \quad a < x < b, t > 0 \tag{8.28a}$$

$$\tau(a;t) = \tau(b;t) = 0, \quad t > 0, \tag{8.28b}$$

$$\tau(x;\ 0) = \tau_0(x), \quad a < x < b \tag{8.28c}$$

with

$$\phi(x;t) = \frac{\partial\mu}{\partial t} - k\frac{\partial^2\mu}{\partial x^2} - k_1\mu(x;t) \tag{8.29}$$

$$\tau_0(x) = \tau_{1,0}(x) + \mu(x;\ 0) \tag{8.30}$$

(see systems (8.1) and (8.2)). Thus, if $U(t) = \tau(x;t)$ and $\phi(t) = \phi(x;t)$ are interpreted as functions from $[0,+\infty)$ into $X = L^2(a,b)$, (8.28) leads to the following initial-value problem

$$\frac{\mathrm{d}}{\mathrm{d}t}U(t) = AU(t) + k_1U(t) + \phi(t), t > 0; \ X\text{-}\lim_{t\to 0+} U(t) = U_0 \tag{8.31}$$

where A is defined by (8.3) and where we assume that $U_0 \in D(A)$ (see problem (8.4)). The initial-value problem (8.31) can be solved by using Theorem 4.8 *if* $\phi(t)$ satisfies ($d1$) or ($d2$) of Theorem 4.8. Note that the above procedure *cannot* be used if $\tau_a(t)$ and/or $\tau_b(t)$ are not differentiable; in this case the theory of maximal monotone (non-linear) operators can be of some help (see for instance Barbu 1976).

9

NEUTRON TRANSPORT

9.1. INTRODUCTION

One-speed neutron transport theory in a homogeneous slab S, bounded by the planes $x = -a$ and $x = a$ $(0 < a < \infty)$, leads to the integrodifferential system (Bell and Glasstone 1970, Chapter 2; Wing 1962, Chapter 8):

$$\frac{\partial}{\partial t} n(x,y;t) = -vy \frac{\partial}{\partial x} n(x,y;t) - v\Sigma n(x,y;t) + \tfrac{1}{2}v\gamma \int_{-1}^{1} n(x,y';t)\,\mathrm{d}y', \quad -a < x < a, \quad -1 \le y \le 1, \quad t > 0 \tag{9.1a}$$

$$n(-a,y;t) = 0 \ \forall y \in (0,1]; \quad n(a,y;t) = 0 \ \forall y \in [-1,0) \tag{9.1b}$$

$$n(x,y;0) = n_0(x,y), \quad -a < x < a, \quad -1 \le y \le 1 \tag{9.1c}$$

where v is a given positive constant, $n_0(x,y)$ is a non-negative function, and Σ and γ are positive and may depend on the unknown $n(x,y;t)$ or on some other physical quantities (see section 9.4). More specifically $v > 0$ is the speed of the neutrons in the slab S (in *one-speed* transport theory, it is assumed that all neutrons have the same speed) and $\Sigma = \Sigma_c + \Sigma_s + \Sigma_f$ where $\Sigma_c > 0, \Sigma_s > 0, \Sigma_f > 0$ are the macroscopic cross-sections for radiative capture, scattering, and fission respectively; $\gamma = \Sigma_s + \nu\Sigma_f$ where $\nu > 0$ is the mean number of neutrons released after a fission. Furthermore, the 'state vector' $n(x,y;t)$ of the physical system under consideration is a neutron density, i.e. $n(x,y;t)\,\mathrm{d}x\mathrm{d}y$ is the (expected) number of neutrons that, at time t, are between x and $x+\mathrm{d}x$ and have velocities $\underline{v}$ such that the cosine of the angle between $\underline{v}$ and the x axis is between y and $y+\mathrm{d}y$ (hence, the variable y takes care of the direction of $\underline{v}$).

The transport equation (9.1a) is a balance equation for the neutrons characterized by the parameters x and y ((x,y) neutrons) and shows that, during the infinitesimal time interval dt, the change $dxdydt(\partial n/\partial t)$ of the number of such neutrons is due to the following.

(*a*) A streaming term $-dxdydt(vy\partial n/\partial x)$ that gives the number of (x,y) neutrons which enter or leave the region between x and $x+dx$ during the dt, without interacting with the atomic nuclei of the materials of the slab S.

(*b*1) A sink term $-dxdydt(v\Sigma_c n)$ showing that (x,y) neutrons are removed during the dt because they are captured by the atomic nuclei of S and disappear for good. In fact, this process leads to an excited compound nucleus that gets rid of the excess energy by the emission of radiation. Note that $dxdydt(v\Sigma_c n)$ is the (expected) number of captures during the dt because $dtv\Sigma_c$ is the probability of a capture event during the dt.

(*b*2) A sink term $-dxdydt(v\Sigma_s n)$ that gives the number of (x,y) neutrons which are scattered by the nuclei of S and re-appear with a different value of y (hence, they are lost for the family of (x,y) neutrons (see (*c*1))).

(*b*3) A sink term $-dxdydt(v\Sigma_f n)$ that gives the number of (x,y) neutrons which are captured during the dt by suitable heavy atomic nuclei of S and produce compound nuclei in excited states. Each of these compound nuclei is highly unstable and immediately splits up into two smaller nuclei with the emission of ν secondary neutrons characterized by different values of y (see (*c*2)).

(*c*1) A source term $dxdydt(\frac{1}{2}v\Sigma_s \int_{-1}^{1} n(x,y';t)dy')$ that gives the contribution to the family of (x,y) neutrons of the scattering phenomenon (which is assumed to be isotropic). In other words, (x,y') neutrons with $y'\in[-1,1]$ are scattered by the nuclei of S (see (*b*2)) and re-appear as (x,y'') neutrons. Since the number of (x,y') neutrons is $dxdy'n(x,y';t)$ and since $(dtv\Sigma_s)$ is the probability of a scattering event during the interval dt, $[dxdy'dtv\Sigma_s n(x,y';t)]$ is the number of (x,y')

neutrons scattered during the dt. Moreover, scattering is assumed to be isotropic and so each final y'' belonging to the interval [-1,1] of 'length' 2 is equally probable, i.e. $[dx\,dy'\,dt\,v\Sigma_s n(x,y';t)](dy''/2)$ is the number of neutrons that reappear as (x,y'') neutrons. Of course, the integral with respect to y' takes care of the scattering events from any 'initial' $y' \in [-1,1]$.

(*c*2) A source term $dx\,dy\,dt(\frac{1}{2}v\nu\Sigma_f \int_{-1}^{1} n(x,y';t)dy')$ that gives the contribution to the family of (x,y) neutrons of fission events induced by (x,y') neutrons with any $y' \in [-1,1]$ (see (*b*3)). The factor ν is due to the fact that ν neutrons are produced on the average after each fission.

The two (9.1b) are boundary conditions showing that the slab S is surrounded by a vacuum or by a 'perfect' neutron absorber because the (9.1b) imply that no neutron can enter the slab from outside. In fact, for instance, the condition $n(-a,y;t) = 0\ \forall y \in (0,1]$ means that, at $x = -a$, the density of neutrons moving from left to right is zero. Finally, (9.1c) is an initial condition which states that the neutron density at $t = 0$ is a known non-negative function.

Remark 9.1.

System (9.1) with some minor modifications is a mathematical model for the transfer of monochromatic radiation in a slab, provided that $n(x,y;t)$ is interpreted as a density of photons. □

9.2. LINEAR NEUTRON TRANSPORT IN $L^2((-a,a)\times(-1,1))$

Let X_2 be the Hilbert space $L^2(\Omega)$ with $\Omega = \{(x,y)\colon -a < x < a, -1 < y < 1\}$ and with inner product and norm

$$(f,g) = \int_{-1}^{1} dy \int_{-a}^{a} f(x,y)\overline{g(x,y)}\,dx$$

$$\|f\|_2 = \sqrt{(f,f)} = \{\int_{-1}^{1} dy \int_{-a}^{a} |f(x,y)|^2 dx\}^{1/2}$$

and let X_2^+ be the closed positive cone of X_2:

$$X_2^+ = \{f: f \in X_2;\ f(x,y) \geq 0 \text{ for a.e. } (x,y) \in \Omega\}.$$

If we define the operators

$$B_2 f = -vy\,\frac{\partial f}{\partial x},\ D(B_2) = \{f: f \in X_2; B_2 f \in X_2; f(-a,y) = 0$$

$$\text{for a.e. } y \in (0,1]; f(a,y) = 0 \text{ for a.e. } y \in [-1,0)\},\ R(B_2) \subset X_2 \tag{9.2}$$

$$J_2 f = \frac{1}{2}\int_{-1}^{1} f(x,y')\,\mathrm{d}y',\quad D(J_2) = X_2,\quad R(B_2) \subset X_2 \tag{9.3}$$

then the abstract version of system (9.1) reads as follows:

$$\frac{\mathrm{d}}{\mathrm{d}t}u(t) = (B_2 - v\Sigma I + v\gamma J_2)u(t),\ t > 0;\ X_2\text{-}\lim_{t \to 0+} u(t) = u_0 \tag{9.4}$$

where $u_0 = n_0(x,y)$ and $u(t)$ is the function from $[0,+\infty)$ into X_2 that maps each $t \geq 0$ into the element $n(x,y;t)$ of X_2. Note that, in (9.2), $\partial f/\partial x$ is a generalized derivative and so $f(x,y)$ is absolutely continuous in $x \in [-a,a]$ for almost every $y \in [-1,1]$, (see section 1.3). Hence, the boundary conditions that are included in the definition of $D(B_2)$ make sense because $f(-a,y)$ and $f(a,y)$ are defined for a.e. $y \in [-1,1]$. We have the following lemma.

Lemma 9.1. (*a*) *$J_2 \in \mathscr{B}(X_2)$ with $\|J_2 f\| \leq \|f\|_2\ \forall f \in X_2$;* (*b*) *$J_2$ maps the cone X_2^+ into itself: $J_2[X_2^+] \subset X_2^+$.*

Proof. (*a*) Definition (9.3) gives for any $f \in X_2$.

$$\|J_2 f\|_2^2 = \int_{-1}^{1}\mathrm{d}y\int_{-a}^{a}\mathrm{d}x\left|\frac{1}{2}\int_{-1}^{1} f(x,y')\,\mathrm{d}y'\right|^2 \leq \frac{1}{4}\int_{-1}^{1}\mathrm{d}y\int_{-a}^{a}\mathrm{d}x\left\{\int_{-1}^{1}|f(x,y')|\,\mathrm{d}y'\right\}^2$$

$$\leq \frac{1}{4}\int_{-1}^{1}\mathrm{d}y\int_{-a}^{a}\mathrm{d}x\left\{\int_{-1}^{1}|f(x,y')|^2\mathrm{d}y'\int_{-1}^{1}\mathrm{d}y'\right\} = \int_{-a}^{a}\mathrm{d}x\int_{-1}^{1}|f(x,y')|^2\mathrm{d}y' = \|f\|_2^2\ .$$

(*b*) If $f \in X_2^+$, then $J_2 f$ is obviously non-negative for a.e.

$(x,y)\in\Omega$ and so $J_2 f\in X_2^+$. □

Lemma 9.2. (a) $B_2\in\mathscr{G}(1,0;X_2)$; *(b)* $\exp(tB_2)[X_2^+]\subset X_2^+$ *for any* $t\geq 0$.

Proof. (a) The operator B_2 is densely defined because $D(B_2)$ contains the set $C_0^\infty(\Omega)$ which is dense in $L^2(\Omega)$ (Kato 1966, p.130). Furthermore, if $g\in X_2$ and $z = \alpha+i\beta\in\mathbb{C}$, the equation

$$(zI-B_2)f = g \tag{9.5}$$

can be written as follows

$$\frac{\partial f}{\partial x} + \frac{z}{vy} f = \frac{1}{vy} g$$

where the 'parameter' y belongs to $[-1,1]$ and is not zero. Hence, we have for almost every $(x,y)\in\Omega$

$$f(x,y) = c(y)\exp\left(-\frac{z}{vy}x\right) + \frac{1}{vy}\int_{-a}^{x}\exp\left\{-\frac{z}{vy}(x-x')\right\}g(x',y)\,dx' \tag{9.6}$$

where $c(y)$ is an arbitrary function of y, i.e. it is a constant with respect to x (see Example 2.18). However, f must belong to $D(B_2)$ and so

$$0 = f(-a,y) = c(y)\exp\left(\frac{z}{vy}a\right)$$

for a.e. $y\in(0,1]$

$$0 = f(\ ,y) = c(y)\exp\left(-\frac{z}{vy}a\right) + \frac{1}{vy}\int_{-a}^{a}\exp\left\{-\frac{z}{vy}(a-x')\right\}g(x',y)\,dx'$$

for a.e. $y\in[-1,0)$. As a consequence, $c(y) = 0$ for a.e. $y\in(0,1]$

and

$$c(y) = -\frac{1}{vy}\int_{-a}^{a}\exp\left(\frac{z}{vy}x'\right)g(x',y)\,dx' \text{ for a.e. } y\ [-1,0)$$

and (9.6) becomes

$$f = f(x,y) = \frac{1}{vy}\int_{-a}^{x} \exp\left\{-\frac{z}{vy}(x-x')\right\}g(x',y)\,dx' \quad \text{for a.e. } y\in(0,1], \tag{9.7a}$$

$$f = f(x,y) = -\frac{1}{vy}\int_{x}^{a} \exp\left\{-\frac{z}{vy}(x-x')\right\}g(x',y)\,dx' \quad \text{for a.e. } y\in[-1,0) \tag{9.7b}$$

By a procedure similar to that of Example 2.18, we have from (9.7a) for any $z = \alpha+i\beta$ with $\alpha > 0$:

$$|f(x,y)| \le \frac{1}{vy}\int_{-a}^{x} \exp\left\{\frac{-\alpha}{2vy}(x-x')\right\}\left[\exp\left\{\frac{-\alpha}{2vy}(x-x')\right\}|g(x',y)|\right]dx'$$

and using Schwartz inequality (1.8)

$$|f(x,y)|^2 \le \frac{1}{v^2y^2}\int_{-a}^{x} \exp\left\{\frac{-\alpha}{vy}(x-x')\right\}dx' \int_{-a}^{x} \exp\left\{\frac{-\alpha}{vy}(x-x')\right\}|g(x',y)|^2 dx'$$

$$= \frac{1}{\alpha vy}\left[1-\exp\left\{\frac{-\alpha}{vy}(x+a)\right\}\right]\int_{-a}^{x} \exp\left\{\frac{-\alpha}{vy}(x-x')\right\}|g(x',y)|^2 dx'$$

$$\le \frac{1}{\alpha vy}\int_{-a}^{x} \exp\left\{\frac{-\alpha}{vy}(x-x')\right\}|g(x',y)|^2 dx', \quad \text{for a.e. } y\in(0,1].$$

Hence

$$\int_{-a}^{a} |f(x,y)|^2 dx \le \frac{1}{\alpha vy}\int_{-a}^{a} dx \int_{-a}^{x} \exp\left\{\frac{-\alpha}{vy}(x-x')\right\}|g(x',y)|^2 dx'$$

$$= \frac{1}{\alpha vy}\int_{-a}^{a} dx'\,|g(x',y)|^2 \int_{x'}^{a} \exp\left\{-\frac{\alpha}{vy}(x-x')\right\}dx$$

$$\le \frac{1}{\alpha^2}\int_{-a}^{a} |g(x',y)|^2 dx', \quad \text{for a.e. } y\in(0,1].$$

Since, in an analogous way, we have from (9.7b)

$$\int_{-a}^{a} |f(x,y)|^2 dx \le \frac{1}{\alpha^2}\int_{-a}^{a} |g(x',y)|^2 dx', \quad \text{for a.e. } y\in[-1,0)$$

integration with respect to y gives

$$\|f\|_2^2 \leq \alpha^{-2}\|g\|_2$$

i.e.

$$\|f\|_2 \leq \alpha^{-1}\|g\|_2 \;, \quad \forall\alpha = \operatorname{Re} z > 0, \; g \in X_2. \tag{9.8}$$

As in Example 2.18, we conclude that, for any $g \in X_2$ and for any $z = \alpha + i\beta$ with $\alpha > 0$, equation (9.5) has a unique solution defined by the two (9.7) and satisfying (9.8). This implies that $f = R(z,B_2)g$ with

$$R(z,B_2) \in \mathscr{B}(X_2), \quad \|R(z,B_2)g\|_2 \leq \alpha^{-1}\|g\|_2, \quad \alpha = \operatorname{Re} z > 0. \tag{9.9}$$

Finally, (9.9) and (2.34) show that

$$-R(z,B_2) = -(zI - B_2)^{-1} \in \mathscr{B}(X_2) \subset \mathscr{C}(X_2)$$

if $\operatorname{Re} z > 0$, and so $-(zI - B_2) \in \mathscr{C}(X_2)$ because of Theorem 2.3. It follows that

$$B_2 = -(zI - B_2) + zI \in \mathscr{C}(X_2)$$

because $zI \in \mathscr{B}(X_2)$ and this concludes the proof of (a).

(b) As in Lemma 8.4, $\exp(tB_2)$ obviously maps X_2^+ into itself if $t = 0$. If $t > 0$ and $g \in X_2^+$, then the (9.7) with $z = n/t$ show that $f(x,y) = R(z,B_2)g \geq 0$ for a.e. $(x,y) \in \Omega$ and so $\{(n/t)R(n/t,B_2)\}g \in X_2^+$ and $\{(n/t)R(n/t,B_2)\}^n g \in X_2^+$, $n = 1,2,\ldots$. Since $B_2 \in \mathscr{G}(1,0;X_2)$ and since X_2^+ is a closed subset of X_2, we conclude that

$$\exp(tB_2)g = X_2\text{-}\lim_{n\to\infty} \left\{\frac{n}{t} R\left(\frac{n}{t},B_2\right)\right\}^n g \in X_2^+$$

for any $t > 0$ and $g \in X_2^+$ and (b) is proved. □

Lemma 9.3. If the cross-sections Σ_c, Σ_s, and Σ_f are constant, then (a) $A_2 = B_2 - v\Sigma I \in \mathscr{G}(1,-v\Sigma;X_2)$;

(*b*) $\exp(tA_2)[X_2^+] \subset X_2^+ \;\forall t \geq 0$; (*c*) $A_2 + v\gamma J_2 \in \mathcal{G}(1, v(\gamma-\Sigma); X_2)$;
(*d*) $\exp[t(A_2+v\gamma J_2)][X_2^+] \subset X_2^+ \;\forall t \geq 0$.

Proof. (*a*) If $z' = \alpha' + i\beta'$, we have

$$(z'I - A_2)^{-1} = \{(z'+v\Sigma)I - B_2\}^{-1}$$

and (9.9) with $z = z'+v\Sigma = (\alpha'+\Sigma)+i\beta'$ gives

$$\|(z'I-A_2)^{-1}g\|_2 \leq (\alpha'+v\Sigma)^{-1}\|g\|_2 \quad \forall\, \alpha'+v\Sigma > 0, \quad g \in X_2. \tag{9.10}$$

Furthermore, $D(A_2)$ is dense in X_2 because

$$D(A_2) = D(B_2) \cap D(I) = D(B_2) \cap X_2 = D(B_2)$$

and $A_2 \in \mathcal{C}(X_2)$ because $B_2 \in \mathcal{C}(X_2)$ and $-v\Sigma I \in \mathcal{B}(X_2)$. Hence, $A_2 \in \mathcal{G}(1, -v\Sigma; X_2)$.

(*b*) Since $-v\Sigma I$ commutes with B_2, (5.22) gives

$$\exp(tA_2) = \exp(-v\Sigma t)\exp(tB_2), \quad t \geq 0$$

and (*b*) is proved because of (*b*) of Lemma 9.2.

(*c*) $$\|v\gamma J_2 f\|_2 = v\gamma\|J_2 f\|_2 \leq v\gamma\|f\|_2 \quad \forall f \in X_2$$

and so (*a*) of Theorem 5.1 gives

$$A_2 + v\gamma J_2 \in \mathcal{G}(1, -v\Sigma + v\gamma; X_2).$$

(*d*) The operator $v\gamma J_2$ maps X_2^+ into itself because of (*b*) of Lemma 9.1 and because $v\gamma > 0$, whereas (*b*) of this Lemma shows that $\exp(tA_2)$ also maps X_2^+ into itself for any $t \geq 0$. Then, (5.5) with $v\gamma J_2$ instead of B proves that $Z^{(n)}(t)[X_2^+] \subset X_2^+$ $\forall t \geq 0$, $n = 0, 1, 2, \ldots$ and so

$$\exp\{t(A_2+v\gamma J_2)\}f = X_2 \text{-}\lim_{n\to\infty} Z^{(n)}(t)f \in X_2^+ \tag{9.11}$$

for any $f \in X_2^+$ and $t \geq 0$ because X_2^+ is a closed subset of X_2. □

Now if the cross-sections Σ_c, Σ_s, and Σ_f are constant, then Σ and γ are constant as well, (9.4) is a *linear* initial-value problem, and its unique strict solution has the form

$$u(t) = \exp\{t(A_2+v\gamma J_2)\}u_0, \quad t \geq 0 \tag{9.12}$$

provided that $u_0 \in D(A_2) = D(B_2)$ because of (*c*) of Lemma 9.3 and because of Theorem 4.7. Furthermore, if $u_0 \in D(A_2) \cap X_2^+$, then (9.12) and (*d*) of Lemma 9.3 show that $u(t) \in X_2^+ \; \forall t \geq 0$, i.e. $n(x,y;t) \geq 0$ for a.e. $(x,y) \in \Omega$ and $t \geq 0$. Finally, if $u_0 \in D(A_2) \cap X_2^+$, we have from (9.12)

$$N(t) = \int_{-1}^{1} dy \int_{-a}^{a} n(x,y;t)\,dx$$

$$\leq \left\{ \int_{-1}^{1} dy \int_{-a}^{a} \{n(x,y;t)\}^2 dx \int_{-1}^{1} dy \int_{-a}^{a} dx \right\}^{1/2} = 2\sqrt{a}\|u(t)\|_2$$

$$\leq 2\sqrt{a}\exp\{v(\gamma-\Sigma)t\}\|u_0\|_2, \quad t \geq 0 \tag{9.13}$$

which gives an upper bound for the *total* number $N(t)$ of neutrons that are in the slab S at time t. Note that $N(t)$ is a *real* function of $t \geq 0$ and *not* a function from $[0,+\infty)$ into the B-space X_2.

Remark 9.2.

The total number of neutrons $N(t)$ is a fundamental quantity in the theory of nuclear reactors because the energy produced by fission per unit time is proportional to $\Sigma_f N(t)$ (if Σ_f is a constant). Thus, (9.13) shows that the energy produced in S per unit time tends to zero as $t \to +\infty$ if $\gamma - \Sigma = (\nu-1)\Sigma_f - \Sigma_c < 0$, i.e. roughly speaking, if the 'neutron-multiplying' power of the materials of S is smaller than their 'neutron-capturing' power. The physical meaning of (9.13) can be better understood if we introduce the operator

$$K_2 f = \int_{-1}^{1} dy \int_{-a}^{a} f(x,y)\,dx, \quad D(K_2) = X_2, \quad R(K_2) \subset \mathbb{C} \tag{9.14}$$

that belongs to $\mathscr{B}(X_2,\mathbb{C})$ (see Exercise 9.4). Then applying K_2

to both sides of the two (9.4) we obtain

$$\frac{\mathrm{d}}{\mathrm{d}t} N(t) = v \{\int_{-1}^{0} y\, n(-a,y;t)\,\mathrm{d}y - \int_{0}^{1} y\, n(a,y;t)\,\mathrm{d}y\} +$$

$$v(\gamma-\Sigma)N(t), \quad t > 0; \quad \lim_{t\to 0+} N(t) = K_2 u_0 \tag{9.15}$$

because

$$K_2 u(t) = N(t)$$

$$K_2 B_2 u(t) = -v \int_{-1}^{1} \mathrm{d}y\{y \int_{-a}^{a} \frac{\partial}{\partial x} n(x,y;t)\,\mathrm{d}x\}$$

with $u(t) \in D(A_2) \cap X_2^+ = D(B_2) \cap X_2^+ \; \forall\, t \geq 0$. Integrating system (9.15) we have for any $t \geq 0$

$$0 \leq N(t) = \exp\{v(\gamma-\Sigma)t\}K_2 u_0 - \Phi(t) \leq \exp\{v(\gamma-\Sigma)t\}K_2 u_0 \tag{9.16}$$

where

$$0 \leq \Phi(t) = v \int_0^t \exp\{v(\gamma-\Sigma)(t-s)\}\{ \int_{-1}^{0} |y| n(-a,y;s)\,\mathrm{d}y +$$

$$\int_0^1 y\, n(a,y;s)\,\mathrm{d}y\}\mathrm{d}s$$

i.e. $\Phi(t)$ gives the number of neutrons escaping from the slab per unit time through the boundary planes $x = -a$ and $x = a$. □

9.3. SPECTRAL PROPERTIES OF THE TRANSPORT OPERATOR $A_2+v\gamma J_2$

If the cross-sections Σ_c, Σ_s, and Σ_f are constant, (9.10) shows that $\{z : z\in\mathbb{C}; \mathrm{Re}\; z > -v\Sigma\} \subset \rho(A_2)$ with $\| R(z,A_2)g\|_2 \leq (\mathrm{Re}\; z+v\Sigma)^{-1}\| g \|_2$. On the other hand, since $\| v\gamma J_2 g\|_2 \leq v\gamma\| g \|_2$, we have

$$\| v\gamma R(z,A_2)J_2\| \leq v\gamma(\mathrm{Re}\; z+v\Sigma)^{-1} < 1 \quad \forall\, \mathrm{Re}\; z > v(\gamma-\Sigma)$$

and so

$$\| R(z,A_2+v\gamma J_2)\| = \| [(zI-A_2)\{I-v\gamma R(z,A_2)J_2\}]^{-1}\|$$

$$= \|\{I-v\gamma R(z,A_2)J_2\}^{-1}R(z,A_2)\| \leq \frac{1}{1-v\gamma(\text{Re } z+v\Sigma)^{-1}} (\text{Re } z+v\Sigma)^{-1}$$

$$= \{\text{Re } z-v(\gamma-\Sigma)\}^{-1}, \quad \text{Re } z > v(\gamma-\Sigma) \tag{9.17}$$

(see Example 2.16) with $z = 1$ and with $v\gamma R(z,A_2)J_2$ instead of A). Hence, $\{z:z\in\mathbb{C};\text{Re } z>v(\gamma-\Sigma)\} \subset \rho(A_2+v\gamma J_2)$. Note that the strip of the complex plane $\{z:z\in\mathbb{C};-v\Sigma < \text{Re } z < v(\gamma-\Sigma)\}$ is contained in $\rho(A_2)$ but not necessarily in $\rho(A_2+v\gamma I_2)$. Also, whether such a strip contains isolated eigenvalues of $A_2+v\gamma J_2$ is a fundamental question as far as the asymptotic behaviour of $u(t)$ as $t\to+\infty$ is concerned (see Example 7.10). We then consider the equation

$$(zI-A_2-v\gamma J_2)f = g, \quad g\in X_2, \quad -v\Sigma < \text{Re } z < v(\gamma-\Sigma) \tag{9.18}$$

i.e.

$$(zI-A_2)f = v\gamma J_2f+g$$

$$f = v\gamma R(z,A_2)J_2f+R(z,A_2)g \tag{9.19}$$

because $z\in\rho(A_2)$ and so $R(z,A_2) = (zI-A_2)^{-1}$ exists and belongs to $\mathscr{B}(X_2)$. If we put

$$\phi = \phi(x) = J_2f, \quad G = G(x) = J_2R(z,A_2)g \tag{9.20}$$

$$H_z = v\gamma J_2R(z,A_2) \tag{9.21}$$

applying the operator J_2 to both sides of (9.19) we obtain

$$\phi = H_z\phi + G. \tag{9.22}$$

Note that (9.22) is an integral equation in the Hilbert space $L^2(-a,a)$ (see Exercise 9.5), and as such it is 'simpler' than (9.18) which is an integrodifferential equation in $L^2(\Omega)$.

Lemma 9.4. If $z = \alpha+i\beta$ is such that $-v\Sigma < \alpha < v(\gamma-\Sigma)$ and $g\in X_2$, then (a) $G\in L^2(-a,a), H_z\in\mathscr{B}(L^2(-a,a))$; (b) if $f\in X_2$ is a solution of (9.18), then $\phi = J_2 f$ belongs to $L^2(-a,a)$ and is a solution of (9.22); conversely, if (9.22) has a unique solution $\phi\in L^2(-a,a)$, then $f = v\gamma R(z,A_2)\phi+R(z,A_2)g$ belongs to X_2 and is the solution of (9.18).

Proof. (*a*) We have

$$\| G;L^2(-a,a)\| = \left\{\int_{-a}^{a} |G(x)|^2 dx\right\}^{1/2} = \left\{\frac{1}{2}\int_{-1}^{1} dy \int_{-a}^{a} |G(x)|^2 dx\right\}^{1/2}$$

$$\leq \frac{1}{\sqrt{2}} \| J_2\| \; \| R(z,A_2)\| \; \| g\|_2 \leq \frac{1}{\sqrt{2}} \frac{1}{\alpha+v\Sigma} \| g\|_2 < \infty$$

and so $G\in L^2(-a,a)$. In an analogous way, if $\phi\in L^2(-a,a)$,

$$\| H_z\phi;L^2(-a,a)\| = \frac{1}{\sqrt{2}} \| H_z\phi\|_2 = \frac{1}{\sqrt{2}} \| v\gamma J_2 R(z,A_2)\phi\|_2$$

$$\leq \frac{1}{\sqrt{2}} \frac{v\gamma}{\alpha+v\gamma} \| \phi\|_2 = \frac{v\gamma}{\alpha+v\gamma} \| \phi;L^2(-a,a)\|$$

and $H_z\in\mathscr{B}(L^2(-a,a))$.

(*b*) If $f\in X_2$ is a solution of (9.18), then $\phi\in L^2(-a,a)$ because

$$\| \phi;L^2(-a,a)\| = \frac{1}{\sqrt{2}} \| \phi\|_2 = \frac{1}{\sqrt{2}} \| J_2 f\|_2 \leq \frac{1}{\sqrt{2}} \| f\|_2 < \infty$$

and it satisfies (9.22) as we have proved when we derived (9.22) from (9.18). Conversely, assume that (9.22) has a unique solution $\phi\in L^2(-a,a)$ and consider the element $f = v\gamma R(z,A_2)\phi+R(z,A_2)g$ that belongs to X_2 as is easy to verify. Then

$$J_2 f = v\gamma J_2 R(z,A_2)\phi+J_2 R(z,A_2)g = H_z\phi+G$$

i.e. $J_2 f = \phi$. Furthermore,

$$(zI-A_2)f = v\gamma(zI-A_2)R(z,A_2)\phi+(zI-A_2)R(z,A_2)g$$

$$(zI-A_2)f = v\gamma J_2 f+g$$

and f satisfies (9.18). □

By using the equivalence of (9.18) to (9.22) (in the sense of (*b*) of Lemma 9.4) the following lemma can be proved.

Lemma 9.5. (a) The strip $\{z:z\in\mathbb{C};-v\Sigma<\mathrm{Re}\ z<v(\gamma-\Sigma)\}$ *contains* n *(real) eigenvalues* $\alpha_1>\alpha_2>\ldots>\alpha_n$ *of the transport operator* $A_2+v\gamma J_2$, *with* $n\geq 1$ *(i.e. at least one eigenvalue* α_1 *of* $A_2+v\gamma J_2$ *exists with* $-v\Sigma<\alpha_1<v(\gamma-\Sigma)$*).*

(b) If D_j *is defined by (7.21) with* α_j *instead of* z_1 *and with* $\Gamma=\{z:z\in\mathbb{C};|z-\alpha_j|=r\}$, $0<r<\min\{(\alpha_j-\alpha_{j+1}),\ j=1,\ n-1\}$, *then* $D_j=0$, $j=1,2,\ldots,n$, *and the eigenvalues* α_j *are all semisimple. (c) If* $z\neq\alpha_j\ \forall j=1,2,\ldots\ n$, *and* $-v\Sigma<\mathrm{Re}\ z<v(\gamma-\Sigma)$, *then* $z\in\rho(A_2+v\gamma J_2)$. □

The proof of Lemma 9.5 involves a detailed study of the properties of the integral operator H_z and it will not be given here (see Wing 1962, Chapter 8). Note that Lemma 9.5 and (7.48) imply that (9.12) can be written as follows

$$u(t)=\sum_{j=1}^{n}\exp(\alpha_j t)P_j u_0+Z_0(t)P_0u_0,\qquad t\geq 0 \tag{9.23}$$

where $Z_0(t)$ is the semigroup generated by $(A_2+v\ J_2)_0$, the resctiction of $(A_2+v\gamma J_2)$ to $M_0=R(P_0)$ (see section 7.6). On the other hand, it can be shown that $\exp(zt)\in\rho(Z_0(t))$ for all $t>0$ if $z\neq\alpha_j\ \forall j=1,2,\ldots,$ and $-v\Sigma<\mathrm{Re}\ z<v(\gamma-\Sigma)$ (Vidav 1970). Hence, as in Example 7.10,

$$\|Z_0(t)\|\leq\hat{M}\exp\{(\beta_1+\varepsilon)t\},\qquad t\geq t_\varepsilon \tag{9.24}$$

where $-v\Sigma<\beta_1<\beta_1+\varepsilon<\alpha_n$, and so the last term on the right-hand side of (9.23) becomes negligible in norm with respect to the first n terms as $t\to+\infty$.

The above results can be summarized as follows.

Theorem 9.1. If the cross sections Σ_c, Σ_s, *and* Σ_f *are constant and* $u_0\in D(A_2)\cap X_2^+$, *then the linear neutron transport problem (9.4) has a unique strict solution* $u(t)$ *that belongs to* $D(A_2) \cap X_2^+ \ \forall t \geq 0$ *and is given by (9.23) with* $Z_0(t)$ *satisfying (9.24).* □

Note that $u(t) \cong \Sigma_{j=1}^{n} \exp(\alpha_j t) P_j u_0$ for 'large' values of t.

9.4. A SEMILINEAR NEUTRON TRANSPORT PROBLEM

If $\tau(x;t)$ is the temperature in the slab S, define the 'average' temperature $\tau(t)$ as

$$\tau(t) = \frac{1}{2a} \int_{-a}^{a} \tau(x;t)\,dx$$

and assume that $\tau(t)$ satisfies the system

$$\frac{d}{dt}\tau(t) = -h\{\tau(t)-\tau_c\}+h_1 \int_{-1}^{1} dy' \int_{-a}^{a} n(x',y';t)\,dx' \quad (9.25a)$$

$$\tau(0) = \tau_0 \quad (9.25b)$$

where $h > 0$ is the coefficient of convective heat transfer, $h_1 > 0$ is the energy-generation coefficient (which is proportional to Σ_f), τ_c is the coolant temperature, and τ_0 is a given temperature.

Equation (9.25a) is a balance for the thermal energy in S and shows that the average temperature $\tau(t)$ changes because of heat transfer to a coolant at temperature τ_c $h[\tau-\tau_c]$, and because of the energy production by fissions $h_1 \int_{-1}^{1} dy' \int_{-a}^{a} n\,dx'$. The coupling between systems (9.25) and (9.1) is given by the integral term of (9.25a) and by the fact that Σ, γ, and h_1 are assumed to be linear functions of τ:

$$\left.\begin{aligned} \Sigma(\tau) &= \Sigma_0[1+c_1\{\tau(t)-\tau_c\}] \\ \gamma(\tau) &= \gamma_0[1+c_2\{\tau(t)-\tau_c\}], \\ h_1(\tau) &= h_{10}[1+c_3\{\tau(t)-\tau_c\}] \end{aligned}\right\} \quad (9.26)$$

where $\Sigma_0 = \Sigma(\tau_c) > 0$, $\gamma_0 = \gamma(\tau_c) > 0$, $h_{10} = h_1(\tau_c) > 0$, and c_1, c_2, c_3 are suitable constant coefficients (Bell and

Gladstone 1970).

Since the basic physical quantities are the total number of neutrons and the real function $\tau(t)$, we shall study system (9.1a,b,c)+(9.25a,b)+(9.26) in the real B-space $X = X_1 \times R^1$ of all ordered couples

$$f = \begin{pmatrix} f_1 \\ f_2 \end{pmatrix}, \quad f_1 \in X_1 = L^1(\Omega), \quad f_2 \in R^1$$

with norm

$$\| f; X \| = h_{10} \| f_1 \|_1 + h | f_2 | = h_{10} \int_{-1}^{1} dy \int_{-a}^{a} | f_1(x,y) | dx + h | f_2 | \quad (9.27)$$

where the factors h_{10} and h are introduced to adjust dimensions. If we define the operators

$$B_1 f_1 = -vy \frac{\partial}{\partial x} f_1, \; D(B_1) = \{ f_1 : f_1 \in X_1; B_1 f_1 \in X_1; f_1(-a,y) = 0$$

$$\text{for a.e. } y \in (0,1]; f(a,y) = 0 \text{ for a.e. } y \in [-1,0) \}, R(B_1) \subset X_1 \quad (9.28a)$$

$$A_1 = B_1 - v\Sigma_0 I, \quad D(A_1) = D(B_1), \quad R(A_1) \subset X_1 \quad (9.28b)$$

$$J_1 f_1 = \tfrac{1}{2} \int_{-1}^{1} f_1(x,y') dy', \; D(J_1) = X_1, \; R(J_1) \subset X_1 \quad (9.29)$$

$$K_1 f_1 = \int_{-1}^{1} dy' \int_{-a}^{a} f_1(x',y') dx', \quad D(K_1) = X_1, \quad R(K_1) \subset R^1 \quad (9.30)$$

then (9.1a,b,c), (9.25a,b), and the assumptions (9.26) lead to the semilinear system

$$\frac{d}{dt} w_1(t) = (A_1 + v\gamma_0 J_1) w_1(t) - c_1 v \Sigma_0 w_2(t) w_1(t) + c_2 v \gamma_0 w_2(t) J_1 w_1(t),$$

$$t > 0 \quad (9.31a)$$

$$\frac{d}{dt} w_2(t) = -h w_2(t) + h_{10} K_1 w_1(t) + c_3 h_{10} w_2(t) K_1 w_1(t), \quad t > 0 \quad (9.31b)$$

$$X_1 - \lim_{t \to 0+} w_1(t) = w_{10}, \quad R^1 - \lim_{t \to 0+} w_2(t) = w_{20}$$

$$(9.31c)$$

where $w_{10} = n_0(x,y)$, $w_{20} = \tau_0 - \tau_c$, $w_1(t) = n(x,y;t)$ is a function from $[0,+\infty)$ into X_1, and $w_2(t) = \tau(t)-\tau_c$ is a real function. System (9.31) can be put into the more compact form

$$\frac{d}{dt} W(t) = (A+J)W(t)+F(W(t)), \qquad t > 0; \qquad X\text{-}\lim_{t\to 0+} W(t) = W_0 \tag{9.32}$$

with

$$W(t) = \begin{pmatrix} w_1(t) \\ w_2(t) \end{pmatrix}, \qquad W_0 = \begin{pmatrix} w_{10} \\ w_{20} \end{pmatrix}$$

$$A = \begin{pmatrix} A_1, & 0 \\ 0, & -hI \end{pmatrix}, \qquad D(A) = D(A_1)\times R^1, \qquad R(A) \subset X \tag{9.33}$$

$$J = \begin{pmatrix} v\gamma_0 J_1, & 0 \\ h_{10}K_1, & 0 \end{pmatrix}, \qquad D(J) = X, \qquad R(J) \subset X \tag{9.34}$$

$$F(f) = \begin{pmatrix} -c_1 v\Sigma_0 f_2 f_1 + c_2 v\gamma_0 f_2 J_1 f_1 \\ c_3 h_{10} f_2 K_1 f_1 \end{pmatrix}, \qquad D(F) = X, \quad R(F) \subset X. \tag{9.35}$$

We have the following lemmas.

Lemma 9.6. (*a*) $B_1 \in \mathscr{G}(1,0;X_1)$; (*b*) $A_1 \in \mathscr{G}(1,-v\Sigma_0;X_1)$ *and* $\exp(tA_1) = \exp(-v\Sigma_0 t)\exp(B_1 t)$ $\forall t \geq 0$; (*c*) $J_1 \in \mathscr{B}(X_1)$ *with* $\| J_1 f_1 \| \leq \| f_1 \|_1$, $K_1 \in \mathscr{B}(X_1, R^1)$ *with* $|K_1 f_1| \leq \| f_1 \|_1$.

Proof. (*a*) If $\mathrm{Re}\, z > 0$, $f_1 = R(z,B_1)g_1$ is still given by (9.7a,b) with f_1 and g_1 instead of f and g and so it is easy to prove that $B_1 \in \mathscr{G}(1,0;X_1)$ by a procedure similar to that of Remark 2.11.

(*b*) Since $A_1 = B_1 - v\Sigma_0 I$ and $v\Sigma_0 I$ commutes with A_1, (*b*) is proved just as (*a*) and (*b*) of Lemma 9.3.

(*c*) follows without difficulty from (9.29), (9.30), and from the definition of $\|\cdot\|_1$. □

Lemma 9.7. (*a*) $A \in \mathcal{G}(1,-\eta_0;X)$ *with* $\eta_0 = \min\{h, v\Sigma_0\}$ *and*

$$\exp(tA) = \begin{pmatrix} \exp(tA_1), & 0 \\ 0 & , \exp(-ht) \end{pmatrix}, \quad \forall t \geq 0;$$

(*b*) $J \in \mathcal{B}(X)$ *with* $\|Jf;X\| \leq \eta_1 \|f;X\|$ $\forall f \in X$, *with* $\eta_1 = h + v\gamma_0$;
(*c*) $A+J \in \mathcal{G}(1,\eta_1-\eta_0;X)$.

Proof. (*a*) The equation $(zI-A)f = g$, i.e. the system

$$(zI-A_1)f_1 = g_1, \qquad (z+h)f_2 = g_2$$

gives

$$f_1 = R(z,A_1)g_1, \qquad f_2 = (z+h)^{-1}g_2, \qquad \forall z > -\eta_0$$

because $z > -\eta_0$ implies $z > -v\Sigma_0$ and $z > -h$ and so $z \in \rho(A_1)$ and $(z+h) > 0$. Hence,

$$R(z,A)g = \begin{pmatrix} R(z,A_1), & 0 \\ 0 & , (z+h)^{-1} \end{pmatrix} \begin{pmatrix} g_1 \\ g_2 \end{pmatrix} \quad \forall g \in X, z > -\eta_0 \tag{9.36}$$

with

$$\begin{aligned} \|R(z,A)g;X\| &= h_{10}\|R(z,A_1)g_1\|_1 + h|(z+h)^{-1}g_2| \\ &\leq h_{10}(z+v\Sigma_0)^{-1}\|g_1\|_1 + h(z+h)^{-1}|g_2| \\ &\leq (z+\eta_0)^{-1}[h_{10}\|g_1\|_1 + h|g_2|] = (z+\eta_0)^{-1}\|g;X\| \end{aligned}$$

and $R(z,A) \in \mathcal{B}(X)$ $\forall z > -\eta_0$. As usual, this implies that $A \in \mathcal{C}(X)$. Furthermore, $D(A)$ is dense in X since $D(A) = D(A_1) \times R^1$ and $D(A_1)$ is dense X_1 because $A_1 \in \mathcal{G}(1,-v\Sigma_0;X_1)$. Finally, (9.36) gives

$$\left\{\frac{n}{t} R\left(\frac{n}{t},A\right)\right\}^n g = \begin{pmatrix} \left\{\frac{n}{t} R\left(\frac{n}{t},A_1\right)\right\}^n, & 0 \\ 0 & , \left\{\frac{n}{t}\left(\frac{n}{t}+h\right)^{-1}\right\}^n \end{pmatrix}\begin{pmatrix} g_1 \\ g_2 \end{pmatrix}$$

and this proves that $\exp(tA)$ is a diagonal metrix with elements $\exp(tA_1)$ and $\exp(-ht)$ because

$$X_1 - \lim_{n\to\infty} \left\{\frac{n}{t} R\left(\frac{n}{t},A_1\right)\right\}^n g_1 = \exp(tA_1)g_1$$

$$R^1 - \lim_{n\to\infty} \left\{\frac{n}{t}\left(\frac{n}{t}+h\right)^{-1}\right\}^n g_2 = R^1 - \lim_{n\to\infty} \left(1+\frac{th}{n}\right)^{-n} g_2 = \exp(-ht)g_2.$$

(b)
$$\|Jf;X\| = h_{10}\|v\gamma_0 J_1 f_1\|_1 + h|h_{10}K_1 f_1|$$

$$\leq h_{10}v\gamma_0\|f_1\|_1 + hh_{10}\|f_1\|_1$$

$$= \eta_1 h_{10}\|f_1\|_1 \leq \eta_1\|f;X\|$$

with $\eta_1 = v\gamma_0+h$.

(c) $A+J\in\mathcal{G}(1,\eta_1-\eta_0;X)$ because of (a) and (b) of this Lemma and because of Theorem 5.1. □

Lemma 9.8. (a) $\|F(\hat{f})-F(f);X\| \leq \eta_2[\|\hat{f};X\|+\|f;X\|]\|\hat{f}-f;X\|$ $\forall \hat{f},f\in D(F) = X$ *with*

$$\eta = \tfrac{1}{2}\left(\frac{|c_1|v\Sigma_0+|c_2|v\gamma_0}{h} + |c_3|\right);$$

(b) *$F(f)$ is Fréchet differentiable at any $f\in X$ and its Fréchet derivative F_f:*

$$F_f g = \begin{pmatrix} -c_1 v\Sigma_0 f_2+c_2 v\gamma_0 f_2 J_1, & -c_1 v\Sigma_0 f_1+c_2 v\gamma_0(J_1 f_1) \\ c_3 h_{10} f_2 K_1 & , \quad c_3 h_{10}(K_1 f_1) \end{pmatrix}\begin{pmatrix} g_1 \\ g_2 \end{pmatrix} \tag{9.37}$$

is such that $\|F_f g;X\| \leq 2\eta_2\|f;X\|\|g;X\|$, $\forall f,g\in X$, *and* $\|F_{\hat{f}}g-F_f g;X\| \to 0$ as $\|\hat{f}-f;X\| \to 0$.

Proof. (*a*) (9.35) gives for any $\hat{f}$, $f \in D(F) = X$:

$$F(\hat{f})-F(f) = \begin{pmatrix} -c_1 v\Sigma_0(\hat{f}_2\hat{f}_1 - f_2 f_1) + c_2 v\gamma_0(\hat{f}_2 J_1 \hat{f}_1 - f_2 J_1 f_1) \\ c_3 h_{10}(\hat{f}_2 K_1 \hat{f}_1 - f_2 K_1 f_1) \end{pmatrix}. \quad (9.38)$$

However,

$$\begin{aligned} \|\hat{f}_2\hat{f}_1 - f_2 f_1\|_1 &= \|\hat{f}_2\hat{f}_1 - \hat{f}_2 f_1 + \hat{f}_2 f_1 - f_2 f_1\|_1 \\ &\leq \|\hat{f}_2(\hat{f}_1 - f_1)\|_1 + \|(\hat{f}_2 - f_2) f_1\|_1 \\ &= |\hat{f}_2|\|\hat{f}_1 - f_1\|_1 + |\hat{f}_2 - f_2|\|f_1\|_1 \end{aligned}$$

and also

$$\begin{aligned} \|\hat{f}_2\hat{f}_1 - f_2 f_1\|_1 &= \|\hat{f}_2\hat{f}_1 - f_2\hat{f}_1 + f_2\hat{f}_1 - f_2 f_1\|_1 \\ &\leq |\hat{f}_2 - f_2|\|\hat{f}_1\|_1 + |f_2|\|\hat{f}_1 - f_1\|_1 \end{aligned}$$

and so

$$2\|\hat{f}_2\hat{f}_1 - f_2 f_1\|_1 \leq [|\hat{f}_2 - f_2|(\|\hat{f}_1\|_1 + \|f_1\|_1) + \|\hat{f}_1 - f_1\|_1(|\hat{f}_2| + |f_2|)].$$

In an analogous way, $2\|\hat{f}_2 J_1 \hat{f}_1 - f_2 J_1 f_1\|_1$ and $2|\hat{f}_2 K_1 \hat{f}_1 - f_2 K_1 f_1|$ are bounded by the same square bracket as above because of (*c*) of Lemma 9.6. Thus,

$$\begin{aligned} \|F(\hat{f}) - F(f);X\| &\leq \tfrac{1}{2}[h_{10}(|c_1|v\Sigma_0 + |c_2|v\gamma_0) + h|c_3|h_{10}][|\hat{f}_2 - f_2| \\ &\quad (\|\hat{f}_1\|_1 + \|f_1\|_1) + \|\hat{f}_1 - f_1\|_1(|\hat{f}_2| + |f_2|)] \\ &= \eta_2\ [h|\hat{f}_2 - f_2|h_{10}(\|\hat{f}_1\|_1 + \|f_1\|_1) + h_{10}\|\hat{f}_1 - f_1\|_1 h(|\hat{f}_2| + |f_2|)] \\ &\leq \eta_2\ \|\hat{f} - f;X\|\ [\|\hat{f};X\| + \|f;X\|] \end{aligned}$$

and (*a*) is proved.

(*b*) Substituting $\hat{f} = f+g$ into (9.38), it is easy to verify that

$$F(f+g)-F(f)]-F_f g = \begin{pmatrix} -c_1 v\Sigma_0 g_1 g_2 + c_2 v\gamma_0 g_2 J_1 g_1 \\ \\ c_3 h_{10} g_2 K_1 g_1 \end{pmatrix}$$

where $F_f g$ is given by (9.37) and the norm of the right-hand side is not larger than $2\eta_2 \| g;X\|^2$ (see section 3.6). Finally, by procedures similar to those of (a), it can be shown that

$$\| F_f g\| \leq 2\eta_2 \| f;X\| \, \| g;X\| , \qquad \forall f,g \in X$$

and that

$$\| F_{\hat{f}} g - F_f g;X\| \to 0 \text{ as } \| \hat{f}-f;X\| \to 0$$

(see Exercise 9.9). □

If D is the open and convex set

$$D = \{f:f \; X;\| f-W_0;X\| < r\}, \qquad r > 0$$

then for any $\hat{f},f \in D$

$$\| f;X\| \leq \| f-W_0;X\| + \| W_0;X\| < r + \| W_0;X\|$$

$$\| \hat{f};X\| < r + \| W_0;X\|$$

and Lemma 9.8 gives

$$\| F(\hat{f})-F(f);X\| \leq 2\eta_2 [r+\| W_0;X\|]\| \hat{f}-f;X\|$$

$$\| F_f g;X\| \leq 2\eta_2 [r+\| W_0;X\|]\| g;X\|$$

for any $\hat{f},f \in D$ and $g \in X$. It follows from the above results and from Theorem 5.4 that the semilinear initial-value problem (9.32) has a unique strict solution $W(t) \in D \cap D(A)$ $\forall t \in [0,\hat{t}]$, with $\hat{t}$ suitably small.

Remark 9.3.

Note that $F(f)$ satisfies (5.63b,c,d); however, we cannot conclude that $W(t)$ exists over an *a priori* given interval $[0,t_0]$ because $W(t)$ does not necessarily satisfy condition (5.63f). □

The strict solution $W(t)$ of system (9.32) is also the continuous solution of the integral equation (5.34):

$$W(t) = Z(t)W_0 + \int_0^t Z(t-s)F(W(s))\mathrm{d}s \qquad t\in[0,\hat{t}] \tag{9.39}$$

with $Z(t) = \exp\{t(A+J)\}$, and so, if $\phi(t) = \|W(t);X\|$,

$$0 \le \phi(t) \le \exp\{(\eta_1-\eta_0)t\}\|W_0;X\| + \eta_2\int_0^t \exp\{(\eta_1-\eta_0)(t-s)\}\{\phi(s)\}^2\mathrm{d}s$$

because of (c) of Lemma 9.7 and of (a) of Lemma 9.8 with $f = \theta_X$ (and so with $F(\theta_X) = \theta_X$) and with $\hat{f} = W(s)$. Now if $\phi_1(t)$ is the solution of the integral equation

$$\phi_1(t) = \exp\{(\eta_1-\eta_0)t\}\|W_0;X\| + \eta_2\int_0^t \exp\{(\eta_1-\eta_0)(t-s)\}\{\phi_1(s)\}^2\mathrm{d}s$$

i.e. if

$$\frac{\mathrm{d}}{\mathrm{d}t}\phi_1(t) = (\eta_1-\eta_0)\phi_1(t)+\eta_2\{\phi_1(t)\}^2, \qquad \phi_1(0) = \|W_0;X\| \tag{9.40}$$

then it is not difficult to prove that $0 \le \phi(t) \le \phi_1(t)$ for any $t \ge 0$ such that $\phi_1(t) < +\infty$. On the other hand, we have from (9.40)

$$\frac{\mathrm{d}}{\mathrm{d}t}\frac{1}{\phi_1(t)} = -(\eta_1-\eta_0)\frac{1}{\phi_1(t)} - \eta_2, \qquad \frac{1}{\phi_1(0)} = \frac{1}{\|W_0;X\|}$$

for any $t \ge 0$ such that $\phi_1(t) > 0$. Hence,

$$\frac{1}{\phi_1(t)} = \left(\frac{1}{\|W_0;X\|} + \frac{\eta_2}{\eta_1-\eta_0}\right)\exp\{-(\eta_1-\eta_0)t\} - \frac{\eta_2}{\eta_1-\eta_0}$$

and so

$$\|W(t)\| \le \phi_1(t) = \frac{(\eta_1-\eta_0)\|W_0;X\|\exp\{(\eta_1-\eta_0)t\}}{(\eta_1-\eta_0)+\eta_2\|W_0;X\|-\eta_2\|W_0;X\|\exp\{(\eta_1-\eta_0)t\}} \tag{9.41}$$

with $0 \le t < \hat{t}_1 = \min\{\hat{t}, t_1\}$, where t_1 is the first positive value of t such that $1/\phi_1(t_1) = 0$. Note that

$$\| W(t);X \| = h_{10} \int_{-1}^{1} dy \int_{-a}^{a} |n(x,y;t)|\,dx + h|\tau(t)-\tau_c|$$

and so (9.41) gives an upper bound for the total number of neutrons and for the average temperature in S.

The above results may be summarized as follows.

Theorem 9.2. If

$$W_0 = \begin{pmatrix} w_{10} \\ \\ w_{20} \end{pmatrix} = \begin{pmatrix} n_0(x,y) \\ \\ \tau_0 - \tau_c \end{pmatrix} \in D(A) = D(A_1) \times R^1$$

then the semilinear problem (9.32), i.e. the abstract version of system (9.1a,b,c)+(9.25a,b)+(9.26), has a unique strict solution $W(t)$ *that belongs to the set* $D \cap D(A)$ *for any* $t \in [0,\hat{t}]$ *with* $\hat{t}$ *suitably small, and it satisfies the inequality (9.41).* □

EXERCISES

9.1. Prove that the operator J_2, defined by (9.3), is a projection.

9.2. Show that $\mathrm{Re}(B_2 f, f) \le 0$ $\forall f \in D(B_2)$ with B_2 defined by (9.2). Hint: $2\mathrm{Re}(B_2 f, f) = (B_2 f, f) + (f, B_2 f)$.

9.3. Study problem (9.4) in the B-space $X_1 = L^1(\Omega)$. Hint: Prove that $J_2 \in \mathscr{B}(X_1)$ with $\| J_2 f \|_1 \le \| f \|_1$ $\forall f \in X_1$, (see Example 7.3), and that $B_2 \in \mathscr{G}(1,0;X_1)$ (see Remark 2.11).

9.4. Prove that the operator K_2 defined by (9.14) belongs to $\mathscr{B}(X_2, \mathbb{C})$, with $|K_2 f| \le 2\sqrt{a} \| f \|_2$.

9.5. Show that the operator H_z defined by (9.21) can be written as follows

$$H_z \phi = \tfrac{1}{2}\gamma \int_{-a}^{a} H(x,x';z)\phi(x')\,dx', \quad \forall \phi \in L^2(-a,a)$$

with

$$H(x,x';z) = \int_1^{+\infty} \exp\left(-\frac{z+v\Sigma}{v}|x-x'|s\right)\frac{ds}{s}, \qquad \mathrm{Re}z+v\Sigma > 0.$$

Hint: It follows from (9.21), (9.3), and (9.7a,b) that

$$H_z\phi = \frac{v\gamma}{2}\left(\int_0^1 dy\left[\frac{1}{vy}\int_{-a}^{x} \exp\left\{-\frac{z}{vy}(x-x')\right\}\phi(x')dx' - \right.\right.$$

$$\left.\left.\int_{-1}^{0} dy\left[\frac{1}{vy}\int_{x}^{a} \exp\left\{-\frac{z}{vy}(x-x')\right\}\phi(x')dx'\right]\right).$$

Then, use the substitution $s = 1/y$ for the first integral and the substitution $s = -1/y$ for the second.

9.6. Study the initial-value problem (9.4) and the procedure leading to (9.16) with $a = +\infty$.

9.7. Assume that $\gamma-\Sigma < 0$ and that a source term $g(t) = g_0 \exp(i\omega t)$ is present on the right-hand side of (9.4). Find the asymptotic behaviour of $u(t)$ as $t\to+\infty$. Hint: See Example 4.11.

9.8. Derive the integral equation (9.39) directly from system (9.31).

9.9. Prove that $\| F_{\hat{f}}g - F_f g; X\| \to 0$ as $\|\hat{f}-f;X\|\to 0$. Hint: Note that $F_f g$ is linear in f.

9.10. Consider the system (9.1a,b,c)+(9.25a,b) and assume that $\Sigma(\tau)$, $\gamma(\tau)$, and $h_1(\tau)$ satisfy the following conditions:
(i) $|d\Sigma(\tau)/d\tau| \le c$, $|d\gamma(\tau)/d\tau| \le \bar{c}$, $|dh_1(\tau)/d\tau| \le \hat{c}\ \forall \tau\in R^1$;
(ii) the derivatives of $\Sigma(\tau)$, $\gamma(\tau)$, and of $h_1(\tau)$ are continuous functions of $\tau\in R^1$.
Prove that the corresponding semilinear problem has a unique strict solution over *any given* $[0,t_0]$. Hint: Write the abstract problem as follows:

$$\frac{d}{dt}w_1(t) = B_1w_1(t) - va_1(w_2(t))w_1(t) + va_2(w_2(t))J_1w_1(t), \qquad t>0$$

$$\frac{d}{dt}w_2(t) = -hw_2(t) + a_3(w_2(t))K_1w_1(t), \qquad t>0$$

$$X_1\text{-}\lim_{t\to 0+} w_1(t) = w_{10}, \qquad R^1\text{-}\lim_{t\to 0+} w_2(t) = w_{20}$$

with $w_1 = n$, $w_2 = \tau-\tau_c$, $a_1(w_2) = \Sigma(\tau)$, $a_2(w_2) = \gamma(\tau)$, $a_3(w_2) = h_1(\tau)$.

9.11. Study the abstract version of system (9.1a)+(9.1c) with 'perfect reflection' boundary conditions:

$$n(-a,y;t) = n(a,y;t), \quad -1 \le y \le 1, \quad t > 0.$$

Hint: Define the operator B_2 as follows:

$$B_2 f = -vy \frac{\partial f}{\partial x}$$

$$D(B_2) = \{f: f\in X_2; B_2 f \in X_2; f(-a,y) = f(a,y) \text{ for a.e. } y\in[-1,1]\}.$$

10
A SEMILINEAR PROBLEM FROM KINETIC THEORY OF VEHICULAR TRAFFIC

10.1. INTRODUCTION

Let the physical system S be the vehicular traffic flowing in a specific direction on a divided multi-lane motorway. Suppose that x is a co-ordinate specifying the position on the motorway, v is the vehicle speed (x is assumed to increase in the direction of the velocity vector $\underline{v}$), and that w is the 'desired' speed, namely the speed that drivers would like to have (Prigogine and Herman 1971). Furthermore let $n(x,v,w;t)$ be the vehicle density, i.e. let $n(x,v,w;t)\mathrm{d}x\,\mathrm{d}v\mathrm{d}w$ be the (expected) number of vehicles that, at time t, are between x and $x+\mathrm{d}x$, have speed between v and $v+\mathrm{d}v$ and whose drivers would like to have a speed between w and $w+\mathrm{d}w$. Then, a recent model for *dilute* traffic leads to the following non-linear system, (Paveri-Fontana 1975):

$$\frac{\partial}{\partial t}\, n(x,v,w;t) = -v\,\frac{\partial}{\partial x}\, n(x,v,w;t) - \frac{\partial}{\partial v}\left\{\frac{w-v}{T}\, n(x,v,w;t)\right\} +$$

$$F(n), \quad x\in(0,a);\quad v,w\in[v_1,v_2];\quad t>0 \tag{10.1a}$$

$$n(x,v,w;0) = n_0(x,v,w), \quad x\in(0,a);\quad v,w\in[v_1,v_2] \tag{10.1b}$$

$$n(0,v,w;t) = n_e(v,w;t),\quad v,w\in[v_1,v_2],\quad t>0 \tag{10.1c}$$

with

$$F(n) = q\{K_2(n)-K_1(n)\} \tag{10.2a}$$

$$K_2(n) = \left\{\int_{v_1}^{v_2} n(x,v,w';t)\,\mathrm{d}w'\right\}\left\{\int_{v}^{v_2} (v'-v)\,n(x,v',w;t)\,\mathrm{d}v'\right\} \tag{10.2b}$$

$$K_1(n) = n(x,v,w;t)\int_{v_1}^{v} (v-v')\left\{\int_{v_1}^{v_2} n(x,v',w';t)\,\mathrm{d}w'\right\}\mathrm{d}v'. \tag{10.2c}$$

In (10.1) and (10.2), $0\le v_1<v_2<+\infty$, $T>0$ is a given relaxation time, a is the length of the motorway, $q\in[0,1]$ is the probability of not passing, and $n_0(x,v,w)$ and $n_e(v,w;t)$ are given non-negative functions.

Equation (10.1a) is a balance for the number of vehicles that, at time t, are characterized by the parameters x,v,w, ((x,v,w) vehicles), and shows that this number changes because of the following factors.

(*a*) A streaming term $-v\partial n/\partial x$ that gives the number of (x,v,w) vehicles which move in or out of the region between x and $x+\mathrm{d}x$ with speed v, without 'interacting' with (x,v',w') vehicles.

(*b*) An acceleration term $-\partial\{n(w-v)/T\}/\partial v$ that gives the number of (x,v,w) vehicles which change their speed v because their drivers desire to have the speed w. Note that a 'small' value of the relaxation time T means that drivers have a 'strong' desire to reach the speed w.

(*c*1) A source term $qK_2(n)$ showing that (x,v',w) vehicles with $v' \geq v$ 'interact' (with probability q) with (x,v,w') vehicles $\forall w' \in [v_1,v_2]$ and so they slow down and become (x,v,w) vehicles. Note that the first factor on the right-hand side of (10.2b) is the total number of vehicles with speed v at x, independently of the speed desired by their drivers.

(*c*2) A sink term $qK_1(n)$ that gives the number of (x,v,w) vehicles that slow down and become vehicles with speed $v' \leq v$.

Finally, (10.1b) is an initial condition, whereas (10.1c) gives the number of vehicles entering the motorway at $x = 0$.

If traffic is *spatially homogeneous*, then $n = n(v,w;t)$ and (10.1a,b) lead to the semilinear system

$$\frac{\partial}{\partial t} n(v,w;t) = -\frac{\partial}{\partial v}\left\{\frac{w-v}{T} n(v,w;t)\right\}+F(n), \qquad v,w\in[v_1,v_2],\ t>0 \tag{10.3a}$$

$$n(v,w;0) = n_0(v,w), \quad v,w \in [v_1,v_2]. \tag{10.3b}$$

Remark 10.1.

A detailed study of system (10.3) is important because the role of desired speeds is enhanced and the structure of the slowing down term $F(n)$ is retained. Quadratic structures similar to that of F are interesting because they are typical of two-body interactions. □

10.2. PRELIMINARY LEMMAS

The method of characteristics applied to equation (10.3a) with $q = 0$, i.e. with $F(n) \equiv 0$, suggests the introduction of the family of operators

$$Z(t)f = \exp\left(\frac{t}{T}\right) f(\tilde{v}(v,w;t),w), \quad t \in (-\infty,+\infty) \tag{10.4}$$

where

$$\tilde{v}(v,w;t) = v+(v-w)\left\{\exp\left(\frac{t}{T}\right)-1\right\} = w-(w-v)\exp\left(\frac{t}{T}\right) \tag{10.5}$$

namely, $Z(t)$ maps each $f = f(v,w)$ into $\exp(t/T)f(\tilde{v},w)$.

Remark 10.2.

It is easy to verify that $n(v,w;t) = \exp(t/T)n_0(\tilde{v},w)$ satisfies (10.3a) with $q = 0$, if $n_0(v',w)$ is partially differentiable with respect to v' (in the classical sense). □

If X is the real B-space $L^1(R^1 \times R^1)$ with norm

$$\|f\| = \int_{-\infty}^{+\infty}\int_{-\infty}^{+\infty} |f(v,w)|\,dv\,dw \tag{10.6}$$

we have the following lemma.

Lemma 10.1. (a) $Z(t) \in \mathscr{B}(X)$ with $\|Z(t)f\| = \|f\|$ $\forall f \in X$, $\forall t \in (-\infty,+\infty)$; (b) the family $\{Z(t), t \in (-\infty,+\infty)\}$ is a group, i.e. $Z(0) = I$, $Z(t)Z(s) = Z(t+s)$, $\forall\, t,s \in (-\infty,+\infty)$.

Proof. (*a*) Definition (10.4) gives

$$\|Z(t)f\| = \exp\frac{t}{T}\int_{-\infty}^{+\infty}\int_{-\infty}^{+\infty}|f(\tilde{v}(v,w;t),w)|\,dv\,dw =$$

$$\exp\left(\frac{t}{T}\right)\int_{-\infty}^{+\infty}\int_{-\infty}^{+\infty}\exp\left(-\frac{t}{T}\right)|f(\tilde{v},w)|\,d\tilde{v}\,dw = \|f\|.$$

(*b*) $Z(0)f = f$ because $\tilde{v}(v,w;0) = v$. On the other hand, we have from (10.5)

$$\begin{aligned}\tilde{v}(\tilde{v}(v,w;s),w;t) &= w-\{w-\tilde{v}(v,w;s)\}\exp\left(\frac{t}{T}\right)\\ &= w-\left[w-\left\{w-(w-v)\exp\left(\frac{s}{T}\right)\right\}\right]\exp\left(\frac{t}{T}\right)\\ &= w-(w-v)\exp\left(\frac{t+s}{T}\right) = \tilde{v}(v,w;t+s)\end{aligned}$$

and so

$$\begin{aligned}Z(t)Z(s)f &= Z(t)\left[\exp\left(\frac{s}{T}\right)f(\tilde{v}(v,w;s),w)\right]\\ &= \exp\left(\frac{t+s}{T}\right)f(\tilde{v}(\tilde{v}(v,w;s),w;t),w) = \exp\left(\frac{t+s}{T}\right)f(\tilde{v}(v,w;t+s),w)\\ &= Z(t+s)f. \quad \square\end{aligned}$$

Now the closed linear subset of X

$$X_0 = \{f: f\in X; f(v,w) = 0 \text{ for a.e. } (v,w)\notin[v_1,v_2]\times[v_1,v_2]\} \tag{10.7}$$

is itself a B-space with norm (10.6) and, if $Z_0(t)$ is the restriction of $Z(t)$ to X_0

$$Z_0(t)f = Z(t)f, \quad \forall f\in X_0, t\geq 0 \tag{10.8}$$

then $Z_0(t)$ can be considered as an operator from X_0 into itself if $t\geq 0$. To see this, let $g = Z_0(t)f$, i.e.

$$g(v,w) = \exp\left(\frac{t}{T}\right)f(\tilde{v}(v,w;t),w)$$

with $f \in X_0$, and consider any $t \geq 0$. Then the following hold.

(i) If $w \notin [v_1, v_2]$, $f(\tilde{v}, w) = 0$ for a.e. $\tilde{v} \in R^1$ and so $g(v,w) = 0$.

(ii) If $w \in [v_1, v_2]$ and $v < v_1$, then $v - w < 0$ and (10.5) gives $\tilde{v} < v < v_1$; similarly, if $w \in [v_1, v_2]$ and $v > v_2$, then $\tilde{v} > v_2$. Hence, if $w \in [v_1, v_2]$ and $v \notin [v_1, v_2]$, then $\tilde{v} \notin [v_1, v_2]$ and so $f(\tilde{v}, w) = 0$, i.e. $g(v,w) = 0$.

We conclude that $g(v,w) = 0$ for a.e. $(v,w) \notin [v_1, v_2] \times [v_1, v_2]$, i.e., $g = Z_0(t) f \in X_0 \ \forall f \in X_0, t \geq 0$.

Lemma 10.2. *(a) The family* $\{Z_0(t), t \geq 0\}$ *is contained in* $\mathscr{B}(X_0)$, *with* $\| Z_0(t) f \| = \| f \| \ \forall f \in X_0, t \geq 0$, *and it is a semi-group; (b)* $Z_0(t) f$ *is a strongly continuous function of* t *for any* $t \geq 0$ *and* $f \in X_0$; *(c)* $Z_0(t)$ *maps* X_0^+ *into itself, where*

$$X_0^+ = \{f : f \in X_0 ; f(v,w) \geq 0 \text{ for a.e. } (v,w) \in [v_1, v_2] \times [v_1, v_2]\}$$

is the closed positive cone of X_0.

Proof. (a) If t is non-negative, $Z_0(t)$ maps X_0 into itself (see above), and $Z_0(t) f = Z(t) f \ \forall f \in X_0$. Then, (a) and (b) of Lemma 10.1 give:

$$\| Z_0(t) f \| = \| Z(t) f \| = \| f \| \ \forall f \in X_0, t \geq 0$$

$$Z_0(0) f = Z(0) f = f$$

$$Z_0(t) Z_0(s) f = Z(t) Z_0(s) f = Z(t) Z(s) f$$

$$= Z(t+s) f = Z_0(t+s) f, \quad \forall f \in X_0, \ t, s \geq 0.$$

Note that t and s must be non-negative because $Z_0(t)$ does not necessarily map X_0 into itself if $t < 0$.

(b) If $f \in X_0$ and $t, t+h \geq 0$, we have from (10.4) and from (10.8):

$$|Z_0(t+h)f - Z_0(t)f| = \left|\exp\left(\frac{t+h}{T}\right) f(\tilde{v}'', w) - \exp\left(\frac{t}{T}\right) f(\tilde{v}', w)\right|$$

$$\leq \exp\left(\frac{t+h}{T}\right) |f(\tilde{v}'', w) - f(\tilde{v}', w)| + \exp\left(\frac{t}{T}\right) \left|\exp\left(\frac{h}{T}\right) - 1\right| |f(\tilde{v}', w)|$$

with $\tilde{v}'' = \tilde{v}(v, w; t+h)$, $\tilde{v}' = \tilde{v}(v, w; t)$. Now if we assume that $f \in C([v_1, v_2] \times [v_1, v_2]) \subset X_0$, then $f(v, w)$ is uniformly continuous and, in particular,

$$|f(v'', w) - f(v', w)| < \varepsilon \quad \forall \, v', w \in [v_1, v_2]$$

if $|v'' - v'| < \mu_0 = \mu_0(\varepsilon)$. On the other hand, (10.5) gives

$$|\tilde{v}'' - \tilde{v}'| = |v - w| \exp\left(\frac{t}{T}\right) \left|\exp\left(\frac{h}{t}\right) - 1\right| \leq \delta \exp\left(\frac{t}{T}\right) \left|\exp\left(\frac{h}{T}\right) - 1\right|$$

with $\delta = v_2 - v_1$, and so

$$|\tilde{v}'' - \tilde{v}'| = |\tilde{v}(v, w; t+h) - \tilde{v}(v, w; t)| < \mu_0(\varepsilon), \quad \forall \, v, w \in [v_1, v_2]$$

provided that $|h| < h_0(\mu_0(\varepsilon), t) = \hat{h}_0(\varepsilon, t)$. We conclude that that

$$|f(\tilde{v}'', w) - f(\tilde{v}', w)| < \varepsilon \quad \forall v, w \in [v_1, v_2]$$

if $|h| < \hat{h}_0(\varepsilon, t)$, and consequently

$$|Z_0(t+h)f - Z_0(t)f| \leq \exp\left(\frac{t+\hat{h}_0}{T}\right) \varepsilon + \exp\left(\frac{t}{T}\right) \left|\exp\left(\frac{h}{t}\right) - 1\right| |f(\tilde{v}', w)|$$

for any $v, w \in [v_1, v_2]$. Hence,

$$\| Z_0(t+h)f - Z_0(t)f \| \leq \exp\left(\frac{t+\hat{h}_0}{T}\right) \varepsilon \int_{v_1}^{v_2} \int_{v_1}^{v_2} \mathrm{d}v \, \mathrm{d}w + \exp\left(\frac{t}{T}\right) \left|\exp\left(\frac{h}{T}\right) - 1\right|$$

$$\times \int_{v_1}^{v_2} \int_{v_1}^{v_2} |f(\tilde{v}', w)| \, \mathrm{d}v \, \mathrm{d}w$$

$$= \delta^2 \exp\left(\frac{t+\hat{h}_0}{T}\right) \varepsilon + \exp\left(\frac{h}{T}\right) - 1| \, \| f \|$$

if $|h| < \hat{h}_0 = \hat{h}_0(\varepsilon, t)$ (see (*a*) of Lemma 10.1), and $\| Z_0(t+h)f - Z_0(t)f \| \to 0$ as $h \to 0$ for any $f \in C([v_1, v_2] \times [v_1, v_2])$. However, $C([v_1, v_2] \times [v_1, v_2])$ is dense in X_0 because it contains $C_0^{\infty}([v_1, v_2] \times [v_1, v_2])$, and (*b*) is proved by the usual extension procedure from a dense subset to the whole space X_0.

Let the operator B_0 be defined by

$$B_0 f = X\text{-}\lim_{h\to 0+} \frac{1}{h} [Z_0(h)f - f] \qquad (10.9)$$

and have domain $D(B_0)$ composed of *all* the elements of X_0 for which the limit (10.9) exists. Then, $B_0 \in \mathscr{G}(1,0;X_0)$ and the semigroup generated by B_0 coincides with $Z_0(t)$:

$$\exp(tB_0)f = Z_0(t)f \;\; \forall f \in X_0, \quad t \geq 0.^{\dagger}$$

Finally, if we define the operator B_1 as follows

$$B_1 f = -\frac{\partial}{\partial v}\left\{\frac{w-v}{T} f(v,w)\right\}$$
$$D(B_1) = \left\{f : f \in X_0;\; \frac{\partial f}{\partial v} \in C([v_1,v_2]\times[v_1,v_2]\right\} \qquad (10.10)$$

where $\partial f/\partial v$ is a classical derivative, it is not difficult to show that the limit (10.9) exists if $f \in D(B_1)$ and that

$$X\text{-}\lim_{h\to 0+} \frac{1}{h}\{Z_0(h)f - f\} = B_1 f, \quad \forall f \in D(B_1)$$

(see Exercise 10.1). Hence, $B_1 f = B_0 f \; \forall f \in D(B_1)$, and $D(B_1) \subset D(B_0)$ because $D(B_0)$ is the set of *all* $f \in X_0$ for which the limit (10.9) exists. Thus, B_0, the generator of the semigroup $Z_0(t)$, is an extension of the operator B_1 which basically is the 'physical' differential operator on the right-hand side of (10.3a).

---

† It can be proved that, if $\{Z_0(t), t \geq 0\} \subset \mathscr{B}(X_0)$ is a semigroup and $Z_0(t)f$ is continuous at $t = 0$ (i.e. $\|Z_0(t)f - f\| \to 0$ as $t \to 0+$, $\forall f \in X_0$), then the linear operator B_0 defined by (10.9) belongs to $\mathscr{G}(M,\beta;X_0)$ with suitable values of M and β. Furthermore, the semigroup generated by B_0 coincides with $Z_0(t)$: $\exp(tB_0) = Z_0(t)$, $\forall t \geq 0$. Note that this can be interpreted as a 'converse part' of Theorems 4.4 and 4.5, and that the definition (10.9) is suggested by (4.31e) at $t = 0$.

Remark 10.3.

A direct definition of the abstract version of the physical operator $-\partial[\{(w-v)/T\}f]/\partial v$, such as

$$Bf = -\frac{\partial}{\partial v}\left[\frac{w-v}{T} f\right]$$

$$D(B) = \{f: f \in X_0; (w-v)f \in X_0; Bf \in X_0\}$$

(the derivative is now in the generalized sense), leads to some technical difficulties as far as the derivation of the resolvent operator $R(z,B)$ is concerned. It is then simpler to start from a physically reasonable semigroup $Z_0(t)$ and take its generator B_0 as the abstract version of the physical differential operator under consideration. □

If we interpret B_0 as the abstract version of the differential operator on the right-hand side of (10.3a), system (10.3a)+(10.3b) leads to the following initial-value problem in the B-space X_0:

$$\frac{\mathrm{d}}{\mathrm{d}t} u(t) = B_0 u(t) + F(u(t)), \quad t > 0; \quad X\text{-}\lim_{t\to 0+} u(t) = u_0 \quad (10.11)$$

with $u(t) = n(v,w;t)$ and $u_0 = n_0(v,w) \in D(B_0)$. In particular, if $q = 0$, i.e. if passing is completely free, then the strict solution of (10.11) has the form

$$u(t) = Z_0(t)u_0, \quad t \geq 0 \quad (10.12a)$$

and, explicitly,

$$n(v,w;t) = \exp\left(\frac{t}{T}\right) n_0(\tilde{v}(v,w;t),w). \quad (10.12b)$$

Note that $|v-w| \leq |v-w|\exp(t/T) = |\tilde{v}-w| \; \forall t \geq 0$, and so the semigroup $Z_0(t)$ tends to reduce the absolute value of the difference between the real speed v and the desired speed w.

10.3. THE OPERATORS F, K_1 AND K_2

The definitions (10.2) show that K_1, K_2, and $F = q[K_2-K_1]$

map X_0 into itself and so they can be interpreted as operators from X_0 into X_0, with domains $D(K_1) = D(K_2) = D(F) = X_0$. The main properties of these operators are summarized in the following Lemma.

Lemma 10.3. If $\delta = v_2 - v_1$, *then* (*a*) $\|K_j(f_1) - K_j(f)\| \leq \delta[\|f_1\| + \|f\|]\|f_1 - f\|$, $\forall f_1, f \in X_0$, $j = 1,2,$; (*b*) $K_j(f)$ *is Fréchet differentiable at any* $f \in X_0$ *and its F derivative* $K_{j,f}$ *is such that* $\|K_{j,f}g\| \leq 2\delta\|g\|\|f\|$, $\|K_{j,f_1}g - K_{j,f}g\| \leq 2\delta\|g\|\|f_1 - f\|$ $\forall f_1, f, g \in X_0$; (*c*) K_1 *and* K_2 *map* X_0^+ *into itself;* (*d*) $\|F(f_1) - F(f)\| \leq 2q\delta[\|f_1\| + \|f\|]\|f_1 - f\|$, $F_f = q[K_{2,f} - K_{1,f}]$, $\|F_f g\| \leq 4q\delta\|g\|\|f\|$, $\|F_{f_1}g - F_f g\| \leq 4q\delta\|g\|\|f_1 - f\|$, $\forall f_1, f, g \in X_0$.

Proof. (*a*) (10.2b) gives for any f_1, $f \in X_0$:

$$|K_2(f_1) - K_2(f)| \leq \int_{v_1}^{v_2} |f_1(v,w')|\,dw' \int_{v}^{v_2} (v'-v)|f_1(v',w) - f(v',w)|\,dv' +$$

$$\int_{v_1}^{v_2} |f_1(v,w') - f(v,w')|\,dw' \int_{v}^{v_2} (v'-v)|f(v',w)|\,dv'$$

$$\leq \delta\left\{ \int_{v_1}^{v_2} |f_1(v,w')|\,dw' \int_{v_1}^{v_2} |f_1(v',w) - f(v',w)|\,dv' + \int_{v_1}^{v_2} |f_1(v,w') - f(v,w')|\,dw' \int_{v_1}^{v_2} |f(v',w)|\,dv' \right\}$$

because $0 \leq v'-v \leq v_2 - v_1 = \delta$, $\forall v' \in [v, v_2]$. Hence,

$$\|K_2(f_1) - K_2(f)\| = \int_{v_1}^{v_2}\int_{v_1}^{v_2} |K_2(f_1) - K_2(f)|\,dv\,dw$$

$$\leq \delta[\|f_1\|\|f_1 - f\| + \|f_1 - f\|\|f\|] = \delta[\|f_1\| + \|f\|]\|f_1 - f\|$$

because K_2 maps X_0 into itself and so the integrals can be taken over $[v_1, v_2]$. Since a similar result holds for K_1, (*a*) is proved.

(*b*) If f and g belong to X_0, we have from (10.2b,c):

$$K_j(f+g)-K_j(f) = K_{j,f}g+K_j(g), \quad j = 1,2,$$

with

$$K_{2,f}g = \int_{v_1}^{v_2} f(v,w')\,dw' \int_{v}^{v_2} (v'-v)g(v',w)\,dv' + \int_{v_1}^{v_2} g(v,w')\,dw' \int_{v}^{v_2} (v'-v)f(v',w)\,dv' \tag{10.13a}$$

$$K_{1,f}g = f(v,w) \int_{v_1}^{v} (v-v') \int_{v_1}^{v_2} g(v',w')\,dw'dv' + g(v,w) \int_{v_1}^{v} (v-v') \int_{v_1}^{v_2} f(v',w')\,dw'dv' \tag{10.13b}$$

and with

$$\|K_j(g)\| \leq \delta\|g\|^2, \quad j = 1,2$$

as follows from (*a*) of this lemma with $f_1 = g$ and with $f = \theta_X$ (and so with $K_j(\theta_X) = \theta_X$). Thus, $K_{j,f}$ is indeed the F derivative of $K_j(f)$ because $\|K_j(g)\|/\|g\| \leq \delta\|g\| \to 0$ as $\|g\| \to 0$ (see section 3.6), and (10.13a,b) give

$$\|K_{j,f}g\| \leq 2\delta\|g\|\|f\|, \quad \|K_{j,f_1}g-K_{j,f}g\| \leq 2\delta\|g\|\|f_1-f\| \tag{10.13c}$$

for any $f_1,f,g\in X_0$ (see Exercise 10.2).

(*c*) follows without difficulty from the definitions (10.2b,c).

(*d*) Since $F(f) = q\{K_2(f)-K_1(f)\}\ \forall f\in X_0$, (*d*) follows from (*a*) and (*b*) of this lemma. □

10.4. THE OPERATORS J AND K_3

Part (*d*) of Lemma 10.3 shows that the non-linear operator F satisfies the assumptions (5.63b,c,d) of Theorem 5.5. To prove that (5.63f) is also satisfied, we define the operators

$$Jf = \int_{-\infty}^{+\infty}\int_{-\infty}^{+\infty} f(v,w)\,dv\,dw = \int_{v_1}^{v_2}\int_{v_1}^{v_2} f(v,w)\,dv\,dw, \qquad D(J) = X_0,\ R(J) \subset R^1 \tag{10.14}$$

$$K_3(f) = (v_2-v_1)f(v,w)Jf - K_1(f), \quad D(K_3) = X_0,\ R(K_3) \subset X_0. \tag{10.15}$$

Note that Jn gives the *total* number of vehicles at time t.

Lemma 10.4. *(a)* $J \in \mathscr{B}(X_0, R^1)$ *with* $|Jf| \le \|f\|\ \forall f \in X_0$; *(b)* $0 \le Jf = \|f\|\ \forall f \in X_0^+$; *(c)* $JZ_0(t)f = Jf\ \forall f \in X_0, t \ge 0$; *(d)* $J[K_2(f) - K_1(f)] = 0\ \forall f \in X_0$.

Proof. *(a)* and *(b)* easily follow from definition (10.14). *(c)* As in the proof of *(a)* of Lemma 10.1, we have from (10.8) and from (10.4)

$$\begin{aligned} JZ_0(t)f &= \exp\left(\frac{t}{T}\right)\int_{-\infty}^{+\infty}\int_{-\infty}^{+\infty} f(\tilde{v}(v,w;t),w)\,dv\,dw \\ &= \exp\left(\frac{t}{T}\right)\int_{-\infty}^{+\infty}\int_{-\infty}^{+\infty} \exp\left(-\frac{t}{T}\right) f(\tilde{v},w)\,d\tilde{v}\,dw = Jf. \end{aligned}$$

(d) (10.2b) and (10.2c) give

$$J\{K_2(f) - K_1(f)\} = \int_{v_1}^{v_2} dv \left\{ \int_{v_1}^{v_2} f(v,w'')\,dw'' \int_{v_1}^{v_2} (v'-v) \int_{v_1}^{v_2} f(v',w'')\,dw''\,dv' \right\}$$

$$= (Jf)\int_{v_1}^{v_2}\int_{v_1}^{v_2} v'f(v',w'')\,dw''\,dv' - \int_{v_1}^{v_2}\int_{v_1}^{v_2} vf(v,w'')\,dw''\,dv\,(Jf) = 0$$

and Lemma 10.4 is proved. □

Remark 10.4.

Since $q\{K_2(n) - K_1(n)\}$ gives the change per unit time of the number of (v,w) vehicles because of slowing-down events (see *(c1)* and *(c2)* of section 10.1), $J[q\{K_2(n) - K_1(n)\}]$ is the change per unit time of the *total* number of vehicles because of slowing-down events. However, $J[q\{K_2(n)-K_1(n)\}] = 0$ and so the total number of vehicles is not affected by such events. □

Lemma 10.5. (*a*) $\| K_3(f_1)-K_3(f) \| \le 2\delta[\| f_1\| + \| f\|]\| f_1-f\|$, $\forall f_1, f \in X_0$; (*b*) $K_3(f) \in X_0^+$, $\forall f \in X_0^+$.

Proof. (*a*) Definition (10.15) gives

$$|K_3(f_1)-K_3(f)| \le |K_1(f_1)-K_1(f)| + \delta|f_1(v,w)||J[f_1-f]| + \delta|f(v,w)-f_1(v,w)||Jf|$$

and so

$$\begin{aligned}\| K_3(f_1)-K_3(f)\| &\le \| K_1(f_1)-K_1(f)\| + \delta\| f_1\| \, |J[f_1-f]| + \delta\| f_1-f\| |Jf| \\ &\le \| K_1(f_1)-K_1(f)\| + \delta\| f_1\| \| f_1-f\| + \delta\| f_1-f\| \| f\|\end{aligned}$$

because of (*a*) of Lemma 10.4. Using (*a*) of Lemma 10.3 with $j = 1$, we finally obtain

$$\| K_3(f_1)-K_3(f)\| \le 2\delta[\| f_1\| + \| f\|]\| f_1-f\| .$$

(*b*) If $f \in X_0^+$, $f(v,w) \ge 0$ for a.e. $(v,w) \in [v_1,v_2]\times[v_1,v_2]$ and (10.2c) gives

$$\begin{aligned}K_1(f) &= f(v,w)\int_{v_1}^{v_2}\int_{v_1}^{v_2}(v-v')f(v',w')\,dw'dv' \\ &\le f(v,w)(v_2-v_1)\int_{v_1}^{v_2}\int_{v_1}^{v_2} f(v',w')\,dw'dv' = (v_2-v_1)f(v,w)Jf\end{aligned}$$

and $K_3(f) \ge 0$ for a.e. $(v,w) \in [v_1,v_2]\times[v_1,v_2]$. □

Finally, if

$$G(f) = q\{K_2(f)+K_3(f)\}, \quad D(G) = X_0 \tag{10.16}$$

we have the following lemma.

Lemma 10.6. (*a*) $\|G(f_1)-G(f)\| \leq 3q\delta[\|f_1\|+\|f\|]\|f_1-f\|, \forall f_1, f\in X_0$; (*b*) $G(f)\in X_0^+$, $\forall f\in X_0^+$; (*c*) $JG(f) = q\delta(Jf)^2$ $\forall f\in X_0$.

Proof. (*a*) Definition (10.16) gives

$$\|G(f_1)-G(f)\| \leq q\delta[\|f_1\|+\|f\|]\|f_1-f\|+2q\delta[\|f_1\|+\|f\|]\|f_1-f\|$$

because of (*a*) of Lemma 10.3 with $j = 2$ and (*a*) of Lemma 10.5. (*b*) follows from (*c*) of Lemma 10.3 and from (*b*) of Lemma 10.5. (*c*) We have from (10.15), (10.16), and from (*d*) of Lemma 10.4:

$$JG(f) = q[JK_2(f)+\delta(Jf)^2-JK_1(f)] = q\delta(Jf)^2 . \quad \square$$

10.5. GLOBAL SOLUTION OF THE ABSTRACT PROBLEM (10.11)

The conditions (5.63a,b,c,d,e) of Theorem 5.5 hold for the initial-value problem (10.11) because (i) $B_0\in \mathcal{G}(1,0;X_0)$ (see section 10.2 after (10.9)) and so (5.63a) is satisfied with $M = 1$ and $\beta = 0$, (ii) F has the properties listed in (*d*) of Lemma 10.3 and so (5.63b,c,d) are satisfied with $\alpha(\rho_1,\rho_2) = 2q\delta[\rho_1+\rho_2]$, $\alpha_1(\rho) = 4q\delta\rho$, (iii) $u_0\in D(B_0)$ by assumption.

The proof that condition (5.63f) is also satisfied will be divided into two parts: first we show that the total number of vehicles $Ju(t)$ is time independent, and then we prove that $u(t)\in X_0^+$ provided that $u_0\in D(B_0)\cap X^+$ (and this implies that $\|u(t)\| = Ju(t) = \eta$ = a constant).

Lemma 10.7. If system (10.11) has a strict solution $u(t)$ *defined over* $[0,t_1]\subset[0,t_0]$, *then* $Ju(t) = Ju_0 = \eta$ $\forall t\in[0,t_1]$.

Proof. If $u(t)$ is a strict solution of (10.11) for $t\in[0,t_1]$, then it is also the unique continuous solution of the integral equation

$$u(t) = Z_0(t)u_0 + \int_0^t Z_0(t-s)F(u(s))\mathrm{d}s \quad t\in[0,t_1] \tag{10.17}$$

(see Exercise 10.3). Applying the operator J to both sides of (10.17), we obtain

$$Ju(t) = Ju_0 + \int_0^t JF(u(s))\mathrm{d}s = Ju_0, \quad \forall\, t\in[0,t_1]$$

because of (a), (c), and (d) of Lemma 10.4. □

Since (10.2a), (10.15), and (10.16) give for any $f\in X_0$

$$F(f) = G(f)-q\{K_3(f)+K_1(f)\} = G(f)-q\delta fJf$$

$u(t)$ is also a strict solution of the initial-value problem

$$\frac{\mathrm{d}}{\mathrm{d}t}\, u(t) = \{B_0-q\delta Ju(t)\}u(t)+G(u(t)), \qquad 0<t\le t_1 \tag{10.18a}$$

$$X\text{-}\lim_{t\to 0+} u(t) = u_0\in D(B_0). \tag{10.18b}$$

However, $Ju(t) = \eta\ \forall t\in[0,t_1]$ and (10.18a) becomes

$$\frac{\mathrm{d}}{\mathrm{d}t}\, u(t) = [B_0-q\delta\eta I]u(t)+G(u(t)), \qquad 0<t\le t_1. \tag{10.19}$$

Now $(-q\delta\eta I)$ commutes with B_0 and so the semigroup $\{Z_1(t) = \exp[t(B_0-q\delta\eta I)], t\ge 0\}$ is such that, for any $t\ge 0$,

$$Z_1(t) = \exp(-q\delta\eta t)Z_0(t)f \tag{10.20a}$$

$$\| Z_1(t)f \| = \exp(-q\delta\eta t)\|Z_0(t)f\| = \exp(-q\delta\eta t)\|f\| \le \|f\|, \quad \forall f\in X_0 \tag{10.20b}$$

$$Z_1(t)f\in X_0^+, \quad \forall\, f\in X_0^+ \tag{10.20c}$$

(see Lemma 10.2). Moreover, $u(t)$ is the unique continuous solution of the integral version of (10.19)+(10.18b):

$$u(t) = Z_1(t)u_0 + \int_0^t Z_1(t-s)G(u(s))\mathrm{d}s, \qquad t\in[0,t_1]. \tag{10.21}$$

Lemma 10.8. If $u_0 \in D(B_0) \cap X_0^+$ *and if system (10.11) has a stric strict solution* $u(t)$ *defined over* $[0,t_1] \subset [0,t_0]$ *then* $u(t) \in X_0^+$.

Proof. The continuous solution of (10.21) can be sought by the usual method of successive approximations:

$$u^{(m)}(t) = Z_1(t)u_0 + \int_0^t Z_1(t-s)G(u^{(m-1)}(s))\,ds, \qquad m = 1,2,\ldots \tag{10.22}$$

with $u^{(0)}(t) = u_0$. Note that $u^{(m)}(t) \in X_0^+ \;\forall m = 0,1,2,\ldots$ because $u_0 \in X_0^+$ and because of (10.20c) and (*b*) of Lemma 10.6. Hence, (*b*) of Lemma 10.4 gives $Ju^{(m)}(t) = \|u^{(m)}(t)\|$ and applying J to both sides of (10.22) we obtain

$$\|u^{(m)}(t)\| = \exp(-q\delta\eta t)JZ_0(t)u_0 + \int_0^t \exp\{-q\delta\eta(t-s)\}JZ_0(t-s)\, G(u^{(m-1)}(s))\,ds$$

$$= \exp(-q\delta\eta t)\eta + q\delta \int_0^t \exp[-q\delta\eta(t-s)]\|u^{(m-1)}(s)\|^2\, ds \tag{10.23}$$

for $m = 1,2,\ldots$ and $t \in [0,t_1]$, because of (10.20a), (*c*) of Lemma 10.4, and (*c*) of Lemma 10.6. If $m = 1$, (10.23) gives

$$\|u^{(1)}(t)\| = \exp(-q\delta\eta t)\eta + q\delta\eta^2 \int_0^t \exp\{-q\delta\eta(t-s)\}ds = \eta$$

because $\|u^{(0)}(s)\|^2 = \|u_0\|^2 = \eta^2$. Iterating this procedure, we have

$$\|u^{(m)}(t)\| = \eta, \quad \forall m = 0,1,2,\ldots, \quad t \in [0,t_1]. \tag{10.24}$$

Finally, (10.20b), (10.22), (10.24), and (*a*) of Lemma 10.6 give

$$\|u^{(m+1)}(t) - u^{(m)}(t)\| \leq \int_0^t \|G(u^{(m)}(s)) - G(u^{(m-1)}(s))\|\,ds$$

$$\leq 3q\delta \int_0^t \{\|u^{(m)}(s)\| + \|u^{(m-1)}(s)\|\}\|u^{(m)}(s) - u^{(m-1)}(s)\|\,ds$$

$$\leq c \int_0^t \| u^{(m)}(s) - u^{(m-1)}(s) \| ds$$

with $c = 6q\delta\eta$. Hence, if $m = 1,2,$

$$\| u^{(2)}(t) - u^{(1)}(t) \| \leq c \int_0^t \{ \| u^{(1)}(s) \| + \| u_0 \| \} ds = 2\eta c t$$

$$\| u^{(3)}(t) - u^{(2)}(t) \| \leq c \int_0^t \| u^{(2)}(s) - u^{(1)}(s) \| ds$$

$$\leq 2\eta c^2 \int_0^t s ds = 2 \frac{\eta c^2 t^2}{2}$$

and, in general,

$$\| u^{(m+1)}(t) - u^{(m)}(t) \| \leq 2\eta \frac{(ct)^m}{m!} \leq 2\eta \frac{(ct_1)^m}{m!}$$

for any $m = 0,1,2,\ldots,$ and $t \in [0,t_1]$. Thus,

$$\| u^{(m+j)}(t) - u^{(m)}(t) \| \leq \sum_{i=m}^{m+j-1} \| u^{(i+1)}(t) - u^{(i)}(t) \|$$

$$\leq 2\eta \sum_{i=m}^{m+j-1} \frac{(ct_1)^i}{i!} \leq 2\eta \sum_{i=m}^{\infty} \frac{(ct_1)^i}{i!} \leq 2\eta \frac{(ct_1)^m}{m!} \exp(ct_1) \tag{10.25}$$

$\forall m,j = 0,1,2,\ldots, t \in [0,t_1]$. Inequality (10.25) shows that $\{u^{(m)}(t), m = 0,1,2,\ldots\}$ is a Cauchy sequence in X_0, uniformly in $t \in [0,t_1]$, because $(ct_1)^m/m! \to 0$ as $m \to +\infty$ (in other words, $\{u^{(m)}, m = 0,1,2,\ldots\}$ is a Cauchy sequence in the space $C([0,t_1];X_0)$). We conclude that $\{u^{(m)}(t), m = 0,1,2,\ldots\}$ is convergent in X_0, uniformly with respect to $t \in [0,t_1]$, and that its limit is the continuous solution of (10.21):

$$u(t) = X\text{-}\lim_{m\to\infty} u^{(m)}(t) \; . \tag{10.26}$$

Since $u^{(m)}(t) \in X_0^+$ and $\| u^{(m)}(t) \| = \eta \; \forall m = 0,1,\ldots,$ (10.26) implies that $u(t) \in X_0^+$ because X_0^+ is a closed subset of X_0 and that $\| u(t) \| = \eta \; \forall t \in [0,t_1]$. □

Lemma 10.8 shows that (5.63f) is satisfied if

$u_0 \in D(B_0) \cap X_0^+$ and so Theorem 5.5 leads to the following.

Theorem 10.1. If $u_0 \in D(B_0) \cap X_0^+$ and $t_0 > 0$ is given a priori, the initial-value problem (10.11) has a unique (global) solution $u(t)$ defined over the whole $[0, t_0]$ and such that, for any $t \in [0, t_0]$, $u(t) \in D(B_0) \cap X_0^+$ with

$$\|u(t)\| = \int_{v_1}^{v_2} \int_{v_1}^{v_2} n(v,w;t)\,dv\,dw$$

$$= \int_{v_1}^{v_2} \int_{v_1}^{v_2} n_0(v,w)\,dv\,dw = \|u_0\| .$$

EXERCISES

10.1. Prove that $\|h^{-1}\{Z_0(h)f - f\} - B_1 f\| \to 0$ as $h \to 0+$, $\forall f \in D(B_1)$ with B_1 defined by (10.10). Hint: If $\tilde{v}(v,w;h) = v + k$ with $k = (v-w) \times \{\exp(h/t) - 1\}$ because of (10.5), we have $f(\tilde{v}(v,w;h), w) - f(v,w) = f(v+k,w) - f(v,w) = kg(v+\rho k, w)$ with $0 \le \rho \le 1$, because

$$g(v,w) = \frac{\partial f(v,w)}{\partial v} \in C([v_1,v_2]\times[v_1,v_2]) .$$

Then

$$\left|\frac{f(v+k,w) - f(v,w)}{h} + \frac{w-v}{T}\, g(v,w)\right| = \left|\frac{k}{h}\, g(v+\rho k,w) + \frac{w-v}{T}\, g(v,w)\right|$$

$$= \left|\frac{w-v}{T}\right| \left|\frac{\exp(h/T)-1}{h/t}\, g(v+\rho k,w) - g(v,w)\right|$$

$$\le \frac{\delta}{T}\left\{\frac{\exp(h/t)-1}{(h/t)}\, |g(v+\rho k,w) - g(v,w)| + \left\{\frac{\exp(h/t)-1}{(h/t)} - 1\right\} |g(v,w)|\right]$$

$$\le \frac{\delta}{T}\left\{\exp\left(\frac{h}{T}\right) |g(v+\rho k,w) - g(v,w)| + \frac{h}{T}\exp\left(\frac{h}{T}\right) |g(v,w)|\right\}$$

because

$$1 \le \frac{\{\exp(h/t)-1\}}{(h/t)} \le \exp\left(\frac{h}{T}\right)$$

and

$$0 \le \frac{\{\exp(h/t)-1\}}{(h/t)} - 1 \le \left(\frac{h}{t}\right) \exp\left(\frac{h}{t}\right), \quad \forall\, h > 0.$$

However, $g(v,w)$ is also uniformly continuous, and so ... .

10.2. Prove inequalities (10.13c). Hint: For instance, (10.13a) gives

$$|K_{2,f}g| \leq \delta\left\{\int_{v_1}^{v_2} f(v,w')\,dw' \int_{v_1}^{v_2} g(v',w)\,dv' + \int_{v_1}^{v_2} g(v,w')\,dw' \int_{v_1}^{v_2} f(v',w)\,dv'\right\}.$$

10.3. Show that the integral equation (10.17) has at most one strongly continuous solution over $[0,t_1]$. Hint: If $u(t)$ and $U(t)$ are continuous solutions of (10.17) over $[0,t_1]$, then $\|u(t)\|$ and $\|U(t)\|$ are continuous functions of $t\in[0,t_1]$ and so $0 \leq \|u(t)\| \leq \hat{n}$, $0 \leq \|U(t)\| \leq \hat{n}$ for any $t\in[0,t_1]$. Then, (10.17) gives

$$\|u(t)-U(t)\| \leq 2q\delta \int_0^t \{\|u(s)\|+\|U(s)\|\}\|u(s)-U(s)\|\,ds$$

$$\leq 4q\delta\hat{n} \int_0^t \|u(s)-U(s)\|\,ds$$

and so, using Gronwall's inequality, ...

10.4. Since the solution of (10.17) depends on the parameter $q\in[0,1]$, re-write (10.17) as follows:

$$u(t;q) = Z_0(t)u_0+q \int_0^t Z_0(t-s)\Phi(u(s;q))\,ds$$

where $\Phi(f) = K_2(f)-K_1(f)$ and $Z_0(t)$ do *not* depend on q. Prove that $u(t;q)$ is strongly differentiable with respect to q and that its strong derivative $u'(t;q) = \partial u(t;q)/\partial q$ satisfies the *linear* equation

$$u'(t;q) = \int_0^t Z_0(t-s)\Phi(u(s;q))\,ds + q \int_0^t Z_0(t-s)\Phi_{u(s;q)}u'(s;q)\,ds$$

where $\Phi_{u(s;q)}$ is the F derivative of $\Phi(f)$ at $f = u(s;q)$. Show that in particular,

$$[u'(t;q)]_{t=0} = \int_0^t Z_0(t-s)\Phi(Z_0(s)u_0)\,ds$$

Hint: Use a procedure similar to that of Lemma 5.4.

10.5. Study the abstract version of system (10.3) under the assumption that the probability q depends on the vehicle density n, so that $F(n) = q(n)\{K_2(n)-K_1(n)\}$. Hint: Study the properties of the operators $H_j(f) = q(f)K_j(f), j = 1,2, f\in X_0$, under the following assumptions: (i) $q = q(f)$ is a function from $D(q) = X_0$ into R^1 with $0 \le q(f) \le 1, \forall f\in X_0$; (ii) $|q(f_1)-q(f)| \le \hat{\alpha}\|f-f_1\| \;\forall\; f_1, f\in X_0$; (iii) $q(f)$ is F differentiable and its F derivative q_f is such that $|q_f g| \le \hat{\alpha}_1\|g\|\,\|f\|$, $|q_{f_1}g-q_f g| \to 0$ as $\|f_1-f\| \to 0, \forall\; f_1, f, g\in X_0$.

11
THE TELEGRAPHIC EQUATION AND THE WAVE EQUATION

11.1. INTRODUCTION

If S is a telegraphic cable (a conductor with distributed parameters, see Tychonov and Samarski 1964), the voltage $\mathcal{V}(x,t)$ and the electric current $\mathcal{I}(x,t)$ at the position x of S and at time t satisfy the system

$$\frac{\partial}{\partial t}\mathcal{V}(x,t) = -\frac{1}{C}\frac{\partial}{\partial x}\mathcal{I}(x,t) - \frac{G}{C}\mathcal{V}(x,t), \qquad t > 0 \tag{11.1a}$$

$$\frac{\partial}{\partial t}\mathcal{I}(x,t) = -\frac{1}{L}\frac{\partial}{\partial x}\mathcal{V}(x,t) - \frac{R}{L}\mathcal{I}(x,t), \qquad t > 0 \tag{11.1b}$$

$$\mathcal{V}(x,0) = \mathcal{V}_0(x), \qquad \mathcal{I}(x,0) = \mathcal{I}_0(x) \tag{11.1c}$$

with $-\infty < x < +\infty$. The constants $C > 0, G \geq 0, L > 0, R \geq 0$ are respectively the capacity, the coefficient of loss due to imperfect insulation, the coefficient of self-induction, and the resistance (all per unit length of S), and $\mathcal{V}_0(x)$ and $\mathcal{I}_0(x)$ are assigned functions.

Remark 11.1.

In (11.1), $-\infty < x < +\infty$, i.e. S is a 'very long' cable. However, if the length of S is $2l < \infty$, then the (11.1) hold for $-l < x < l$ and must be supplemented with some kind of boundary condition such as, for instance,

$$\mathcal{V}(-l,t) = \mathcal{V}^-(t), \quad \mathcal{V}(l,t) = R_r\,\mathcal{I}(l,t)$$

where $\mathcal{V}^-(t)$ is a given function of t (the message that is being transmitted) and R_r is the resistance of the receiver. □

Differentiating (11.1a) with respect to t and (11.1b) with respect to x, we obtain

$$\frac{\partial^2\mathcal{V}}{\partial t^2} = -\frac{1}{C}\frac{\partial^2\mathcal{I}}{\partial x\partial t} - \frac{G}{C}\frac{\partial\mathcal{V}}{\partial t} = -\frac{1}{C}\left[-\frac{1}{L}\frac{\partial^2\mathcal{V}}{\partial x^2} - \frac{R}{L}\frac{\partial\mathcal{I}}{\partial x}\right] - \frac{G}{C}\frac{\partial\mathcal{V}}{\partial t}$$

and using (11.1a) to eliminate $\partial \mathscr{I}/\partial x$:

$$\frac{\partial^2 \mathscr{V}}{\partial t^2} = -\frac{RC+LG}{LC}\frac{\partial \mathscr{V}}{\partial t} + \frac{1}{LC}\frac{\partial^2}{\partial x^2} - \frac{RG}{LC}\mathscr{V} \tag{11.2a}$$

which is called the 'telegraphic' equation because it governs the propagation of a signal in the cable S. The initial conditions to be used together with (11.2a) can be derived from (11.1c) and (11.1a):

$$\mathscr{V}(x,0) = \mathscr{V}_0(x) \tag{11.2b}$$

$$\left[\frac{\partial \mathscr{V}}{\partial t}(x,t)\right]_{t=0} = -\frac{1}{C}\frac{\mathrm{d}\mathscr{I}_0}{\mathrm{d}x}(x) - \frac{G}{C}\mathscr{V}_0(x) = \mathscr{V}_1(x). \tag{11.2c}$$

If we put

$$u_1(x;t) = \mathscr{V}(x,t), \qquad u_2(x;t) = \sqrt{\left(\frac{L}{C}\right)}\,\mathscr{I}(x,t) \tag{11.3}$$

then systems (11.1) and (11.2) become

$$\frac{\partial}{\partial t}u_1(x;t) = -v\frac{\partial}{\partial x}u_2(x;t) - a_1 u_1(x;t), \qquad t > 0 \tag{11.4a}$$

$$\frac{\partial}{\partial t}u_2(x;t) = -v\frac{\partial}{\partial x}u_1(x;t) - a_2 u_2(x;t), \qquad t > 0 \tag{11.4b}$$

$$u_1(x;0) = u_{10}(x) = \mathscr{V}_0(x), \qquad u_2(x;0) = u_{20}(x) = \sqrt{\left(\frac{L}{C}\right)}\,\mathscr{I}_0(X) \tag{11.4c}$$

and

$$\frac{\partial^2}{\partial t^2}u_1(x;t) = -(a_1+a_2)\frac{\partial}{\partial t}u_1(x;t) + v^2\frac{\partial^2}{\partial x^2}u_1(x;t) - a_1 a_2 u_1(x;t), \qquad t > 0 \tag{11.5a}$$

$$u_1(x;0) = \mathscr{V}_0(x) \tag{11.5b}$$

$$\left[\frac{\partial}{\partial t}u_1(x;t)\right]_{t=0} = \mathscr{V}_1(x) \tag{11.5c}$$

with $-\infty < x < +\infty$ and with

$$v = \frac{1}{\sqrt{(LC)}} > 0, \quad a_1 = \frac{G}{C} \geq 0, \quad a_2 = \frac{R}{L} \geq 0. \tag{11.6}$$

Remark 11.2.

Maxwell's equations in a uniform material lead to systems similar to (11.4) and to (11.5) under the assumption of plane symmetry (see Exercise 11.6). □

11.2. PRELIMINARY LEMMAS

Since $\int_{-\infty}^{+\infty} \mathscr{I}^2(x,t)\,dx$ is a relevant physical quantity (see section 11.3), we introduce the Hilbert space $X_0 = L^2(R^1)$ with the usual inner product and norm:

$$(f_1,g_1)_0 = \int_{-\infty}^{+\infty} f_1(x)\overline{g_1(x)}\,dx, \qquad \|f_1\|_0 = \left\{\int_{-\infty}^{+\infty} |f_1(x)|^2 dx\right\}^{1/2}$$

and the Hilbert space $X = X_0 \times X_0$ of all ordered couples of elements of X_0 with inner product and norm

$$(f,g) = (f_1,g_1)_0 + (f_2,g_2)_0, \qquad \|f\| = [\|f_1\|_0^2 + \|f_2\|_0^2]^{1/2}$$

where

$$g = \begin{pmatrix} f_1 \\ f_2 \end{pmatrix} \quad \text{and} \quad g = \begin{pmatrix} g_1 \\ g_2 \end{pmatrix}.$$

We also define the operators

$$B_0 f_1 = -v\frac{d}{dx} f_1, \quad D(B_0) = \left\{f_1 : f_1 \in X_0;\ \frac{df_1}{dx} \in X_0\right\}$$

$$R(B_0) \subset X_0 \tag{11.7}$$

$$Bf = \begin{pmatrix} 0, & B_0 \\ B_0, & 0 \end{pmatrix} \begin{pmatrix} f_1 \\ f_2 \end{pmatrix}$$

$$D(B) = D(B_0) \times D(B_0) = \left\{f : f = \begin{pmatrix} f_1 \\ f_2 \end{pmatrix};\ f_1, f_2 \in D(B_0)\right\}, \qquad R(B) \subset X \tag{11.8}$$

$$Jf = -\begin{pmatrix} a_1, & 0 \\ 0, & a_2 \end{pmatrix}\begin{pmatrix} f_1 \\ f_2 \end{pmatrix}, \qquad D(J) = X, R(J) \subset X \tag{11.9}$$

(see (11.4a) and (11.4b)).

Lemma 11.1. *(a)* $B_0 \in \mathcal{G}'(1,0;X_0)$; *(b) if* $\{Z_0(t) = \exp(tB_0)$ $t \in (-\infty,+\infty)\}$ *is the group generated by* B_0, *then* $\|Z_0(t)f_1\|_0 = \|f_1\|_0, \forall f_1 \in X_0, t \in (-\infty,+\infty)$; *(c)* $Z_0(t)f_1 = f_1(x-vt), \forall f_1 \in X_0$, $t \in (-\infty,+\infty)$, *i.e.* $Z_0(t)$ *maps each* $f_1 = f_1(x) \in X_0$ *into* $f_1(x-vt)$.

Proof. *(a)* By a procedure similar to that of Example 2.18, the general solution of the equation

$$(zI-B_0)f_1 = g_1, \quad g_1 \in X_0, \quad z = \text{a real parameter}$$

can be put into the form

$$f_1 = f_1(x) = c_1\exp\left(-\frac{z}{v}x\right) + \frac{1}{v}\int_{-\infty}^{x} \exp\left\{-\frac{z}{v}(x-y)\right\}g_1(y)\,dy \tag{11.10a}$$

or into the form

$$f_1 = f_1(x) = c_2\exp\left(-\frac{z}{v}x\right) - \frac{1}{v}\int_{x}^{+\infty} \exp\left\{-\frac{z}{v}(x-y)\right\}g_1(y)\,dy \tag{11.10b}$$

where c_1 and c_2 are arbitrary constants. Now, as will be made clear in the following, the integral term of (11.10a) defines an element of X_0 if $z > 0$, whereas the integral term of (11.10b) is an element of X_0 if $z < 0$. Since $\exp(-zx/v)$ diverges as $x \to -\infty$ if $z > 0$ and as $x \to +\infty$ if $z < 0$, the constants c_1 and c_2 are to be taken equal to zero because f_1 must belong to $D(B_0)$ and so to X_0. Hence, the two (11.10) give

$$f_1 = f_1(x) = \frac{1}{v}\int_{-\infty}^{x} \exp\left\{-\frac{z}{v}(x-y)\right\}g_1(y)\,dy, \quad z > 0 \tag{11.11a}$$

$$f_1 = f_1(x) = -\frac{1}{v}\int_{x}^{+\infty} \exp\left\{-\frac{z}{v}(x-y)\right\}g_1(y)\,dy, \quad z < 0. \tag{11.11b}$$

As in Example 2.18, we have from (11.11a)

$$|f_1(x)|^2 \leq \frac{1}{v^2} \int_{-\infty}^{x} \exp\left\{-\frac{z}{v}(x-y)\right\}dy \int_{-\infty}^{x} \exp\left\{-\frac{z}{v}(x-y)\right\} |g_1(y)|^2 dy$$

$$= \frac{1}{zv} \int_{-\infty}^{x} \exp\left\{-\frac{z}{v}(x-y)\right\}|g_1(y)|^2 dy$$

$$\|f_1\|_0^2 = \int_{-\infty}^{+\infty} |f_1(x)|^2 dx \leq \frac{1}{zv} \int_{-\infty}^{+\infty} dy|g_1(y)|^2 \int_{y}^{+\infty} \exp\left\{-\frac{z}{v}(x-y)\right\}dx$$

$$= \frac{1}{z^2} \|g_1\|_0^2, \qquad z > 0 .$$

Since a similar result can be derived from (11.11b) if $z > 0$, we conclude that

$$\|f_1\|_0 = \|R(z,B_0)g_1\|_0 \leq |z|^{-1}\|g_1\|_0, \quad \forall\, z \neq 0 \tag{11.12}$$

where $R(z,B_0)g_1$ is given by (11.11a) if $z > 0$ and by (11.11b) if $z < 0$. Hence, $B_0 \in \mathcal{G}'(1,0;X_0)$ (see (4.17a,b,c″) with $M = 1$, $\beta = 0$) because (11.12) implies that $(zI-B_0)^{-1} \in \mathcal{B}(X_0) \subset \mathcal{C}(X_0)$ for any $z \neq 0$ and so, as usual, $B_0 \in \mathcal{C}(X_0)$, and because $D(B_0)$ is dense in X_0 since $D(B_0) \supset C_0^\infty(R^1)$.

(*b*) See Example 4.9.

(*c*) If $w_0 \in D(B_0)$, the unique strict solution of the problem

$$\frac{dw}{dt} = B_0 w, \qquad X_0\text{-}\lim_{t\to 0} w(t) = w_0 \tag{11.13}$$

has the form

$$w(t) = Z_0(t)w_0, \qquad t \in (-\infty,+\infty)$$

because $B_0 \in \mathcal{G}'(1,0;X_0)$ (see Theorem 4.6). Now, if $w_0 = \phi \in C_0^\infty(R^1)$, then $\phi \in D(B_0)$,

$$\phi(x) = \int_{-\infty}^{x} \phi'(y)\,dy, \quad \forall\, x \in R^1$$

and for any $x_1 \in R^1$, $h < 0$:

$$|\frac{1}{h}\{\phi(x_1-vh)-\phi(x_1)\}+v\phi'(x_1)|^2 = |\frac{1}{h}\int_{x_1}^{x_1-vh}\{\phi'(y)-\phi'(x_1)\}dy|^2$$

$$= |\frac{1}{h}\int_0^{-vh}\{\phi'(x_1+y_1)-\phi'(x_1)\}dy_1|^2$$

$$\leq \frac{1}{|h|^2}\int_0^{v|h|}dy_1\int_0^{v|h|}|\phi'(x_1+y_1)-\phi'(x_1)|^2dy_1$$

$$= \frac{v}{|h|}\int_0^{v|h|}|\phi'(x_1+y_1)-\phi'(x_1)|^2dy_1. \tag{11.14}$$

However, $\phi\in C_0^\infty(R^1)$ implies that $\phi'\in C_0(R^1)$ and so a suitable finite interval $[-\lambda,\lambda]$ exists such that $\phi'(x) \equiv 0$ if $x\notin[-\lambda,\lambda]$ and $|\phi'(x_1+\delta)-\phi'(x_1)| < \varepsilon, \forall x_1\in R^1$ if $|\delta| < \delta_0 = \delta_0(\varepsilon)$. Thus, if we put

$$\hat{w}(t) = \phi(x-vt), \quad x\in R^1, \quad t\in(-\infty,+\infty)$$

we obtain from (11.14) with $x_1 = x-vt$

$$\|\frac{1}{h}\{\hat{w}(t+h)-\hat{w}(t)\}-B_0\hat{w}(t)\|_0^2 = \int_{-\infty}^{+\infty}\frac{1}{h}[\phi(x-vt-vh)-\phi(x-vt)]+v\phi'(x-vt)|^2dx$$

$$\leq \frac{v}{|h|}\int_{-\infty}^{+\infty}dx_1\int_0^{v|h|}|\phi'(x_1+y_1)-\phi'(x_1)|^2dy_1$$

$$= \frac{v}{|h|}\int_0^{v|h|}dy_1\int_{-\infty}^{+\infty}|\phi'(x_1+y_1)-\phi'(x_1)|^2dx_1$$

$$= \frac{v}{|h|}\int_0^{v|h|}dy_1\int_{-\lambda-v|h|}^{\lambda}|\phi'(x_1+y_1)-\phi'(x_1)|^2dx_1$$

because $\phi'(x_1) \equiv 0$ if $x_1\notin[-\lambda,\lambda], \phi'(x_1+y_1) \equiv 0$ if $x_1\notin[-\lambda-v|h|,\lambda]$ for any $y_1\in[0,v|h|]$. Hence,

$$\|\frac{1}{h}\{\hat{w}(t+h)-\hat{w}(t)\}-B_0\hat{w}(t)\|_0^2 \leq \frac{v}{|h|}\varepsilon^2v|h|(2\lambda+v|h|)$$

$$\leq v^2(2\lambda+\delta_0)\varepsilon^2$$

provided that $|h| < \delta_0/v$. Hence, $\hat{w}(t)$ is strongly differentiable with $d\hat{w}/dt = B_0\hat{w}$ because a similar inequality can be derived if $h > 0$. Since $\hat{w}(0) = \phi(x)$, we conclude that $\hat{w}(t)$ is the strict solution of problem (11.13) with $w_0 = \phi$ and so

$$\phi(x-vt) = \hat{w}(t) = Z_0(t)\phi \tag{11.15}$$

i.e. (c) is proved if $f_1 = \phi \in C_0^\infty(R^1)$. However, $C_0^\infty(R^1)$ is dense in X_0 and $\|Z_0(t)f_1\|_0 = \|f_1\|_0, \forall f_1 \in X_0$, and (11.15) can be extended to the whole space X_0 by the usual procedure. □

Remark 11.3.

> $Z_0(t)f_1 = f_1(x-vt)$ represents a 'wave' that propagates toward the right (in the positive x direction) with speed v and without distortion. To understand this, consider a new co-ordinate system defined by the relation $x_1 = x-vt$ that moves toward the right with speed v (with respect to the old co-ordinate system). In such a new system $Z_0(t)f_1 = f_1(x_1)$, i.e. the graph of $Z_0(t)f_1$ is time independent. It follows that in the old system the graph of $Z_0(t)f_1$ translates rigidly toward the right with speed v. □

Lemma 11.2. (a) $B \in \mathcal{G}'(1,0;X)$; (b) *if* $\{Z(t) = \exp(tB), t \in (-\infty,+\infty)\}$ *is the group generated by* B, *then* $\|Z(t)f\| = \|f\|$ $\forall f \in X$, $t \in (-\infty,+\infty)$; (c) *for any* $f \in X$ *and* $t \in (-\infty,+\infty)$

$$Z(t)f = \frac{1}{2}\begin{pmatrix} Z_0(t)+Z_0(-t), & Z_0(t)-Z_0(-t) \\ Z_0(t)-Z_0(-t), & Z_0(t)+Z_0(-t) \end{pmatrix}\begin{pmatrix} f_1 \\ f_2 \end{pmatrix}. \tag{11.16}$$

Proof. (a) The equation $(zI-B)f = g$ can be written as follows:

$$zf_1-B_0f_2 = g_1, \quad zf_2-B_0f_1 = f_2$$

and so

$$z(f_1+f_2)-B_0(f_1+f_2) = g_1+g_2, \quad -z(f_2-f_1)-B_0(f_2-f_1) = g_1-g_2$$

However, any real $z \neq 0$ belongs to $\rho(B_0)$ because of (11.12) and the above system gives

$$f_1+f_2 = R(z,B_0)[g_1+g_2], \qquad f_2-f_1 = R(-z,B_0)[g_1-g_2] \tag{11.17}$$

for any $z \neq 0$. Hence,

$$f_1 = \tfrac{1}{2}[\{R(z,B_0)-R(-z,B_0)\}g_1+\{R(z,B_0)+R(-z,B_0)\}g_2]$$

$$f_2 = \tfrac{1}{2}[\{R(z,B_0)+R(-z,B_0)\}g_1+\{R(z,B_0)-R(-z,B_0)\}g_2]$$

i.e.

$$f = R(z,B)g = \frac{1}{2}\begin{pmatrix} R(z,B_0)-R(-z,B_0), & R(z,B_0)+R(-z,B_0) \\ R(z,B_0)+R(-z,B_0), & R(z,B_0)-R(-z,B_0) \end{pmatrix}\begin{pmatrix} g_1 \\ g_2 \end{pmatrix} \tag{11.18}$$

for any $g \in X$ and $z \neq 0$. On the other hand, we have from (11.17) and from (11.12)

$$\|f\|^2 = \|f_1\|_0^2+\|f_2\|_0^2 = \tfrac{1}{2}(\|f_1+f_2\|_0^2+ \|f_2-f_1\|_0^2)$$

$$\leq \frac{1}{2z^2}\|g_1+g_2\|_0^2 + \frac{1}{2z^2}\|g_1-g_2\|_0^2 = \frac{1}{z^2}[\|g_1\|_0^2+\|g_2\|_0^2] = \frac{1}{z^2}\|g\|^2$$

(see Exercise 11.3), and so

$$\|R(z,B)g\| \leq |z|^{-1}\|g\|, \quad \forall g \in X, \quad z \neq 0. \tag{11.19}$$

We conclude that $B \in \mathcal{G}'(1,0;X)$ because (11.19) implies that $B \in \mathcal{C}(X)$ (see (*a*) of Lemma 11.1) and because $D(B)$ is dense in X ($D(B)$ contains $C_0^\infty(R^1)\times C_0^\infty(R^1)$ which is dense in $X = X_0 \times X_0$).

(*b*) See Example 4.9.

(*c*) (11.16) is trivial if $t = 0$ because $Z_0(0) = I$. If $t \neq 0$ and if $V = V(t/n) = (n/t)R(n/t,B_0) = (I-t/n\,B_0)^{-1}, W = V(-t/n) = (I+ t/n\,B_0)^{-1}$, then we have from (11.18)

$$(I-\frac{t}{n}B)^{-1} = \frac{1}{2}\begin{pmatrix} V+W, & V-W \\ V-W, & V+W \end{pmatrix}.$$

Hence,

$$(I-\frac{t}{n}B)^{-2} = \frac{1}{4}\begin{pmatrix} V+W, & V-W \\ V-W, & V+W \end{pmatrix}\begin{pmatrix} V+W, & V-W \\ V-W, & V+W \end{pmatrix} = \frac{1}{2}\begin{pmatrix} V^2+W^2, & V^2-W^2 \\ V^2-W^2, & V^2+W^2 \end{pmatrix}$$

and, in general,

$$(I-\frac{t}{n}B)^{-n} = \frac{1}{2}\begin{pmatrix} V^n+W^n, & V^n-W^n \\ V^n-W^n, & V^n+W^n \end{pmatrix}, \qquad n = 1,2,\ldots$$

which proves (11.15) because $X_0\text{-}\lim_{n\to\infty} V^n(t/n)g_0 = Z_0(t)g_0$ and $X_0\text{-}\lim_{n\to\infty} W^n(t/n)g_0 = Z_0(-t)g_0$ for any $g_0 \in X_0$. □

Note that (*c*) of Lemma 11.1 and (11.16) give

$$Z(t)f = \frac{1}{2}\begin{pmatrix} Z_0(t)[f_1+f_2] \\ Z_0(t)[f_1+f_2] \end{pmatrix} + \frac{1}{2}\begin{pmatrix} Z_0(-t)[f_1-f_2] \\ Z_0(-t)[-f_1+f_2] \end{pmatrix}$$

i.e.

$$Z(t)f = \frac{1}{2}\begin{pmatrix} 1 \\ 1 \end{pmatrix}\{f_1(x-vt) + f_2(x-vt)\} + \frac{1}{2}\begin{pmatrix} 1 \\ -1 \end{pmatrix}\{f_1(x+vt)-f_2(x+vt)\}. \tag{11.20}$$

Lemma 11.3. (*a*) $J \in \mathcal{B}(X)$, $\|Jf\| \le a\|f\|$ $\forall f \in X$, *with* $a = \max\{a_1,a_2\}$; (*b*) *if*

$$A = B+J, \quad D(A) = D(B), \quad R(A) \subset X \tag{11.21}$$

then $A \in \mathcal{G}'(1,a;X)$ *and the group* $\{\exp(tA), t \in (-\infty,+\infty)\}$ *is such that* $\|\exp(tA)f\| \le \exp(-bt)\|f\|$ $\forall f \in X, t \ge 0$ *with* $b = \min\{a_1,a_2\}$.

Proof. (*a*) We have from definition (11.9):

$$\| Jf \|^2 = a_1^2 \| f_1 \|_0^2 + a_2^2 \| f_2 \|_0^2 \le a^2 [\| f_1 \|_0^2 + \| f_2 \|_0^2] = a^2 \| f \|^2 .$$

(*b*) $B \in \mathcal{G}'(1,0;X)$ and $\| J \| \le a$ imply that $A \in \mathcal{G}'(1,a;X)$ becaus of Theorem 5.1 (this holds for generators of groups as well; see Remark 5.1). Now,

$$A + aI = B + J_1, \quad J_1 = \begin{pmatrix} a - a_1, & 0 \\ 0 & , a - a_2 \end{pmatrix} I \tag{11.22}$$

where $\| J_1 f \| \le |a_1 - a_2| \| f \|, \forall f \in X$, (see the proof of (*a*) of this lemma). Since aI commutes with A and $(B+J_1) \in \mathcal{G}'(1, |a_1 - a_2|;X)$, we have for any $t \in (-\infty, +\infty)$

$$\exp(at)\exp(tA) = \exp\{t(B+J_1)\}$$

i.e.

$$\exp(tA) = \exp(-at)\exp\{t(B+J_1)\} . \tag{11.23}$$

However, $(B+J_1) \in \mathcal{G}'(1, |a_1 - a_2|;X)$ and so

$$\| \exp(tA) f \| \le \exp(-at)\exp(|a_1 - a_2| t) \| f \| = \exp(-bt) \| f \|,$$

$$\forall f \in X, t \ge 0$$

because $a = a_1$ and $|a_1 - a_2| = a_1 - a_2$ if $a_1 > a_2$, whereas $a = a_2$ and $|a_1 - a_2| = a_2 - a_1$ if $a_2 > a_1$. If $a_1 = a_2$ (and so $a_1 = a_2 = a$), then $J_1 = 0$ and (11.23) gives for any $t \in (-\infty, +\infty)$

$$\exp(tA) = \exp(-at)\exp(tB) = \exp(-at)Z(t) \tag{11.24}$$

$$\| \exp(tA) f \| = \exp(-at) \| f \|, \quad \forall f \in X, \tag{11.25}$$

because of (*b*) of Lemma 11.2. □

11.3. THE ABSTRACT VERSION OF THE TELEGRAPHIC SYSTEM (11.4)

Using the definitions (11.8) and (11.9), the abstract version of the 'telegraphic system' (11.4) can be written as follows:

$$\frac{\mathrm{d}}{\mathrm{d}t}\, u(t) = (B+J)u(t), \quad t > 0 \tag{11.26a}$$

$$X\text{-}\lim_{t\to 0+} u(t) = u_0 \tag{11.26b}$$

where

$$u(t) = \begin{pmatrix} u_1(t) \\ \\ u_2(t) \end{pmatrix}, \quad u_0 = \begin{pmatrix} u_{10} \\ \\ u_{20} \end{pmatrix}$$

and where $u_j(t) = u_j(x;t), j = 1,2$, are now to be interpreted as elements of X_0, depending on the parameter t, and so

$$\| u_j(t) \|_0 = \left\{ \int_{-\infty}^{+\infty} |u_j(x;t)|^2 \mathrm{d}x \right\}^{1/2} .$$

Since $A = B+J\in \mathcal{G}\,'(1,a;X)$ because of Lemma 11.3, A also belongs to $\mathcal{G}\,(1,a;X)$ and the unique strict solution of the initial-value problem (11.26) has the form

$$u(t) = \exp(tA)u_0 \tag{11.27}$$

provided that $u_0\in D(A)$.

Remark 11.4.

$A\in \mathcal{G}\,'(1,a;X)$ and so $u(t) = \exp(tA)u_0$, with $u_0\in D(A)$, satisfies (11. a) for any $t\in(-\infty,+\infty)$ and (11.26b) as $t\to 0$ (see Remark 4.8). Also, $u(t)$ has a continuous derivative because $\mathrm{d}u/\mathrm{d}t = \exp(tA)Au_0$, $\forall t\in(-\infty,+\infty)$. Finally, since $\exp(-tA)\exp(tA) = I$, (11.27) gives $u_0 = \exp(-tA)u(t)$, which shows that there is a one-to-one correspondence between the initial state vector u_0 and the state vector $u(t)$ at time t. □

It follows from (11.27) and from (*b*) of Lemma 11.3 that

$$\|u(t)\| = \{\|u_1(t)\|_0^2+\|u_2(t)\|_0^2\}^{1/2} \le \exp(-bt)\|u_0\|, \quad t \ge 0. \quad (11.28)$$

Inequality (11.28) has a relevant physical meaning because it shows that the total energy E dissipated by the Joule effect in the cable S is bounded by $(aC\|u_0\|^2)\exp(-2bt), \forall t \ge 0$. In fact

$$E = G \int_{-\infty}^{+\infty} \mathscr{V}^2(x,t)\,dx + R \int_{-\infty}^{+\infty} \mathscr{I}^2(x,t)\,dx$$

and so

$$E = G\|u_1(t)\|_0^2 + R\frac{C}{L}\|u_2(t)\|_0^2 = C\{a_1\|u_1(t)\|_0^2 + a_2\|u_2(t)\|_0^2\}$$

$$\le Ca\|u(t)\|^2 \le Ca\|u_0\|^2\exp(-2bt).$$

Now if we assume that in particular $a_1 = a_2 (=a \ge 0)$, then (11.24), (11.25), and (11.27) lead to

$$u(t) = \exp(-at)Z(t)u_0 \quad (11.29)$$

$$\|u(t)\| = \exp(-at)\|u_0\|, \quad \forall t \ge 0$$

$$E = Ca\|u(t)\|^2 = Ca\|u_0\|^2\exp(-2at), \quad \forall t \ge 0.$$

Moreover, (11.29), (11.20), (11.4c), and (11.3) give

$$\mathscr{V}(x,t) = \frac{\exp(-at)}{2}[\{\mathscr{V}_0(x-vt)+\sqrt{\left(\frac{L}{C}\right)}\mathscr{I}_0(x-vt)\}+\{\mathscr{V}_0(x+vt)-\sqrt{\left(\frac{L}{C}\right)}\mathscr{I}_0(x+vt)\}] \quad (11.30)$$

and a similar relation for $\mathscr{I}(x,t)$. Hence, under the assumption $a_1 = a_2$, (11.30) shows that $\mathscr{V}(x,t)$ is the superposition of two waves that propagate with speed v (one toward the right and the other toward the left) and *without changing their shape* but being attenuated because of the factor $\exp(-at)$ (see Remark 11.3). This explains why a 'good' telegraphic cable is built with $a_1 \simeq a_2$ (see Exercise 11.5). The case $a_1 = a_2 = 0$ is of particular interest in connection with Maxwell's equations in a non-conducting material (see

Exercise 11.6 with $\sigma = 0$).

The above results are summarized by the following theorem.

Theorem 11.1. If $u_{10} = \mathscr{V}_0$ *and* $u_{20} = \sqrt{(L/C)}\,\mathscr{I}_0$ *belong to* $D(B_0)$, *the telegraphic system (11.26) has a unique strict solution* $u(t)$ *that is given by (11.27) and satisfies (11.26a) at any* $t \in (-\infty, +\infty)$ *and (11.26b) as* $t \to 0$. *Moreover,* $\|u(t)\| \leq \|u_0\| \exp(-bt)\ \forall t \geq 0$ *with* $b = \min\{b_1, b_2\}$, *and* $\|u(t)\| = \exp(-at)\ \|u_0\|\ \forall t \geq 0$ *if* $a_1 = a_2 = a$.

11.4. THE TELEGRAPHIC EQUATION AND THE WAVE EQUATION

Using the definition (11.7), the abstract version of the telegraphic equation (11.5a) and of the initial conditions (11.5b,c) can be written as follows.

$$\frac{d^2}{dt^2} U_1(t) = -(a_1+a_2)\frac{d}{dt} U_1(t) + B_0^2 U_1(t) - a_1 a_2 U_1(t), \qquad t > 0 \tag{11.31a}$$

$$X_0\text{-}\lim_{t\to 0+} U_1(t) = U_{10}, \qquad X_0\text{-}\lim_{t\to 0+}\left\{\frac{d}{dt} U_1(t)\right\} = \tilde{U}_{10} \tag{11.31b}$$

where, as in section 11.3, $U_1(t) = u_1(x;t)$ is to be interpreted as an element of X_0 depending on the parameter t, and where $U_{10} = \mathscr{V}_0$, $\tilde{U}_{10} = \mathscr{V}_1$.

We say that $U_1(t)$ is a strict solution of the *second-order* initial-value problem (11.31) if

(i) $U_1(t)$ is continuous and belongs to $D(B_0^2) = \{f_1 : f_1 \in D(B_0);\ B_0 f_1 \in D(B_0)\}, \forall t \geq 0$;

(ii) $dU_1(t)/dt$ exists and is continuous, $dU_1(t)/dt \in D(B_0)$, and $B_0\{dU_1(t)/dt\}$ is continuous, $\forall t \geq 0$;

(iii) $d^2 U_1(t)/dt^2$ exists and is continuous, $\forall t \geq 0$;

(iv) $U_1(t)$ satisfies equations (11.31)[†].

[†]The continuity of $U_1(t)$ follows from the existence of dU_1/dt and the continuity of dU_1/dt from the existence of d^2U_1/dt^2. In fact, for example, $\|U_1(t+h)-U_1(t)\|_0 \leq |h|\,\|h^{-1}\{U_1(t+h)-U_1(t)\}-dU_1(t)/dt\|_0 + |h|\,\|dU_1(t)/dt\|_0 \to 0$ as $h \to 0$.

To prove that system (11.31) is equivalent in some sense to (11.26), we need the results listed in the following Lemmas.

Lemma 11.4. If $U_1(t)$ is a strict solution of (11.31) and if $U_{10} \in D(B_0)$, then (a) $B_0 U_1(t)$ is continuous $\forall\, t \geq 0$; (b) $B_0^2 U_1(t)$ is continuous $\forall\, t \geq 0$.

Proof. (*a*) If U_1 is a strict solution of (11.31), then $dU_1(s)/ds$ is continuous and belongs to $D(B_0)$, and $B_0\{dU_1(s)/ds\}$ is continuous for any $s \geq 0$. Then,

$$U_1(t) = U_{10} + \int_0^t \frac{d}{ds} U_1(s)\,ds$$

and Theorem 3.4 gives

$$B_0 U_1(t) = B_0 U_{10} + \int_0^t B_0\left\{\frac{d}{ds} U_1(s)\right\}ds$$

because $B_0 \in \mathscr{C}(X_0)$, and (*a*) is proved.

(*b*) follows from (11.31a) and from the fact that $U_1(t)$, $dU_1(t)/dt$ and $d^2U_1(t)/dt^2$ are continuous. □

Lemma 11.5. Assume that $w(t)$ is continuous and belongs to $D(B_0)$ $\forall\, t \geq 0$, and that $dw(t)/dt$ exists, belongs to $D(B_0)$ with $B_0\{dw(t)/dt\}$ continuous $\forall\, t \geq 0$. Then,

$$\frac{d}{dt}\{B_0 w(t)\} = B_0 \frac{d}{dt} w(t), \qquad \forall\, t \geq 0.$$

Proof. As in the proof of (*a*) of Lemma 11.4, we have

$$B_0 w(t) = B_0 w(0) + \int_0^t B_0\left\{\frac{d}{ds} w(s)\right\}ds$$

and so

$$\frac{d}{dt}\{B_0 w(t)\} = B_0 \frac{d}{dt} w(t). \quad □$$

Now assume that $U_1(t)$ is a strict solution of (11.31) and that $U_{10} \in D(B_0) \cap R(B_0)$, $\tilde{U}_{10} \in D(B_0) \cap R(B_0)$. Then, we have from (11.31a):

$$\frac{d}{dt}\left\{\frac{d}{dt}U_1(t)+a_1U_1(t)\right\} = -a_2\left\{\frac{d}{dt}U_1(t)+a_1U_1(t)\right\}+B_0^2U_1(t)$$

$$\frac{d}{dt}\left[\exp(a_2t)\left\{\frac{d}{dt}U_1(t)+a_1U_1(t)\right\}\right] = \exp(a_2t)B_0^2U_1(t)$$

$$\exp(a_2t)\left\{\frac{d}{dt}U_1(t)+a_1U_1(t)\right\}-\left\{\tilde{U}_{10}+a_1U_{10}\right\}$$
$$= \int_0^t \exp(a_2s)B_0^2U_1(s)\,ds$$

where we recall that $B_0^2U_1(s)$ is continuous $\forall\, s \geq 0$ because of (*b*) of Lemma 11.4. Since $\tilde{U}_{10}$ and U_{10} belong to $R(B_0)$ by assumption, then $\tilde{U}_{10}+a_1U_{10}\in R(B_0)$ and an element $U_{20}\in D(B_0)$ exists such that $\tilde{U}_{10}+a_1U_{10} = B_0U_{20}$. Note that $U_{20}\in D(B_0^2)$ since $B_0U_{20}\in D(B_0)$ because both $\tilde{U}_{10}$ and U_{10} belong to $D(B_0)$. On the other hand, $B_0\in \mathscr{C}(X_0)$, $B_0U_1(t)$ is continuous and belongs to $D(B_0)$, and $B_0^2U_1(t) = B_0\{B_0U_1(t)\}$ is continuous because of Lemma 11.4. Then, Theorem 3.4 gives

$$B_0\int_0^t \exp(a_2s)B_0U_1(s)\,ds = \int_0^t \exp(a_2s)B_0^2U_1(s)\,ds$$

and so

$$\frac{d}{dt}U_1(t)+a_1U_1(t) = \exp(-a_2t)B_0\{U_{20}+\int_0^t \exp(a_2s)B_0U_1(s)\,ds\}.$$

Thus, if we put

$$U_2(t) = \{U_{20}+\int_0^t \exp(a_2s)B_0U_1(s)\,ds\}\exp(-a_2t) \tag{11.32}$$

we obtain

$$\frac{d}{dt}U_1(t) = B_0U_2(t)-a_1U_1(t), \quad t>0 \tag{11.33a}$$

$$\frac{d}{dt}U_2(t) = B_0U_1(t)-a_2U_2(t), \quad t>0 \tag{11.33b}$$

$$X_0\text{-}\lim_{t\to 0+} U_1(t) = U_{10}, \quad X_0\text{-}\lim_{t\to 0+} U_2(t) = U_{20} \tag{11.33c}$$

where $U_{10} \in D(B_0) \cap R(B_0) \subset D(B_0)$ and $U_{20} \in D(B_0^2) \subset D(B_0)$.

Since system (11.33) is formally identical with (11.26), we have

$$U(t) = \begin{pmatrix} U_1(t) \\ U_2(t) \end{pmatrix} = \exp(tA) \begin{pmatrix} U_{10} \\ U_{20} \end{pmatrix} \tag{11.34}$$

and $U(t)$ satisfies (11.33a,b) for any $t \in (-\infty, +\infty)$ and (11.33c) as $t \to 0$. Thus, if $U_1(t)$ is a strict solution of (11.31) with U_{10} and $\tilde{U}_{10}$ in $D(B_0) \cap R(B_0)$ and if $U_2(t)$ is defined by (11.32), then the vector $U(t)$ defined by (11.34) satisfies system (11.33).

Conversely, assume that $U(t)$ is given by (11.34) and that U_{10} and U_{20} belong to $D(B_0^2)$ (hence, $U(t)$ is the strict solution of system (11.33)). It follows that

$$U_0 = \begin{pmatrix} U_{10} \\ U_{20} \end{pmatrix} \in D(A^2)$$

because $D(A^2) = D(B^2) = D(B_0^2) \times D(B_0^2)$ (see Exercise 11.7), and (11.34) gives for any $t \in (-\infty, +\infty)$

$$A^2 U(t) = \exp(tA) A^2 U_0 \tag{11.35}$$

$$\begin{pmatrix} \mathrm{d}^j U_1(t)/\mathrm{d}t^j \\ \mathrm{d}^j U_2(t)/\mathrm{d}t^j \end{pmatrix} = \frac{\mathrm{d}^j}{\mathrm{d}t^j} U(t) = \exp(tA) A^j U_0, \quad j = 1,2. \tag{11.36}$$

Equation (11.35) shows that $U(t) \in D(A^2)$, i.e. $U_1(t)$ and $U_2(t)$ belong to $D(B_0^2)$, whereas (11.36) proves that $U_1(t)$ and $U_2(t)$ have continuous first and second derivatives. Moreover, we have from (11.36) with $j = 1$

$$B \frac{\mathrm{d}U}{\mathrm{d}t} = (A - J) \frac{\mathrm{d}U}{\mathrm{d}t} = \exp(tA) A^2 U_0 - J \exp(tA) A U_0$$

and so $\mathrm{d}U_1/\mathrm{d}t$ and $\mathrm{d}U_2/\mathrm{d}t$ belong to $D(B_0)$ with $B_0[\mathrm{d}U_1/\mathrm{d}t]$ and $B_0[\mathrm{d}U_2/\mathrm{d}t]$ continuous in t.

Using these results, we obtain from (11.33a) and from Lemma 11.5 (with U_2 instead of w)

$$\frac{d^2}{dt^2} U_1 = B_0 \frac{d}{dt} U_2 - a_1 \frac{d}{dt} U_1.$$

On the other hand, since (11.33b) and (11.33a) give

$$B_0 \frac{d}{dt} U_2 = B_0^2 U_1 - a_2 B_0 U_2 = B_0^2 U_1 - a_2 [\frac{d}{dt} U_1 + a_1 U_1]$$

we have

$$\frac{d^2}{dt^2} U_1 = B_0^2 U_1 - (a_1 + a_2) \frac{d}{dt} U_1 - a_1 a_2 U_1$$

i.e. the first component of $U(t)$ satisfies (11.31a) for any $t \in (-\infty, +\infty)$. Finally, the first of (11.33c) coincides with the first of (11.31b), whereas the first component of (11.36) with $j = 1$ and $t = 0$ gives

$$\left[\frac{d}{dt} U_1\right]_{t=0} = B_0 U_{20} - a_1 U_{10}.$$

We conclude that the first component of $U(t)$ is a strict solution of (11.31), with $\tilde{U}_{10} = B_0 U_{20} - a_1 U_{10}$. Hence, we have the following theorem.

Theorem 11.2. If U_{10} and $\tilde{U}_{10}$ belong to $D(B_0) \cap R(B_0)$, then the strict solution $U_1(t)$ of (11.31) and the function $U_2(t)$ defined by (11.32) satisfy system (11.33) with $B_0 U_{20} = \tilde{U}_{10} + a_1 U_{10}$. Conversely, if U_{10} and U_{20} belong to $D(B_0^2)$, then the first component $U_1(t)$ of the strict solution $U(t)$ of system (11.33) is also the strict solution of (11.31) with $\tilde{U}_{10} = B_0 U_{20} - a_1 U_{10}$. Furthermore, $U_1(t)$ satisfies (11.31a) at any $t \in (-\infty, +\infty)$ and the two (11.31b) as $t \to 0$. □

In particular, if $a_1 = a_2 = a$, we have from (11.34), (11.24), and (11.16) for any $t \in (-\infty, +\infty)$

$$U_1(t) = \frac{\exp(-at)}{2} [\{Z_0(t) + Z_0(-t)\} U_{10} + \{Z_0(t) - Z_0(-t)\} U_{20}] \quad (11.37)$$

where $Z_0(t)f_1 = f_1(x-vt)$ because of (b) of Lemma 11.1. Finally, if $a_1 = a_2 = a = 0$, (11.31a) becomes the abstract version of the wave equation (see Exercise 11.8)

$$\frac{d^2}{dt^2} U_1(t) = B_0^2 U_0(t)$$

and (11.37) gives

$$U_1(t) = \tfrac{1}{2} Z_0(t)[U_{10}+U_{20}] + \tfrac{1}{2} Z_0(-t)[U_{10}-U_{20}].$$

Remark 11.5.

Theorem 11.2 explains how the equivalence between (11.31) and (11.33) (or (11.26)) must be interpreted. Note that (11.33) has a unique strict solution if U_{10} and U_{20} belong to $D(B_0)$ and not necessarily to $D(B_0^2)$. □

EXERCISES

11.1. Derive an equation for $\mathscr{I}(x,t)$, similar to (11.2a).

11.2. Show that (11.12) holds with $z_1 = \text{Re}\, z$ instead of z, for any $z = z_1+iz_2$ with $z_1 \neq 0$.

11.3. Prove (a) of Lemma 11.1 under the assumption that X_0 is the Sobolev space $L^{s,2}(R^1)$. Hint: By a procedure similar to that of Example 2.15, define B_0 as follows:

$$B_0 f_1 = \mathscr{F}^{-1}[ivy\mathscr{F}f_1], \quad D(B_0) = L^{s+1,2}(R^1), \quad R(B_0) \subset L^{s,2}(R^1).$$

11.4. Prove the identity

$$(f_1,f_1)_0+(f_2,f_2)_0 = \tfrac{1}{2}(f_1+f_2,f_1+f_2)_0+\tfrac{1}{2}(f_1-f_2,f_1-f_2)_0$$

where f_1 and f_2 are any two elements of the Hilbert space X_0.

11.5. If $a_2 = a_1+\varepsilon$ with $\varepsilon > 0$, then $a = a_2$ and the operator J_1 defined by (11.22) becomes

$$J_1 = \varepsilon \begin{pmatrix} 0, & 0 \\ 0, & 1 \end{pmatrix} I .$$

Show that $\exp(tA)$ can be expanded in powers of ε. Hint: See the Example 5.3.

11.6. Maxwell's equations in a uniform material S can be written

(*a*) $\operatorname{div} \underline{E} = 0$, (*b*) $\operatorname{div} \underline{H} = 0$

(*c*) $\operatorname{curl} \underline{E} = -\mu \frac{\partial}{\partial t} \underline{H}$, (*d*) $\operatorname{curl} \underline{H} = \varepsilon \frac{\partial}{\partial t} \underline{E} + \sigma \underline{E}$

under the assumption that there is no charge in S and no current other than that given by Ohm's law. In (*a*)-(*d*), $\mu > 0$, $\varepsilon > 0$, and $\sigma \geq 0$ are given constants, $\underline{E} = (E_1, E_2, E_3)$ is the electric field, $\underline{H} = (H_1, H_2, H_3)$ is the magnetic field, and the components of $\underline{E}$ and of $\underline{H}$ depend on the position in S and on time.

Under the assumption that E_j and H_j, $j = 1,2,3$, depend only on x and t (plane symmetry), prove that each of the two couples (E_2, H_3) and (E_3, H_2) satisfies a system similar to (11.1a,b).

11.7. Prove that $D(A^2) = D(B^2) = D(B_0^2) \times D(B_0^2)$, where A, B, and B_0 are defined by (11.21), (11.8), and (11.7). Hint: Write the explicit expression of A^2 and of B^2.

11.8. Consider Maxwell's equations of Exercise 11.6 and assume that $\sigma = 0$ and that E_j, H_j, $j = 1,2,3$, depend only on x and t. Prove that each of the components E_2, E_3, H_2, H_3 satisfies the wave equations

$$\frac{\partial^2}{\partial t^2} \gamma(x,t) = v^2 \frac{\partial^2}{\partial x^2} \gamma(x,t), \qquad v = \frac{1}{\sqrt{(\varepsilon\mu)}} .$$

12
A PROBLEM FROM QUANTUM MECHANICS

12.1. INTRODUCTION

The state vector $\Psi(x,t)$ of a particle S (e.g. an electron) in a one-dimensional box (a potential well of infinite depth) extending from $x = -a$ to $x = a$ satisfies the Schrödinger equation (Shiff 1968)

$$\frac{\partial}{\partial t}\Psi(x,t) = \mathrm{i}\,\frac{h}{4\pi m}\,\frac{\partial^2}{\partial x^2}\Psi(x,t), \qquad -a < x < a, \qquad t > 0 \tag{12.1a}$$

and the initial and boundary conditions

$$\Psi(x,0) = \Psi_0(x), \qquad -a < x < a \tag{12.1b}$$

$$\Psi(-a,t) = \Psi(a,t) = 0, \qquad t > 0. \tag{12.1c}$$

In the (12.1) m is the mass of S, $h > 0$ is the Planck's constant, the initial state vector $\Psi_0(x)$ is a given function and

$$\{\int_{-a}^{a} |\Psi(y,t)|^2 \mathrm{d}y\}^{-1}\ |\Psi(x,t)|^2 \mathrm{d}x$$

is the probability of finding the particle S between x and $x+\mathrm{d}x$ at time t.

If X is the Hilbert space $L^2(-a,a)$ with the usual inner product $(\cdot,\cdot)$ and norm $\|\cdot\| = \|\cdot\|_2$, we define the operator A as follows:

$$Af = k\,\frac{\mathrm{d}^2 f}{\mathrm{d}x^2}$$

$$D(A) = \{f : f \in X;\ \frac{\mathrm{d}^2 f}{\mathrm{d}x^2} \in X;\ f(-a) = f(a) = 0\}, \qquad R(A) \subset X \tag{12.2}$$

where $\mathrm{d}^2 f/\mathrm{d}x^2$ is a generalized derivative and $k = h/4\pi m$. Then, the abstract version of system (12.1) reads as follows:

$$\frac{\mathrm{d}}{\mathrm{d}t}u(t) = \mathrm{i}Au(t), \quad t > 0; \qquad X\text{-}\lim_{t\to 0+} u(t) = u_0 \tag{12.3}$$

where, as usual, $u(t) = \Psi(x,t)$ is now an element of X depending on the parameter t, and $u_0 = \Psi_0$ is assumed to belong to $D(iA) = D(A)$ and to be such that $\|u_0\| = 1$ (i.e. u_0 is normalized). The condition

$$\|u_0\|^2 = \int_{-a}^{a} |\Psi_0(x)|^2 dx = 1$$

means that, at $t = 0$, the probability of finding S somewhere between $-a$ and a is equal to 1.

Since A is a self-adjoint operator (see Example 2.14), $iA \in \mathscr{G}'(1,0;X)$ (see Example 4.8) and the unique strict solution of problem (12.3) has the form

$$u(t) = \exp(tiA)u_0, \quad u_0 \in D(A). \tag{12.4}$$

Note that (12.4) satisfies the first of (12.3) at any $t \in (-\infty,+\infty)$ and the second as $t \to 0$, because iA generates a group. Furthermore,

$$\|u(t)\|^2 = \int_{-a}^{a} |\Psi(x,t)|^2 dx = \|u_0\|^2 = 1, \quad \forall\, t \in (-\infty,+\infty) \tag{12.5}$$

because

$$\|u(t)\| = \|\exp(tiA)u_0\| \le \|u_0\|$$

$$\|u_0\| = \|\exp(-itA)u(t)\| \le \|u(t)\|$$

and so $\|u(t)\| = \|u_0\|$ (see Example 4.9). Equation (12.5) shows that at *any* time t the probability of finding S somewhere between $-a$ and a is equal to 1 (the particle S is 'confined for ever' in the box).

12.2. SPECTRAL PROPERTIES OF iA

As in the Example 2.19 (with $-a$ instead of a, with a instead of b, and with $\delta = 2a$), the equation

$$(zI-A)f = 0, \quad f \in D(A) \tag{12.6}$$

leads to the system

$$\frac{d^2}{dx^2} f - \frac{z}{k} f = 0, \quad -a < x < a; \qquad f(-a) = f(a) = 0.$$

It follows that (12.6) has the non-trivial solution

$$f = f^{(n)} = c^{(n)} \sin\left(\pi n \frac{x+a}{2a}\right)$$

(with $c^{(n)} = 2i \exp(-\pi n i/2) c_1^{(n)}$, see (2.86)), if and only if

$$z = z_n = -\frac{k\pi^2}{4a^2} n^2 = -\frac{\pi h}{16ma^2} n^2, \qquad n = 1,2,\ldots . \tag{12.7}$$

Since $\{z : z \in \mathbb{C}; z \neq z_n \ \forall n = 1,2,\ldots\} = \rho(A)$ (see Example 2.19), we conclude that $R_\sigma(A)$ and $C_\sigma(A)$ are empty and that $\{z : z = z_n, n = 1,2,\ldots\} = P_\sigma(A)$. Each of the z_n is a simple eigenvalue of A (see Example 7.6), and the corresponding normalized eigenfunction is given by

$$f^{(n)}(x) = \frac{1}{\sqrt{a}} \sin\left(\pi n \frac{x+a}{2a}\right), \quad n = 1,2,\ldots . \tag{12.8}$$

Remark 12.1.

The eigenvalues z_n have a remarkable physical meaning because

$$E_n = -\frac{h}{2\pi} z_n = \frac{h^2}{32ma^2} n^2, \ n = 1,2,\ldots \tag{12.9}$$

are the energy levels of the particle S, namely, the possible values of the total energy of S. Furthermore, if the energy of S is E_n, then it is known from quantum mechanics that the state vector of S is

$$u^{(n)}(t) = f^{(n)} \exp\left(-i \frac{2\pi}{h} E_n t\right)$$

(which is the strict solution of (12.3) with $u_0 = f^{(n)}$). Correspondingly, S is said to be in a pure state (or in a stationary state) because only the nth eigenfunction and eigenvalue are involved (see also (12.11)). □

Multiplying (12.6) by $i = \sqrt{(-1)}$ and taking into account that $D(iA) = D(A)$, we conclude that $\{\eta : \eta = iz_n, n = 1,2,\ldots\}$

$= P_\sigma(\mathrm{i}A)$ and that $f^{(n)}$ is the normalized eigenfunction corresponding to the simple eigenvalue $\mathrm{i}z_n$ of $\mathrm{i}A$. It follows that $\exp(\mathrm{i}z_n t)$ is an eigenvalue of the group $\exp(t\mathrm{i}A)$ for any $t \neq 0$ and $\exp(t\mathrm{i}A)f^{(n)} = \exp(\mathrm{i}z_n t)f^{(n)}$ (see Example 7.10 and section 8.2). On the other hand, each $f \in X$ can be written

$$f = X\text{-}\lim_{r\to\infty} \sum_{n=1}^{r} (f,f^{(n)})f^{(n)} = \sum_{n=1}^{\infty} (f,f^{(n)})f^{(n)} \tag{12.10}$$

(see section 8.2), and so (12.4) gives

$$u(t) = \sum_{n=1}^{\infty} (u_0,f^{(n)})\exp(t\mathrm{i}A)f^{(n)} = \sum_{n=1}^{\infty} (u_0,f^{(n)})\exp(\mathrm{i}z_n t)f^{(n)}$$

i.e.

$$u(t) = \sum_{n=1}^{\infty} \exp\left(-\mathrm{i}\,\frac{2\pi}{h}\,E_n t\right)(u_0,f^{(n)})f^{(n)} \tag{12.11}$$

because of (12.9). Relation (12.11) shows that the state of the particle S at time t is the superposition of (in general) infinite states corresponding to different values of the energy.

12.3. BOUNDED PERTURBATIONS

Assume that $B \in \mathscr{B}(X)$ and ε is a (small) real parameter, and consider the *perturbed* problem

$$\frac{\mathrm{d}}{\mathrm{d}t}\,w(t) = \mathrm{i}(A+\varepsilon B)w(t), \quad t > 0; \quad X\text{-}\lim_{t\to 0} w(t) = u_0. \tag{12.12}$$

For instance, if an electric field $\underline{F} = (\varepsilon F(x),0,0)$ is present in the box $-a < x < a$ and S is an electron, then

$$Bw(t) = \left\{-\frac{2\pi e}{h}\int_{-a}^{x} F(x')\,\mathrm{d}x'\right\}w(t)$$

where $-e$ is the charge of S, and $B \in \mathscr{B}(X)$ if $F \in L^\infty(-a,a)$ (see Exercise 12.1).

Since $\mathrm{i}(A+\varepsilon B) \in \mathscr{G}'(1,|\varepsilon|\,\|B\|;X)$ because of Theorem 5.1 (see Remark 5.1), the solution of (12.12) has the form

$$w(t) = \exp\{t\mathrm{i}(A+\varepsilon B)\}u_0, \quad u_0 \in D(\mathrm{i}A+\mathrm{i}\varepsilon B) = D(A). \tag{12.13}$$

The perturbed group $\{\exp[ti(A+\varepsilon B)], t\in(-\infty,+\infty)\}$ can be built as in Exercise 5.3:

$$\exp\{ti(A+\varepsilon B)\}u_0 = \sum_{j=0}^{\infty} (i\varepsilon)^j \hat{Z}^{(j)}(t)u_0 \tag{12.14}$$

with

$$\hat{Z}^{(j+1)}(t)u_0 = \int_0^t \exp\{(t-s)iA\}B\hat{Z}^{(j)}(s)u_0 ds, \quad j = 0,1,2,\ldots$$

$$\hat{Z}^{(0)}(t)u_0 = \exp(itA)u_0,$$

(see (5.20) with iA instead of A, B instead of B_0, and with $z = i\varepsilon$). However, the elements $u_0\in D(A)$ and $Bf^{(n)}\in X$ can be written

$$u_0 = \sum_{n=1}^{\infty} (u_0,f^{(n)})f^{(n)}$$

$$Bf^{(n)} = \sum_{r=1}^{\infty} b_{n,r}\, f^{(r)}, \quad b_{n,r} = (Bf^{(n)},f^{(r)})$$

and so we obtain

$$\hat{Z}^{(1)}(t)u_0 = \sum_{n=1}^{\infty} (u_0,f^{(n)}) \int_0^t \exp\{(t-s)iA\}\exp(iz_n s)Bf^{(n)}ds$$

because

$$\hat{Z}^{(0)}(s)u_0 = \sum_{n=1}^{\infty} (u_0,f^{(n)})\exp(iz_n s)f^{(n)}.$$

Hence,

$$\hat{Z}^{(1)}(t)u_0 = \sum_{n,r=1}^{\infty} (u_0,f^{(n)})b_{n,r}\left[\int_0^t \exp\{iz_r(t-s)\}\exp(iz_n s)ds\right]f^{(r)}$$

$$= \sum_{n=1}^{\infty} \{(u_0,f^{(n)})b_{n,n}t\,\exp(iz_n t)\}f^{(n)} +$$

$$\sum_{\substack{n,r=1\\ n\neq r}}^{\infty} \left\{(u_0,f^{(n)})b_{n,r}\,\frac{\exp(iz_n t)-\exp(iz_r t)}{i(z_n-z_r)}\right\} f^{(r)}. \tag{12.15}$$

The explicit expressions of $\hat{Z}^{(j)}(t)u_0, j = 2,3,\ldots$, can be derived by a similar procedure.

In particular, if the initial state of S is the eigenfunction $f^{(\nu)}$ corresponding to the eigenvalue $iz_\nu = -i(2\pi/h)E_\nu$ of the operator iA, then $u_0 = f^{(\nu)}$ and (12.15) gives

$$\hat{Z}^{(1)}(t)u_0 = \{b_{\nu,\nu}t\exp(iz_\nu t)\}f^{(\nu)} + \sum_{\substack{r=1\\ r\neq\nu}}^{\infty} \left\{b_{\nu,r} \times \frac{\exp(iz_\nu t)-\exp(iz_r t)}{i(z_\nu - z_r)}\right\} f^{(r)}$$

because $(u_0, f^{(n)}) = (f^{(\nu)}, f^{(n)})$ and so $(u_0, f^{(n)}) = 0$ if $n \neq \nu$ and $(u_0, f^{(\nu)}) = \|f^{(\nu)}\|^2 = 1$. Thus, (12.13) becomes

$$w(t) = \sum_{r=1}^{\infty} \lambda_{\nu,r} f^{(r)} + w_1(t;\varepsilon) \tag{12.16}$$

with

$$\lambda_{\nu,\nu} = 1 + i\varepsilon b_{\nu,\nu} t\exp(iz_\nu t)$$

$$\lambda_{\nu,r} = \varepsilon b_{\nu,r} \frac{\exp(iz_\nu t)-\exp(iz_r t)}{z_\nu - z_r}, \qquad \nu \neq r$$

and with

$$\|w_1(t;\varepsilon)\| = \left\|\sum_{j=2}^{\infty} (i\varepsilon)^j \hat{Z}^{(j)}(t)u_0\right\| \leq \sum_{j=2}^{\infty} |\varepsilon|^j \frac{(\|B\||t|)^j}{j!} \|u_0\|$$

$$|\varepsilon|^2 \exp(|\varepsilon t|\|B\|)\|u_0\|$$

because $\|\hat{Z}^{(j)}(t)u_0\| \leq \|u_0\| (\|B\||t|)^j/j!$. Relation (12.16) shows that, with an error of the order of $|\varepsilon|^2$,

$$w(t) \cong \sum_{r=1}^{\infty} \lambda_{\nu,r} f^{(r)}$$

i.e. the state of S at any $t > 0$ is a superposition of all the states $f^{(r)}, r = 1,2,\ldots$. Moreover, it is known from quantum mechanics that the probability of finding S in the state of energy E_r with $r \neq \nu$ at time t is given by (with an error of the order of $|\varepsilon|^3$)

$$|\lambda_{\nu,r}|^2 = 2\varepsilon^2 |b_{\nu,r}|^2 \frac{1-\cos\{(z_\nu - z_r)t\}}{(z_\nu - z_r)^2}$$

$$= t^2\varepsilon^2 |b_{\nu,r}|^2 \left\{\frac{\sin(\Delta_{\nu,r})}{\Delta_{\nu,r}}\right\}^2$$

with $\Delta_{\nu,r} = (z_\nu - z_r)t/2$. Note that the function $\{(\sin y)/y\}^2$ has an absolute maximum equal to 1 as $y \to 0$, whereas the other maxima are much smaller than 1. Then, the probability $|\lambda_{\nu,r}|^2$ has appreciable values only if $|\Delta_{\nu,r}|$ is small, i.e. if $r = \nu \pm 1$.

EXERCISES

12.1. Prove that the operator

$$Bf = [-\frac{2\pi e}{h} \int_{-a}^{x} F(x')\,dx']f$$

belongs to $\mathscr{B}(X)$ if $F \in L^\infty(-a,a)$. Hint:

$$|Bf| \le \frac{4\pi a e}{h} \|F\|_\infty |f(x)| \text{ for a.e. } x \in (-a,a).$$

12.2. Find the explicit expression of $\hat{Z}^{(2)}(t)u_0$.

12.3. Prove that $i(A+\varepsilon B) \in \mathscr{G}'(1,0;X)$, where B is defined as in Exercise 12.1 and ε is a real parameter. Hint: Show that $(A+\varepsilon B)$ is self-adjoint, (see Examples 2.14 and 4.8).

12.4. Study the initial-value problem

$$\frac{dw(t)}{dt} = i\{A+\varepsilon B \exp(i\omega t)\}w(t), \quad t > 0; \qquad X\text{-}\lim_{t\to 0} w(t) = u_0 = f^{(\nu)} \tag{12.17}$$

with B defined as in Exercise 12.1, ε real, and $\omega > 0$. Hint: Using Theorem 5.2, transform (12.17) into the integral equation

$$w(t) = \exp(tiA)u_0 + i\varepsilon \int_0^t \exp\{(t-s)iA\}B \exp(i\omega s)w(s)\,ds. \tag{12.18}$$

Then, solve (12.18) by the usual method of successive approximations.

13
A PROBLEM FROM STOCHASTIC POPULATION THEORY

13.1. INTRODUCTION

Let S be a population of bacteria in a culture and indicate by $P(n,t)$ the probability that, at time t, S is composed of n individuals. Under suitable assumptions (see for example Ludwig 1974), $P(n,t)$ satisfies the system

$$\frac{\partial}{\partial t} P(n,t) = -(p+q)nP(n,t)+p[n-1]P(n-1,t) +$$

$$q[n+1]P(n+1,t), \quad t > 0, \quad n = 0,1,2,..., \tag{13.1a}$$

$$P(n,0) = P_0(n), \quad n = 0,1,2,... \tag{13.1b}$$

where $P(-1,t) \equiv 0$ and the $P_0(n)$ are given so that

$$0 \le P_0(n) \le 1, \quad \sum_{n=0}^{\infty} P_0(n) = 1. \tag{13.2}$$

In (13.1a) the non-negative constants p and q are respectively the probabilities per unit time interval of an individual giving birth or dying.

Equation (13.1a) is, in some sense, a balance equation for the probability $P(n,t)$. To understand this, note that $(p\mathrm{d}t)mP(m,t)$ and $(q\mathrm{d}t)mP(m,t)$ respectively are the probabilities of birth events and of death events during the time interval $\mathrm{d}t$ in a population of m individuals. Also, if a birth (death) occurs, the population changes from m to $m+1$ ($m-1$) individuals. Thus, for instance, the probability $P(n,t)$ of having a population of n individuals at time t is increased by birth events in a population of $n-1$ individuals (see the term $p[n-1]P(n-1,t)$ on the right-hand side of (13.1a)) and is decreased by birth events in a population of n individuals (see the term $-pnP(n,t)$).

Finally, (13.1b) and (13.2) show that the initial probabilities $P_0(n)$ necessarily belong to $[0,1]$ and that their sum (i.e. the initial probability of a population of *any* number of individuals) is equal to unity.

13.2. THE ABSTRACT PROBLEM

The conditions (13.2) suggest introducing the real B-space $X = l^1$ of all summable sequences of real numbers with norm

$$\|f\| = \|f\|_1 = \sum_{n=0}^{\infty} |f_n|, \qquad f = \begin{pmatrix} f_0 \\ f_1 \\ \cdots \end{pmatrix}$$

(see Example 1.2), and the closed positive cone of X:

$$X^+ = \{f: f \in X; f_n \geq 0 \ \forall n = 0,1,2,\ldots\}.$$

If we define the operators

$$[Hf]_n = (p+q)nf_n, \quad n = 0,1,\ldots, \quad D(H) = D = \{f: f \in X; \sum_{n=0}^{\infty} n|f_n| < \infty\} \tag{13.3}$$

$$[Kf]_n = p(n-1)f_{n-1} + q(n+1)f_{n+1}, \quad n = 1,2,\ldots$$

$$[Kf]_0 = qf_1, \quad D(K) = D \tag{13.4}$$

$$[Af]_n = -[Hf]_n + [Kf]_n, \quad D(A) = \{f: f \in X; \sum_{n=0}^{\infty} |[Af]_n| < \infty\} \tag{13.5}$$

where the symbols $[g]_n$ or g_n indicate the $(n+1)$th component of the element $g \in X$, then the abstract version of (13.1) reads as follows

$$\frac{\mathrm{d}}{\mathrm{d}t} u(t) = Au(t), \quad t > 0; \qquad X\text{-}\lim_{t\to 0+} u(t) = u_0 \tag{13.6}$$

with

$$u(t) = \begin{pmatrix} P(0,t) \\ P(1,t) \\ \cdots\cdots \end{pmatrix}, \qquad u_0 = \begin{pmatrix} P_0(0) \\ P_0(1) \\ \cdots\cdots \end{pmatrix}.$$

For a reason that will become clear later on, we also

consider the 'approximating' initial-value problem

$$\frac{d}{dt} w(t;r) = A_r w(t;r), \quad t > 0; \quad X\text{-}\lim_{t\to 0+} w(t;r) = u_0 \tag{13.7}$$

where $r \in [0,1)$ is a real parameter, and

$$A_r = -H + rK, \quad D(A_r) = D. \tag{13.8}$$

Note that $D \subset D(A)$ because if $f \in D$ then

$$\sum_{n=0}^{\infty} |[Af]_n| \le \sum_{n=0}^{\infty} \{(p+q)n|f_n| + p(n-1)|f_{n-1}| + q(n+1)|f_{n+1}|\}$$

$$= 2(p+q) \sum_{m=0}^{\infty} m|f_m| < \infty$$

$(f_{-1} = 0)$, and so $f \in D(A)$. Furthermore, if $f \in D$, we have

$$\| Af - A_r f\| = (1-r)\| Kf \| \to 0 \text{ as } r \to 1- \tag{13.9}$$

that justifies the locution 'approximating' for problem (13.7).

13.3. PRELIMINARY LEMMAS

The following lemmas will be used to study the initial value problems (13.6) and (13.7).[†]

Lemma 13.1. (*a*) *H is densely defined;* (*b*) *for every $z > 0$ $(zI+H)^{-1} \in \mathscr{B}(X)$ with $\|(zI+H)^{-1} g\| \le z^{-1}\|g\|$;* (*c*) *$H$ maps $D^+ = D \cap X^+$ into X^+, $(zI+H)^{-1}$ maps X^+ into itself for any $z > 0$;* (*d*) *$H(zI+H)^{-1} \in \mathscr{B}(X)$, $\|H(zI+H)^{-1} g\| \le \|g\|$, $\forall g \in X$, $z > 0$.*

Proof. (*a*) The set D_0, composed of all elements of X with a *finite* number of non-zero components, is dense in X (see Exercise 13.1). It follows that D is dense in X because $D \supset D_0$.

(*b*) The equation $(zI+H)f = g$ with $g \in X$ and $z > 0$ leads to the system

[†] Most of the proofs of this section and of section 13.4 are taken from Kato (1954).

$$zf_n+(p+q)nf_n = g_n, \quad n = 0,1,2,\ldots .$$

Thus,

$$f_n = \{z+(p+q)n\}^{-1}g_n, \quad n = 0,1,2,\ldots$$

and $f \in D = D(H)$ because

$$\sum_{n=0}^{\infty} n|f_n| = \sum_{n=0}^{\infty} \frac{n}{z+(p+q)n}|g_n| \le \frac{1}{p+q}\sum_{n=0}^{\infty}|g_n|$$

$$= (p+q)^{-1}\|g\| < \infty .$$

Moreover,

$$\|f\| = \|(zI+H)^{-1}g\| = \sum_{n=0}^{\infty}\frac{1}{z+(p+q)n}|g_n| \le z^{-1}\|g\| .$$

(*c*) If $g \in X^+$, then $g_n \ge 0 \; \forall n = 0,1,\ldots$ and so

$$[(zI+H)^{-1}g]_n = \{z+(p+q)n\}^{-1}g_n \ge 0,$$

i.e. $(zI+H)^{-1}g \in X^+$. Similarly, if $g \in D \cap X^+$, then $[Hg]_n = (p+q)ng_n \ge 0$ for any $n = 0,1,\ldots$, and $Hg \in X^+$.

(*d*) Since the range of $(zI+H)^{-1}$ is $D(H) = D$ for any $z > 0$, the operator $H(zI+H)^{-1}$ has domain X and $H(zI+H)^{-1}g = g-z(zI+H)^{-1}g$ for any $g \in X$ and $z > 0$ (see Exercise 2.14). Now, if $\phi \in X^+$,

$$[H(zI+H)^{-1}\phi]_n = \phi_n - \frac{z}{z+(p+q)n}\phi_n = \frac{(p+q)n}{z+(p+q)n}\phi_n \ge 0$$

$$\|H(zI+H)^{-1}\phi\| = \sum_{n=0}^{\infty}\frac{(p+q)n}{z+(p+q)n}\phi_n \le \sum_{n=0}^{\infty}\phi_n = \|\phi\|.$$

Hence, $H(zI+H)^{-1}$ maps X^+ into itself and $\|H(zI+H)^{-1}\phi\| \le \|\phi\|$, $\forall z > 0$, $\phi \in X^+$. Finally, if $f \in X$, then $f = f^+ - f^-$ with

$$f_n^+ = f_n,\; f_n^- = 0 \text{ if } f_n \ge 0; \quad f_n^+ = 0,\; f_n^- = -f_n \text{ if } f_n < 0$$

i.e. with $f^+ \in X^+$ and $f^- \in X^+$. It follows that

$$\|H(zI+H)^{-1}f\| \le \|H(zI+H)^{-1}f^{+}\| + \|H(zI+H)^{-1}f^{-}\|$$

$$\le \|f^{+}\| + \|f^{-}\| = \|f\|$$

and (d) is proved. □

Lemma 13.2. (*a*) *K maps $D^{+} = D \cap X^{+}$ into X^{+}*; (*b*) $\|Kf\| \le \|Hf\|$, $\forall f \in D$; (*c*) $\|Kf\| = \|Hf\|$, $\forall f \in D^{+}$.

Proof. (*a*) follows from (13.4) with $f_n \ge 0\ \forall n = 0,1,\ldots$.
(*b*) Relations (13.3) and (13.4) (with $f_{-1} = 0$) give

$$\|Kf\| \le \sum_{n=0}^{\infty} \{p(n-1)|f_{n-1}| + q(n+1)|f_{n+1}|\}$$

$$= p\sum_{m=0}^{\infty} m|f_m| + q\sum_{m=1}^{\infty} m|f_m| = \|Hf\|, \quad \forall f \in D .$$

(*c*) If $f \in D^{+}$, then $f_n \ge 0\ \forall n = 0, 1, \ldots$, and so

$$\|Kf\| = p\sum_{m=0}^{\infty} mf_m + q\sum_{m=1}^{\infty} mf_m = \|Hf\| . \quad \square$$

Lemma 13.3. For any $z > 0$, the linear operator $F(z) = K(zI+H)^{-1}$ has the following properties: (*a*) $F(z) \in \mathscr{B}(X)$, $\|F(z)f\| \le \|f\|$, $\forall f \in X$; (*b*) *$F(z)$ maps X^{+} into itself.*

Proof. (*a*) The domain of $F(z)$ is the whole space X because $(zI+H)^{-1}$ has domain X and range $D(H) = D(K)$. Moreover, (*b*) of Lemma 13.2 and (*d*) of Lemma 13.1 give

$$\|F(z)f\| = \|K\{(zI+H)^{-1}f\}\| \le \|H\{(zI+H)^{-1}f\}\| \le \|f\| .$$

(*b*) follows from (*c*) of Lemma 13.1 and from (*a*) of Lemma 13.2. □

13.4. STRICT SOLUTION OF THE APPROXIMATING PROBLEM (13.7)

The approximating problem (13.7) with $0 \le r < 1$ has a unique strict solution $w(t;r) \in D^{+}\ \forall t \ge 0$ if $u_0 \in D^{+}$, because $A_r \in \mathscr{G}(1,0;X)$ and the semigroup $\{Z_r(t) = \exp(tA_r),\ t \ge 0\}$ maps D^{+} into itself. To see this, note first that the inverse operator $\{I - rF(z)\}^{-1}$ exists for any $z > 0$ and $r \in [0,1)$ because

$\| rF(z)f\| = r\| F(z)f\| \le r\| f\| < \| f\|$. Furthermore,

$$\{I-rF(z)\}^{-1} = \sum_{j=0}^{\infty} r^j \{F(z)\}^j$$

$$\| \{I-rF(z)\}^{-1} f\| \le (1-r)^{-1} \| f\|$$

(see (2.62) of Example 2.16 with $z = 1$ and with $rF(z)$ instead of A). As a consequence, we have for any $z > 0$ and $0 \le r < 1$

$$R(z,A_r) = (zI+H-rK)^{-1} = [\{I-rF(z)\}(zI+H)]^{-1}$$

and so

$$R(z,A_r) = (zI+H)^{-1}\{I-rF(z)\}^{-1} = (zI+H)^{-1} \sum_{j=0}^{\infty} r^j \{F(z)\}^j \quad (13.10)$$

with

$$\| R(z,A_r)f\| \le z^{-1}(1-r)^{-1} \| f\| , \quad \forall f \in X . \quad (13.11)$$

Relations (13.10) and (13.11) show that $R(z,A_r) \in \mathscr{B}(X)$, whereas the second part of (13.10) proves that $R(z,A_r)f \in X^+ \; \forall f \in X^+$ because of (c) of Lemma 13.1 and (b) of Lemma 13.3. Inequality (13.11) can be improved as follows. If $\phi \in X^+$, then $\Phi = R(z,A_r)\phi \in D(A_r) \cap X^+ = D^+$ and we have $r\| K\Phi\| = r\| H\Phi\| \le \| H\Phi\| \le z\| \Phi\| + \| H\Phi\| = \| (zI+H)\Phi\|$ because Φ and $H\Phi$ belong to X^+. Hence,

$$\| (zI-A_r)\Phi\| = \| (zI+H)\Phi - rK\Phi\| \ge \left| \| (zI+H)\Phi\| - r\| K\Phi\| \right|$$

$$= \| (zI+H)\Phi\| - r\| K\Phi\| = z\| \Phi\| + \| H\Phi\| - r\| K\Phi\| \ge z\| \Phi\|$$

and so

$$\| R(z,A_r)\phi\| \le z^{-1} \| \phi\| , \quad \forall z > 0, \quad \phi \in X^+ .$$

Since $f = f^+ - f^-$ with f^+ and f^- in X^+ (see (d) of Lemma 13.1), the above inequality holds for any $f \in X$:

$$\|R(z,A_r)f\| \le z^{-1}\|f\|, \quad \forall z > 0, \quad f \in X. \tag{13.12}$$

Finally, A_r is densely defined because $D(A_r) = D \supset D_0$ and $A_r \in \mathscr{C}(X)$ because $(zI-A_r)^{-1} \in \mathscr{B}(X) \subset \mathscr{C}(X)$ for $z > 0$. We conclude that $A_r \in \mathscr{G}(1,0;X)$ $\forall r \in [0,1)$ and that the semigroup $\{Z_r(t), t \ge 0\}$ maps D^+ into itself because $Z_r(t)[D] \subset D$ and

$$Z_r(t)g = X\text{-}\lim_{j\to 0} \left\{\frac{j}{t} R\left(\frac{j}{t}, A_r\right)\right\}^j g \in X^+, \quad \forall g \in X^+ .$$

Using the above results, we can state the following theorem.

Theorem 13.1. If $0 \le r < 1$, then the approximating initial-value problem (13.7) has the unique strict solution

$$w(t;r) = Z_r(t)u_0 \in D, \quad t \ge 0, \quad \textit{if } u_0 \in D \tag{13.13a}$$

$$w(t;r) = Z_r(t)u_0 \in D^+, \quad t \ge 0, \quad \textit{if } u_0 \in D \cap X^+ = D^+. \; \square \tag{13.13b}$$

13.5. A PROPERTY OF THE STRICT SOLUTION OF THE APPROXIMATING PROBLEM

Consider the following initial-value problem:

$$\frac{d}{dt}v(t;r) = A_r v(t;r) + \{2rpB-(p+q)I\}v(t;r), \quad t > 0 \tag{13.14a}$$

$$X - \lim_{t\to 0+} v(t;r) = v_0 \tag{13.14b}$$

with

$$[v_0]_n = (n+1)[u_0]_{n+1}, \quad n = 0,1,2,\ldots \tag{13.15}$$

$$[Bf]_n = f_{n-1}, \quad n = 1,2,\ldots, \quad [Bf]_0 = 0 \tag{13.16}$$

Eqn (13.14a) can be *formally* derived from (13.7) multiplying the $(n+2)$th component of the first of (13.7) by $(n+1)$ and letting $[v(t;r)]_n = (n+1)[w(t;r)]_{n+1}$. However, (13.14) will be considered as an *independent* initial-value problem for reasons that will become clear later on.

Lemma 13.4. (a) $B \in \mathscr{B}(X)$ *with* $\|Bf\| \le \|f\|$ $\forall f \in X$, $B[X^+] \subset X^+$ *and* $\|Bg\| = \|g\|$ $\forall g \in X^+$; (b) $\Lambda_r = A_r + 2rpB - (p+q)I \in \mathscr{G}(1, 2rp-p-q; X)$ *and the semigroup* $\{\exp(t\Lambda_r), t \ge 0\}$ *maps* X^+ *into itself.*

Proof. (a) See Exercise 13.2.

(b) Since $A_r \in \mathscr{G}(1,0;X)$ and $-(p+q)I$ commutes with A_r, $A_r - (p+q)I \in \mathscr{G}(1,-p-q;X)$ (see Exercise 5.2). It follows that $\Lambda_r \in \mathscr{G}(1,-p-q+2rp;X)$ because $\|2rpB\| \le 2rp$. Finally, $\exp(t\Lambda_r)g \in X^+$ $\forall g \in X^+, t \to 0$, because

$$\exp\{t(A_r - (p+q)I)\} = \exp\{-(p+q)t\}Z_r(t)$$

and $2rpB$ maps X^+ into itself (see the proof of (d) of Lemma 9.3). □

The strict solution of (13.14) is such that, for any $t \ge 0$,

$$v(t;r) = \exp(t\Lambda_r)v_0 \in D \quad \text{if } v_0 \in D(\Lambda_r) = D(A_r) = D \tag{13.17a}$$

$$v(t;r) = \exp(t\Lambda_r)v_0 \in D \cap X^+ = D^+ \quad \text{if } v_0 \in D^+ \tag{13.17b}$$

$$\|v(t;r)\| \le \exp\{(2rp-p-q)t\}\|v_0\| \le \exp\{(p-q)t\}\|v_0\| \tag{13.17c}$$

because of (b) of Lemma 13.4. Note that the condition $v_0 \in D$ is equivalent to the assumption $u_0 \in D(H^2)$ because $D(H^2) = \{f : f \in X; \Sigma_{m=0}^{\infty} m^2 |f_m| < \infty\}$ (see 13.15)).

$$[Yf]_n = f_{n+1}, \quad [Jf]_n = \frac{1}{n+1} f_n, \quad D(Y) = D(J) = X \tag{13.18}$$

then it is easy to prove that Y and J belong to $\mathscr{B}(X)$, with $\|Yf\| \le \|f\|$ and $\|Jf\| \le \|f\|$, and map D into itself and X^+ into itself. Moreover, we have for any $f \in D$:

$$YHf = HYf + (p+q)Yf, \quad YKf = KYf + pBYf + qYYf \tag{13.19}$$

$$\left.\begin{aligned} JHf &= HJf, \quad JKf = KJf-pBJf+2pJBJf+qYJf \\ JBf &= BJf-JBJf \end{aligned}\right\} \qquad (13.20)$$

(see Exercise 13.3). If we put

$$\delta(t;r) = Jv(t;r)-Yw(t;r), \quad t \geq 0 \qquad (13.21)$$

i.e.

$$[\delta(t;r)]_n = \frac{1}{n+1}[v(t;r)]_n-[w(t;r)]_{n+1}, \quad n = 0,1,2,\ldots$$

(see the remark on equation(13.14a), after (13.16) of this section,then we obtain from (13.7) and (13.14)

$$\frac{\mathrm{d}}{\mathrm{d}t}\delta(t;r) = \{A_r+rpB+rqY-(p+q)I\}\delta(t;r), \quad t > 0 \qquad (13.22a)$$

$$X\text{-}\lim_{t\to 0+}\delta(t;r) = \theta_X \qquad (13.22b)$$

because of (13.19), (13.20), and (13.15). Note that $Y\mathrm{d}w/\mathrm{d}t = \mathrm{d}(Yw)/\mathrm{d}t$ and $J\mathrm{d}v/\mathrm{d}t = \mathrm{d}(Jv)/\mathrm{d}t$ because $Y, J\in\mathscr{B}(X)$. However,

$$\Delta_r = A_r+rpB+rqY-(p+q)I\in\mathscr{G}(1,-(1-r)(p+q);X)$$

(see the proof of (*b*) of Lemma 13.4), and so

$$\delta(t;r) = \exp(t\Delta_r)\theta_X \equiv \theta_X, \quad \forall t \geq 0.$$

Hence, we have for any $t \geq 0$ and $0 \leq r < 1$:

$$Jv(t;r) \equiv Yw(t;r)\in D \quad \text{if } u_0\in D(H^2) \qquad (13.23a)$$

$$Jv(t;r) \equiv Yw(t;r)\in D^+ \quad \text{if } u_0\in D(H^2)\cap X^+, \qquad (13.23b)$$

because J and Y map D into itself and X^+ into itself and because of (13.13a,b) and (13.17a,b).

Going back to the approximating initial-value problem (13.7), assume that $u_0\in D(H^2)\cap X^+$. Then, $v_0\in D^+$ and (13.23b) gives for any $m = 0,1,2,\ldots$

$$0 \leq [v(t;r)]_m = (m+1)[w(t;r)]_{m+1} = \frac{1}{p+q}[Hw(t;r)]_{m+1}$$

namely

$$Hw(t;r) = (p+q)Bv(t;r) \tag{13.24}$$

and so $Hw(t;r)$ is continuous in $t \geq 0$ with

$$\|Kw(t;r)\| = \|Hw(t;r)\| \leq (p+q)\|v(t;r)\| \leq (p+q)\|v_0\|\exp\{(p-q)t\}$$

$$\leq (p+q)\|v_0\|\exp\{(p+q)t\}, \quad \forall t \geq 0, \quad 0 \leq r < 1 \tag{13.25}$$

because of (c) of Lemma 13.2, (a) of Lemma 13.4, and (13.17c).

Finally, if

$$\mathcal{N}f = \sum_{n=0}^{\infty} f_n, \quad D(\mathcal{N}) = X, \; R(\mathcal{N}) \subset R^1 \tag{13.26}$$

then $\mathcal{N} \in \mathcal{B}(X, R^1)$ with $\|\mathcal{N}f\| \leq \|f\|$, and $\mathcal{N}g \geq 0$ with $\mathcal{N}g = \|g\|$ if $g \in X^+$. Moreover, (13.7) with $u_0 \in D^+$ gives

$$\frac{d}{dt}\mathcal{N}w(t;r) = -\mathcal{N}Hw(t;r) + r\mathcal{N}Kw(t;r)$$

i.e.

$$\frac{d}{dt}\|w(t;r)\| = -\|Hw(t;r)\| + r\|Kw(t;r)\| = -(1-r)\|Hw(t;r)\|$$

because Hw and Kw belong to X^+. However, $\|w(t;r)\| \leq \|u_0\|$ because of (13.13a) and so, if $u_0 \in D(H^2) \cap X^+$, we obtain for any $t \geq 0$ and $0 \leq r < 1$

$$0 \leq \|u_0\| - \|w(t;r)\| = (1-r)\int_0^t \|Hw(s;r)\|\,ds$$

$$\leq (1-r)\|v_0\|[\exp\{(p+q)t\}-1]$$

because of (13.25). Hence, if $u_0 \in D(H^2) \cap X^+$, we have

$$\lim_{r \to 1-} \|w(t;r)\| = \|u_0\| \tag{13.27}$$

uniformly with respect to t in any finite interval $[0,t_0]$.

13.6. STRICT SOLUTION OF PROBLEM (13.6)

The discussion that follows explains why the problem (13.7) approximates (13.6). If $g \in X^+$ and $z > 0$ are given, then $\gamma(r) = R(z,A_r)g$ belongs to X^+, it is a 'non-decreasing' function of r because (13.10) gives for any $0 \le r \le r_1 < 1$

$$\gamma(r_1) - \gamma(r) = (zI+H)^{-1} \sum_{j=0}^{\infty} (r_1^j - r^j)\{F(z)\}^j g \in X^+$$

and $\|\gamma(r)\| \le z^{-1}\|g\|$ because of (13.12). Hence, $\gamma(r)$ is a non-decreasing and bounded function of $r \in [0,1)$ and so the X-limit of $\gamma(r)$ as $r \to 1-$ exists (see Exercise 13.4). Now if we put $\gamma = X\text{-}\lim_{r\to 1-} \gamma(r)$, the element γ belongs to X^+ with $\|\gamma\| \le z^{-1}\|g\|$, it is uniquely determined by g and z, and obviously depends linearly on g. In other words, $\gamma = R(z)g$ and $R(z)$ is a linear operator with $\|R(z)g\| \le z^{-1}\|g\|$ $\forall g \in X^+$, $z > 0$. Since $f = f^+ - f^-$ with f^+ and f^- in X^+, we finally have for any $f \in X$ and $z > 0$

$$R(z)f = X\text{-}\lim_{r\to 1-} R(z,A_r)f \ , \quad \|R(z)f\| \le z^{-1}\|f\| . \tag{13.28}$$

Furthermore,

$$\lim_{\alpha\to 0+} \|(I-\alpha A_r)^{-1}f - f\| = 0 \ \textit{uniformly in } r \in [0,1) \tag{13.29}$$

because we have for any $g \in D(A_r) = D$

$$\|(I-\alpha A_r)^{-1}g - g\| = \alpha\|(I-\alpha A_r)^{-1}A_r g\| \le \alpha\|A_r g\|$$

$$\le \alpha\{\|Hg\| + r\|Kg\|\} \le 2\alpha\|Hg\| \to 0 \text{ as } \alpha \to 0+, \quad \forall r \in [0,1]$$

since $(I-\alpha A_r)^{-1} = \alpha^{-1}R(\alpha^{-1},A_r)$, see (13.12) with $z = \alpha^{-1}$. The above result can be extended to the whole X by the usual procedure because D is dense in X, and so (13.29) is proved. Relations (13.28) and (13.29) imply that an operator G exists, such that

$$R(z,G) = R(z), \quad z > 0 \tag{13.30}$$

$$G \in \mathcal{G}(1,0;X) \tag{13.31}$$

(see Theorem 2.17 of Kato (1966, p.503) that generalizes Theorem 6.1 of section 6). Thus, if $Z(t) = \exp(tG)$, Theorem 6.1 gives

$$X\text{-}\lim_{r\to 1-} Z_r(t)f = Z(t)f, \quad \forall t \geq 0, \quad f \in X \tag{13.32}$$

(uniformly in t in each finite interval $[0,t_0]$). The operator G and the semigroup $\{Z(t), t \geq 0\}$ also have the following properties.

Lemma 13.5. (a) $D \subset D(G) \subset D(A)$; (b) *G is a restriction of A: $Gf = Af\ \forall f \in D(G)$;* (c) $Z(t)g \in X^+$, $\|Z(t)g\| = \|g\|$, $\forall g \in X^+, t \geq 0$.

Proof. (a), (b) See Kato (1954).

(c) Relation (13.32) with $f = g \in X^+$ shows that $Z(t)$ maps X^+ into itself because $Z_r(t)$ has this property for any $r \in [0,1)$. Moreover, if $u_0 \in D(H^2) \cap X^+$, (13.27) gives

$$\|u_0\| = \lim_{r\to 1-} \|w(t;r)\| = \lim_{r\to 1-} \|Z_r(t)u_0\|$$

and $\|Z(t)u_0\| = \|u_0\|$ because of (13.32). Since $D(H^2)$ is dense in X because $D(H^2) \supset D_0$, then $D(H^2) \cap X^+$ is dense in X^+ and (d) can be obtained by the usual extension procedure. □

Now consider the initial-value problem

$$\frac{d}{dt} W(t) = GW(t), \quad t > 0; \qquad X\text{-}\lim_{t\to 0+} W(t) = u_0 \tag{13.33}$$

and assume that $u_0 \in D^+$ $(= D \cap X^+ \subset D(G) \cap X^+)$. Then, the strict solution of (13.33)

$$W(t) = Z(t)u_0, \quad t \geq 0 \tag{13.34}$$

belongs to $D(G) \cap X^+$ and is such that

$$X\text{-}\lim_{r\to 1-} w(t;r) = W(t), \text{ uniformly in } t\in[0,t_0] \tag{13.35}$$

and

$$\|W(t)\| = \|u_0\| \quad \forall\, t \geq 0 \tag{13.36}$$

because of (13.13b), (13.32), and (*c*) of Lemma 13.5. However, G is a restriction of A, i.e. $GW(t) = AW(t)\ \forall\, t \geq 0$ and so $W(t)$ also satisfies (13.6). Hence, we have the following theorem.

Theorem 13.2. If $u_0\in D^+$, the initial-value problem (13.6) has the strict solution $u(t) = W(t)$ defined by (13.34). Such a solution is the X limit as $r\to 1-$ of the strict solution of the approximating problem (13.7) and $\|u(t)\| = \|u_0\|\ \forall\, t \geq 0$ (so that $\|u(t)\| \equiv 1$ if $\|u_0\| = 1$, see (13.2)). □

13.7. THE EQUATION FOR THE FIRST MOMENT $\langle n\rangle(t)$ OF THE BACTERIA POPULATION

If $u_0\in D^+$, the initial-value problem (13.6) has the strict solution $u(t) = W(t)\in D(G)\cap X^+$ because of Theorem 13.2. The first moment of the bacteria population S is defined by

$$\langle n\rangle(t) = \sum_{n=0}^{\infty} n[W(t)]_n = \sum_{m=0}^{\infty} (m+1)[W(t)]_{m+1} \tag{13.37}$$

(provided that $\langle n\rangle(t)$ is *finite*) and it is the expected number (average number) of individuals of S. An equation for $\langle n\rangle(t)$ can be derived as follows. If

$$\Lambda = G+2pB-(p+q)I, \quad D(\Lambda) = D(G) \tag{13.38}$$

then we have the following lemma.

Lemma 13.6. (a) $\Lambda\in \mathcal{G}(1,p-q;X)$; (b) the semigroup $\{\exp(t\Lambda), t \geq 0\}$ maps X^+ into itself and $\|\exp(t\Lambda)g\| = \exp\{(p-q)t\}\|g\|\ \forall\, g\in X^+, t \geq 0$; (c) $X\text{-}\lim_{r\to 1-} \exp(t\Lambda_r)f = \exp(t\Lambda)f$, $\forall\, f\in X, t \geq 0$, uniformly in t in each finite interval

$[0,t_0]$.

Proof. (*a*) $G-(p+q)I\in \mathscr{G}(1,-p-q;X)$ because $G\in \mathscr{G}(1,0;X)$ and $-(p+q)I$ commutes with G. Hence,

$$\Lambda = \{G-(p+q)I\}+2pB\in \mathscr{G}(1,-p-q+2p;X)$$

because $\|2pB\| \leq 2p$.

(*b*) If $g\in X^+$,

$$\exp\{t(G-(p+q)I)\}g = \exp\{-(p+q)t\}Z(t)g\in X^+$$

and

$$\|\exp\{t(G-(p+q)I)\}g\| = \exp\{-(p+q)t\}\|g\|$$

because of (*c*) of Lemma 13.5. On the other hand, we have as in Example 5.3

$$\exp(t\Lambda)g = \sum_{j=0}^{\infty} (2p)^j \hat{Z}^{(j)}(t)g$$

with

$$\hat{Z}^{(0)}(t)g = \exp\{t(G-(p+q)I)\}g\in X^+$$

$$\hat{Z}^{(j+1)}(t)g = \int_0^t \hat{Z}^{(0)}(t-s)B\hat{Z}^{(j)}(s)g\,ds\in X^+, j = 0,1,2,\ldots .$$

Thus, $\exp(t\Lambda)g\in X^+$ and

$$\|\hat{Z}^{(0)}(t)g\| = \exp\{-(p+q)t\}\|g\|$$

$$\|\hat{Z}^{(1)}(t)g\| = \int_0^t \exp\{-(p+q)(t-s)\}\exp\{-(p+q)s\}\|g\|\,ds$$

$$= \exp\{-(p+q)t\}t\|g\|$$

because $\|Bg\| = \|g\|$ (see (*a*) of Lemma 13.4). Since in general

$$\|\hat{Z}^{(j)}(t)g\| = \exp\{-(p+q)t\}\,\frac{t^j}{j!}\,\|g\|$$

we obtain

$$\|\exp(t\Lambda)g\| = \sum_{j=0}^{\infty} \frac{(2pt)^j}{j!}\exp\{-(p+q)t\}\|g\| = \exp\{(p-q)t\}\|g\|.$$

(c) follows from (13.32) and from the fact that X-lim $2prBf = 2pBf$ as $r\to 1-$ (see Theorem 6.3). □

Now assume that $u_0\, D(H^2)\cap X^+$ ($\subset D(H)\cap X^+ = D^+$). Then, $v_0 \in D^+$ because of (13.15), and (13.23b), (13.13b), and (13.17b) give for any $t \geq 0$ and $r\in[0,1)$

$$J\,\exp(t\Lambda_r)v_0 = Y\exp(tA_r)u_0 \in X^+ .$$

On passing to the X limit as $r\to 1-$, we obtain

$$J\,\exp(t\Lambda)v_0 = Y\,\exp(tG)u_0 \in X^+,\ \forall t \geq 0$$

i.e.

$$JV(t) = YW(t) \in X^+,\quad \forall\, t \geq 0 \tag{13.39}$$

with $V(t) = \exp(t\Lambda)v_0 \in X^+$. Relation (13.39) implies that

$$[V(t)]_m = (m+1)[W(t)]_{m+1},\quad m = 0,1,\ldots,\quad t \geq 0,$$

and so

$$\langle n\rangle(t) = \sum_{m=0}^{\infty}(m+1)[W(t)]_{m+1} = \|V(t)\| < \infty .$$

Since

$$\|V(t)\| = \|\exp(t\Lambda)v_0\| = \exp\{(p-q)t\}\|v_0\|$$

$$= \exp\{(p-q)t\}\sum_{m=0}^{\infty}(m+1)[u_0]_{m+1} = \exp\{(p-q)t\}\langle n\rangle_0$$

we obtain

$$\langle n \rangle(t) = \exp\{(p-q)t\}\langle n \rangle_0, \quad t \geq 0$$

namely

$$\frac{d}{dt}\langle n \rangle(t) = p\langle n \rangle(t) - q\langle n \rangle(t) \tag{13.40a}$$

$$\lim_{t \to 0+} \langle n \rangle(t) = \langle n \rangle_0 \,. \tag{13.40b}$$

System (13.40) governs the evolution of the expected number of individuals in the culture S. The 'physical' meaning of (13.40a) is quite simple; the change of the expected number of individuals during the time interval dt is due to $(p\,dt)\langle n \rangle(t)$ (births) and to $-(q\,dt)\langle n \rangle(t)$, (deaths).

Remark 13.1.

System (13.40) was derived under the assumption

$$u_0 = \begin{pmatrix} P_0(0) \\ P_1(0) \\ \cdots \end{pmatrix} \in D(H^2) \cap X^+$$

i.e.

$$P_0(n) \geq 0 \;\; \forall n = 0,1,\ldots, \qquad \sum_{n=1}^{\infty} n^2 P_0(n) < \infty \tag{13.41}$$

(thus, (13.41) holds if in particular $u_0 \in D_0 \cap X^+$). The assumption (13.41) can be better understood if we derive (13.40a) from (13.1a) (in a heuristic way) as follows. Multiplying both sides of (13.1a) by n and summing over n, we have

$$\frac{d}{dt}\sum_{n=1}^{r} nP(n,t) = p\sum_{m=1}^{r-1} mP(m,t) - q\sum_{m=1}^{r+1} mP(m,t) -$$

$$pr^2 P(r,t) + q(r+1)^2 P(r+1,t) \tag{13.42}$$

and (13.40a) follows from (13.42) if we assume that (i) $\lim_{r\to\infty}$ and d/dt commute, (ii) $r^2 P(r,t) \to 0$ as $r \to \infty$. Note that condition (ii) is certainly satisfied if $\Sigma_{n=1}^{\infty} n^2 P(n,t) < \infty$ (see (13.41)). □

EXERCISES

13.1. Prove that the set D_0 defined in (*a*) of Lemma 13.1 is dense in $X = l^1$. Hint: If $f \in X$ and $\varepsilon > 0$ are given, then $\Sigma_{n=n_0}^{\infty} |f_n| < \varepsilon$ with $n_0 = n_0(\varepsilon)$, and consider the element g such that $g_m = f_m$ $\forall m = 0,1,\ldots,n_0-1$, $g_m = 0 \forall m \geq m_0$.

13.2. Prove (*a*) of Lemma 13.4. Hint: use the definition (13.16).

13.3. Prove formulas (13.19) and (13.20). Hint: For instance,

$$[YHf]_n = [Hf]_{n+1} = (p+q)(n+1)f_{n+1} = (p+q)n[Yf]_n + (p+q)[Yf]_n = [HYf]_n + (p+q)[Yf]_n$$

and so $YHf = HYf + (p+q)Yf$, $\forall f \in D$.

13.4. Prove that the X-$\lim \gamma(r)$ as $r \to 1-$ exists, where $\gamma(r) = R(z, A_r)g$ with $g \in X^+$ and $z > 0$ assigned (see section 13.6). Hint: If $\gamma_n(r), n = 0,1,\ldots,$ are the components of $\gamma(r)$, then $\gamma_n(r) \geq 0$ and $\gamma_n(r_1) - \gamma_n(r) \geq 0$ $\forall n = 0,1,\ldots, 0 \leq r \leq r_1 < 1$, because $\gamma(r)$ and $\gamma(r_1) - \gamma(r)$ belong to X^+. Hence, $\gamma_n(r)$ is a non-decreasing and bounded function of $r \in [0,1)$ for each n because

$$0 \leq \gamma_n(r) \leq \sum_{j=0}^{n} \gamma_j(r) \leq \|\gamma(r)\| \leq z^{-1}\|g\|, \quad \forall r \in [0,1).$$

It follows that $\gamma_n \in R^1$ exists such that

$$0 \leq \lim_{r \to 1-} \gamma_n(r) = \gamma_n = \sup\{\gamma_n(r), 0 \leq r < 1\} \leq z^{-1}\|g\|.$$

Now, if

$$\gamma = \begin{pmatrix} \gamma_0 \\ \gamma_1 \\ \ldots \end{pmatrix}$$

then $\gamma_n \geq 0$ and the inequality $0 \leq \Sigma_{j=0}^{N} \gamma_j(r) \leq z^{-1}\|g\|$ (see above) implies that $0 \leq \Sigma_{j=0}^{N} \gamma_j \leq z^{-1}\|g\|$ and so $0 \leq \|\gamma\| \leq z^{-1}\|g\| < \infty$, i.e. $\gamma \in X^+$. Finally, to show that $\|\gamma - \gamma(r)\| \to 0$ as $r \to 1-$, note that

$$0 \leq \sum_{j=\nu+1}^{\infty} \{\gamma_j - \gamma_j(r)\} \leq \sum_{j=\nu+1}^{\infty} \gamma_j < \varepsilon$$

where $\nu = \nu(\varepsilon)$ is a suitable integer, because $\gamma \in X^+$, i.e. the series $\Sigma_{j=0}^{\infty} \gamma_j$ is convergent. Then, $\|\gamma - \gamma(r)\| = \ldots$.

13.5. If 'immigration' is present, (13.1a) becomes

$$\frac{\partial}{\partial t} P(n,t) = -(p+q)nP(n,t) + p[n-1]P(n-1,t) + q[n+1]P(n+1,t) + \eta\{P(n-1,t) - P(n,t)\}$$

where $\eta \geq 0$ is the probability per unit time that a bacterium is introduced in the culture from outside. Discuss the system composed of the above equation and of (13.1b). Hint: the first of (13.6) becomes

$$\frac{d}{dt} u(t) = Au(t) + \eta\{B-I\}u(t), \quad t > 0.$$

13.6. Study system (13.1) under the assumption that $p = p(n)$ and $q = q(n)$ with $0 \leq p_1 \leq p(n) \leq p_2 < \infty$ and $0 \leq q_1 \leq q(n) \leq q_2 < \infty$ $\forall n = 0,1,2,\ldots$.

BIBLIOGRAPHY

This list of books and articles is by no means complete. They are intended to provide the reader with a 'library' suitable for a further study of semigroups and of equations of evolution in B-spaces.

ADAMS, R.A. (1975). *Sobolev spaces.* Academic Press, New York.

BARBU, V. (1976). *Nonlinear semigroups and differential equations in Banach spaces.* Noordhoff International, Leyden.

BELL, G.I. and GLASSTONE, S. (1970). *Nuclear reactor theory.* Van Nostrand Reinhold, New York.

CARSLAW, H.S. and JAEGER, J.C. (1959). *Conduction of heat in solids.* Clarendon Press, Oxford.

HILLE, E. and PHILLIPS, R.S. (1957). *Functional analysis and semigroups.* American Math.Soc.Colloq.Publ., Providence, R.I.

KATO, T. (1954). *J.Math.Soc.Japan* 6, 1-15.

——— (1966). *Perturbation theory for linear operators.* Springer, New York.

LUDWIG, D. (1974). *Stochastic population theories.* Springer, Berlin.

MARTIN, R.H. Jr. (1976). *Nonlinear operators and differential equation in Banach spaces.* Wiley, New York.

MURRAY, J.D. (1978). *Lectures on non-linear differential-equation models in biology.* Clarendon Press, Oxford.

PAVERI-FONTANA, S.L. (1975). *Transp.Res.* 9, 225-35.

PRIGOGINE, I. and HERMAN, R. (1971). *Kinetic theory of vehicular traffic.* American Elsevier, New York.

ROYDEN, H.L. (1963). *Real analysis.* McMillan, New York.

SCHIFF, L.I. (1968). *Quantum mechanics.* McGraw-Hill, New York.

SEGAL, I. (1963). *Ann.Math.* 78, 339-63.

TAYLOR, A.E. (1958). *Introduction to functional analysis.* Toppan, Tokyo.

TITCHMARSH, E.C. (1948). *Introduction to the theory of Fourier integrals*. Clarendon Press, Oxford.

TROTTER, H.F. (1958). *Pacific J.Math* 8, 887-919.

TYCHONOV, A.N. and SAMARSKI, A.A. (1964). *Partial differential equations of mathematical physics*. Holden-Day, San Francisco.

VIDAV, I. (1970). *J.Math.Anal.Applic.* 30, 264-79.

WING, G.M. (1962). *An introduction to transport theory*. Wiley, New York.

SUBJECT INDEX

Absolutely continuous functions, 24
Algebraic eigenspace, 256
Algebraic multiplicity, 256, 257

Banach space, 10
Bounded convergence theorem, 203
Bounded operator, 40
Bounded perturbation, 179

Cauchy's integral formula, 122
Cauchy sequence, 9
Chain-rule for Fréchet derivatives, 127
Closed subset, 20
Commutation property, 185
Complete normed space, 9
Cone, 288
Cone property, 31
Continuity, strong, 101
Continuous spectrum, 75
Contraction mapping theorem, 57
Convection operator, 83
  and the heat diffusion operator, 190, 217
  and the streaming term in neutron transport, 296, 298, 309
  and the streaming term in vehicular traffic, 320
  and the telegraphic equation, 340
  as a generator of a semigroup, 141

Densely defined operator, 48
Dense subset, 28
Derivative
  Fréchet, 123
  strong, 106
Differentiation under the integral sign, 114
Direct sum, 246
Discretization of the time variable, 164
Dissipative operator, 144

Eigenfunction, 74
  normalized, 75
  orthonormal family, 283

Eigenspace
  algebraic, 256
  geometric, 256

Eigenvalue, 74
  isolated, 256
  semisimple, 258
  simple, 257
Energy method, 144
Equivalent norms, 12
Extension, 38
Extension of a densely defined bounded operator, 48

First moment, 377
Fourier coefficient, 233
Fourier-Plancherel transform, 34
Fourier transform, 32
Frechet derivative, 123
Fredholm integral operator, 46, 68
Functional, linear and bounded, 44

Galerkin method, 233
Generalized derivatives, 22
Generator of a strongly continuous semigroup, 325
Geometric eigenspace, 256
Geometric multiplicity, 256
Graph of an operator, 60
Gronwall's inequality, 119
Group generated by a bounded linear operator, 130
Group generated by a closed operator, 160, 321, 341, 344, 346, 358
$\mathcal{G}(1,0;X)$, 152
$\mathcal{G}(M,0;X)$, 158
$\mathcal{G}(M,\beta;X)$, 158
$\mathcal{G}'(M,\beta;X)$, 158

Heat diffusion operator, 68, 70, 285
  and heat conduction in a rigid body, 280, 284
  and oscillating heat sources, 175
  and the convection operator, 190, 217
  and the Schrödinger equation, 357
  as a generator of a semigroup, 141, 282
  perturbation, 178, 182, 217, 285
  spectral properties, 87, 93, 263, 273
  with non-zero boundary conditions, 292
Hilbert space, 12
Holomorphic functions, 120
Hölder inequality, 27
Homogeneous initial-value problem, 129, 162, 224

Imbedding, 29
Initial-value problem, 2, 114, 129, 134, 162, 224
  and heat conduction, 282, 285
  and neutron transport, 298, 310
  and quantum mechanics, 357
  second order, 350
  semilinear, 191
  and stochastic population theory, 366
  and vehicular traffic, 326
Inner product, 10
Inverse operator, 38, 39

Laurent expansion of the resolvent, 255
Lebesgue integral, 18
Lebesgue measure, 16
Linear operator, 39
  closed, 62
Linear space, 7
Linear subset, 21
Linearly dependent elements, 234
Lipschitz operator, 51
Locally integrable function, 21

Maxwell's equations, 340, 356
Mean-value theorem, 108
Multiplicity
  algebraic, 256, 257
  geometric, 256

Neutron transport, 296
  linear, 297
  nonlinear, 308
  one-group diffusion approximation, 281
  spectral properties, 304
  with temperature dependent cross sections, 308
Non-homogeneous initial-value problem, 129, 162, 167
Nonzero boundary conditions, 292
Norm, 8
  of an operator, 40
Normed space, 8

Open sphere, 20
Open subset, 20
Operator, 38
  bounded, 40
  closed, 60
  closed and linear, 62
  densely defined, 48
  dissipative, 144
  inverse, 38, 39
  linear, 39
  Lipschitz, 51
  norm of, 40
  self-adjoint, 66
  strictly contractive, 57
  symmetric, 66
Oscillating heat sources, 175
Oscillating sources, 171

Parseval relation, 33
Periodic solutions, 172
Perturbation
  bounded, 179
relatively bounded, 190
  time-dependent and bounded, 187
Point spectrum, 74
Positive cone, 288
Positive solutions, 288, 303, 331
Projection, 245

Relatively bounded perturbation, 190
Residual spectrum, 75
Resolvent
  operator, 76, 122
  of a bounded operator, 77
  of $A+B$ with A closed and B bounded, 94
  of a self-adjoint operator, 95
  of iA with A self-adjoint, 97
  set, 76, 122
Restriction, 38
Riemann integral, strong, 110

Schrödinger equation, 357
Schwarz inequality, 11
Self-adjoint operator, 66
Semigroup generated by a closed operator, 152, 158, 159
Semilinear initial-value problem, 191
  global solution, 205
  local solution, 193
Sequence
  of approximating problems, 224, 228, 241
  of Banach spaces, 225
  Cauchy, 9
  of semigroups, 219
Simple function, 17
Solution
  global, 205, 331
  local, 193, 205
  mild, 196
  positive, 188, 303, 331
  strict, 129, 162, 193, 350
Spaces
  $\mathscr{B}(D,Y)$, 41
  $C([a,b])$, 15
  $C^j([a,b])$, 35

$C_B^j(\Omega)$, 30
$C^j(\overline{\Omega})$, 36
$C_0(\Omega)$, 22
$C^m(\Omega)$, 29
$C([t_1,t_2];X)$, 105
$C^1([t_1,t_2];X)$, 127
$\mathscr{C}(X,Y)$, 61
Lip(D,Y), 51
l^p, 13
$L^1_{\text{loc}}(\Omega)$, 21
$L^p(a,b)$, 16
$L^{s,2}(R^n)$, 34
$W^{m,p}(\Omega)$, 25
$W_0^{m,p}(\Omega)$, 28
Spectral properties of the semigroup, 273
Spectral radius, 79
Spectral representation of a closed operator, 265
of the semigroup, 269
Spectrum, 75
continuous, 75
isolated point of, 249
point, 74
residual, 75
State vector, 1
Stochastic population theory, 365
Support of a function, 22
Symmetric operator, 22

Telegraphic equation, 339, 350
Time-dependent bounded perturbation, 187

Uniform boundedness theorem, 217

Vector space, 7
Vehicular traffic, 319

Wave equation, 355